Design-Build Subsurface Projects

Design-Build Subsurface Projects

Editor

Marcel Talbot

scitus academics

Design-Build Subsurface Projects
Edited by **Marcel Talbot**

ISBN: 978-1-68117-455-6
Library of Congress Control Number: 2016937623

SCITUS Academics LLC,
www.scitusacademics.com
Box No. 4766, 616 Corporate Way,
Suite 2, Valley Cottage,
NY 10989

Notice

Reasonable efforts have been made to publish reliable data and views articulated in the chapters are those of the individual contributors, and not necessarily those of the editors or publishers. Editors or publishers are not responsible for the accuracy of the information in the published chapters or consequences of their use. The publisher believes no responsibility for any damage or grievance to the persons or property arising out of the use of any materials, instructions, methods or thoughts in the book. The editors and the publisher have attempted to trace the copyright holders of all material reproduced in this publication and apologize to copyright holders if permission has not been obtained. If any copyright holder has not been acknowledged, please write to us so we may rectify.

Printed in United States of America on Acid Free Paper

Preface

With its opportunities for cost containment and substantial risk transfer, Design-Build is increasingly becoming the delivery method of choice for owners with challenging funding limitations. But deciding to use the Design-Build system for underground projects is one thing; successfully implementing it is quite another. Subsurface engineers must design a successful well completion system by selecting equipment that is adequate for both downhole environments and applications. Considerations must be given to the various functions under which the completion equipment must operate and the effects any changes in temperatures or differential pressure will have on the equipment. The completion system must also be efficient and cost effective to achieve maximum production and financial goals. Another factor in the selection of specific completion equipment is the production rates of the well. The typical job duties of a Subsurface engineer include managing the interface between the reservoir and the well, including perforations, sand control, artificial lift, downhole flow control, and downhole monitoring equipment.

This book, Design-Build Subsurface Projects, provides a straight-forward, comprehensive look at how to make Design-Build work on complicated projects involving tunnels, highways, dams, and deep foundations. This book includes extensive information about procurement, as well as risk allocation issues, which are significantly different from the Design-Bid-Build approach and serves as an indispensable resource for owners, engineers, construction managers, contractors, and others involved in the design and construction of subsurface projects.

Table of Contents

Chapter 1	An overview of impact of subsurface drainage project studies on salinity management in developing countries 1
Chapter 2	Thermal effect of a borehole thermal energy store on the subsurface 23
Chapter 3	Detection of Surface and Subsurface Cracks in Metallic and Non-Metallic Materials Using a Complementary Split-Ring Resonator 45
Chapter 4	A Mobile Acoustic Subsurface Sensing (MASS) System for Rapid Roadway Assessment 69
Chapter 5	The Search for Sustainable Subsurface Habitats on Mars, and the Sampling of Impact Ejecta 91
Chapter 6	Realistic Subsurface Anomaly Discrimination Using Electromagnetic Induction and an SVM Classifier 121
Chapter 7	Audio-frequency magnetotelluric imaging of the Hijima fault, Yamasaki fault system, southwest Japan 147
Chapter 8	An experimental investigation of geochromatography during secondary migration of petroleum performed under subsurface conditions with a real rock 169

Chapter 9 **Neutron production in the lunar subsurface from alpha particles in galactic cosmic rays** **189**

Chapter 10 **2D structural and thermal models in southeastern Norway based on the recently drilled Årvollskogen borehole and 2D density, magnetic and thermal modelling** **215**

Chapter 11 **Magnetotelluric and temperature monitoring after the 2011 sub-Plinian eruptions of Shinmoe-dake volcano** **255**

Index **283**

CHAPTER 1

An overview of impact of subsurface drainage project studies on salinity management in developing countries

Priyanka Tiwari, Arun Goel

Department of Civil Engineering, National Institute of Technology (Institute of National Importance), Kurukshetra,Haryana 136119, India

ABSTRACT

Subsurface drainage has been used for more than a century to keep water table at a desired level of salinity and waterlogging control. This paper has been focused on the impact assessment of pilot studies in India and some other countries from 1969 to 2014 . This review article may prove quite useful in deciding the installation of subsurface drainage project depending on main design parameters, such as drain depth and drain spacing, installation area and type of used outlet. A number of pilot studies have been taken up in past to solve the problems of soil salinity and waterlogging in India. The general guidelines that arise on the behalf of this review paper are to adapt drain depth >1.2 m and spacing depending on soil texture classification, i.e., 100–150 m for light-textured soils, 50–100 m for medium-textured soils and 30–50 m heavy-textured soils, for better result obtained from the problem areas in Indian soil and climatic conditions. An attempt has been made in the manner of literature survey to highlight the salient features of these studies, and it is hopeful to go a long way in selecting design parameters for subsurface drainage problems in the future with similar soil, water table and climatic conditions.

KEYWORDS

Subsurface drainage; Salinity; Waterlogging; and groundwater table

1. INTRODUCTION

Land and water are two basic natural resources. Due to rapid population growth and fast industrialization, these resources are facing immense pressure and are depleting day by day. According to the FAO land and plant nutrition management service (1994), over 6 % of the world's land is affected by either salinity or sodicity. Much of the world's land is not cultivated; however, a significant proportion of cultivated land is salt-affected, and of the current 230 million ha of irrigated land, 45 million ha is salt-affected (19.5 %), and of the 1500 million ha under dry land agriculture, 32 million ha is salt-affected to varying degrees, i.e., 2.1 % (Hefny et al. 2013). World's large irrigated regions with serious salinity problems are Yellow River Plain in China, San Joaquin Valley in California, KaraKum Canal project in Turkmenistan, Indus Plain in Pakistan, Tigris–Euphrates Plain in Iraq, Murray–Murrumbidgee Area in Australia and lower Nile Valley in Egypt, which need a serious attention of researchers (Ghassemi et al. 1995).

Subsurface drainage is considered as a most suitable approach for groundwater balance and land and water management practices containing the groundwater table at a suitable level (Luthin 1978; Gates and Grismer1989). Agricultural subsurface drainage is a process of removal of excess groundwater from the crop root zone system which promotes safe environment for efficient crop growth and for better health in rural and urban areas. Subsurface drainage lowers the high water tables, and the main causes of the rise in water table are precipitation, excess irrigation, leaching water, seeps from higher land or irrigation canal and ditches and groundwater under artesian pressure. This technique has gained international acceptance. Subsurface agricultural drainage provides agronomical and environmental benefits in terms of improved crop yield, improved soil trafficability, field operations and reduction in sediment and phosphorus losses from agricultural fields (Kornecki et al. 2001).

Subsurface drainage has been found to be the only solution for providing land reclamation on a long-term basis when salts are present in the soil and groundwater. Subsurface drainage has been provided in 75–80 % of irrigated area in Egypt and 25–30 % irrigated area in western USA Goel and Tiwari (2013). Comprehensive reclamation programs involving provision of subsurface drainage in irrigated areas have been embarked in a big way in Pakistan over the past 50 years, while these started much earlier in India during late 1920s (Thatte and Kulkarni 2000), and FAO (1999) reported that India has 2.5 million ha waterlogged land and 3.1 million ha salinity-affected area, while this hazard on the state level in India is extended, i.e., Uttar Pradesh has salinity-affected area over 1 million hectares, 1 million hectares in Gujarat, 0.5 million ha in Punjab, 0.2 million ha in Haryana and a smaller area in Rajasthan, and significant less proportionate area in Tamil Nadu in the south is also affected. At the time of nineties, subsurface drainage installed <10 % of the total irrigated area Zimmer et al. (2002). But currently <0.02 % irrigated area in India is provided with subsurface drainage Tiwari (2011), and Maharashtra state has minimum salt-affected area to the extent of 0.6 M ha (Sethi et al. 2010).Still India is facing varying degree of salinity problems such as saline soil, costal saline and alkalinity (Patel et al. 2002; Mandal and Sharma 2011).

Some past review papers and past remarkable studies on subsurface drainage and salinity management were discussed in this session. The purpose of this literature survey is that the findings and recommendations of these studies would be useful to take as guidelines to plan a new subsurface drainage system in an efficient, effective, economic and ecofriendly manner for future in India and other countries.

Gupta (2002, 2003) has summarized past 100-year Indian efforts to control salinity and waterlogging problems, which were conducted under the supervision of Central Soil Salinity Research Institute (CSSRI) from 1972 to 2002. He has covered several pre- and post-independence pilot studies and their salient findings from 1873 to 1975. He has provided general drainage design guidelines and operationalizes the subsurface drainage system for Indian conditions.

Kaledhonkar et al. (2009) have summarized preliminary studies conducted in India from 1980 to 2008. He has been enlightening on the 18 Haryana, 3 Gujarat, 2 Maharashtra, 2 Andhra Pradesh, 1 Karnataka and 1 Rajasthan pilot studies of their saline areas in the listed states, and he made efforts to focus the subsurface drainage operational and subsurface drainage effluent management problems. They discussed the drainage design parameters, crop production, cropping intensity and cost investment for the subsurface drainage installations but did not considered the maintenance of SSD.

Several researchers have suggested different strategies for salinity management in developing countries from 1969 to 2014. Special issues in SSD are discussed/highlighted, and specific methodologies for design of SSD are summarized in this paper. In the end, some of the general guidelines have been proposed which can be kept in mind while designing any SSD project in future.

2. SUBSURFACE DRAINAGE PILOT OR PROJECT STUDIES

2.1 India

The requirement of salinity management was firstly reported by Punjab Govt. in 1865 to Governor General. Mr. Robertson and started India's first drainage project in 1873 to reclaim the land from waterlogging, and people cure from malaria disease. History of northwest India's salinity has to reclaim through subsurface drainage installation and keep functioning it till achieving the aim of installation (Hanumanthaiah and Satyanarayana 2003). But they always worried about drain discharge disposal. Till few years ago, the drain discharge was disposed in canals. And also they tried to reuse the saline drain water in crop production, but the problem in those days was the pumping out of large quantity of drain discharge. So nowadays, the drainage installation is costly and seen more problematic in disposal point of view. A number of pilot scales subsurface drainage projects undertaken by CSSRI Karnal (Haryana), India, during 1980s have slowly paved the way for communication of large-scale mechanically installed drainage projects in Haryana, Rajasthan, Gujarat, Punjab, Andhra Pradesh, Maharashtra and Karnataka. India is divided into ten climatic zones which played very important role in SSD installation because any region is more affected by their weather conditions. On the basis of the climatic zones, experimental sites are through the India map (Fig. 1).

About 35,000 ha waterlogged saline soils have been reclaimed in these states with significant improvement in crop intensity, yield, land value and farmers income. India's major pilot studies literature survey has been discussed in Table 1. Here we are discussing the important factor of the drainage design criteria with their major outcomes. The following table (Table 1) have described about India's 56 successful pilot studies with their study area, installation year, operational duration, necessary drainage design parameters and their major conclusions.

On the basis of the Table 1, the pilot studies following guidelines are framed which would be helpful in drainage design subsurface drainage system in India. It is expected that these salient points would be useful in new wherein similar soil, water table and climatic conditions existing in India.

2.2 Basic guidelines for Indian subsurface drainage design

Although Drainage manual (1978) and Drainage design factors (1986) have recommended general guidelines for installation procedure and drainage design parameters for more effective performance yet on the basis of pilot studies undertaken in the past 50 years in India, few more guidelines have emerged which are mentioned below.

- Before subjecting any drainage project into a problematic area, the primary data (technical, socioeconomic, geo-hydrological climate) and secondary (water quality) have to be collected, which are relevant to the particular area.

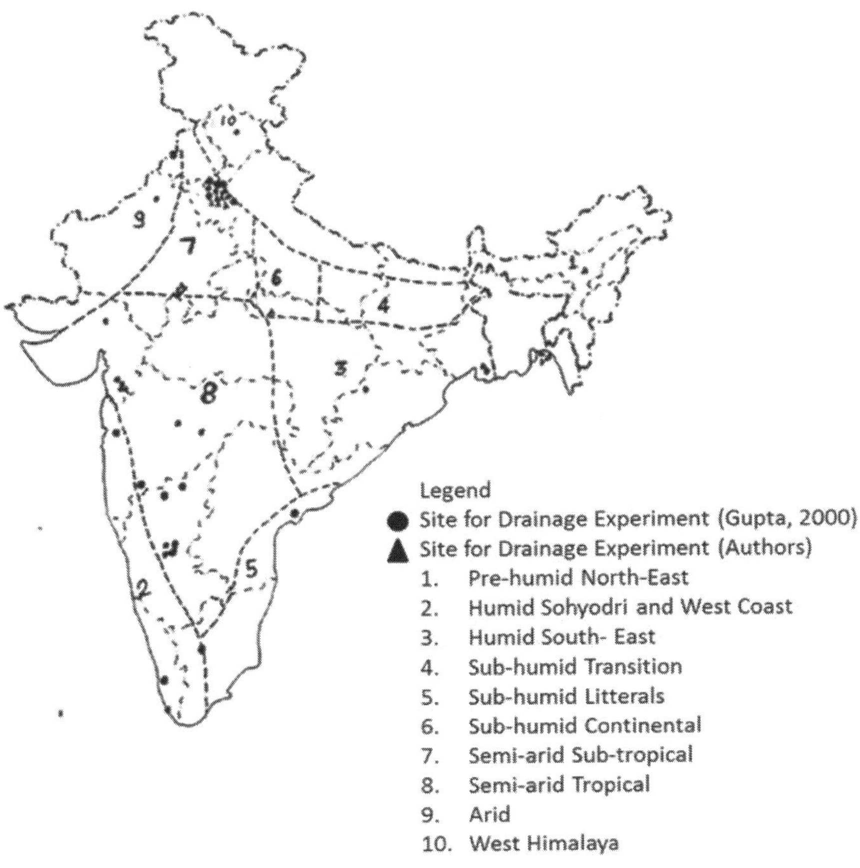

Figure 1 Climatic zone wise distribution of SSD studies in India

Table 1. Pilot studies on subsurface drainage in India

S.no.	References	Study area description for SSD Installation		Drainage design parameters				Major conclusion
		Place and area (ha)	Year of installation	Drain depth (m)	Drain spacing (m)	Type of drain	Type of outlet and envelope material	
Andhra Pradesh								
1	Channabasiah (1972)	Siruguppa	1972	1.2	12	Open drain	Gravity	Significant increase in crop yield is observed
2	Devadattam and Chandra (1995)	Machilipatnam	1995	–	–	Tile drain	–	Increased rice crop yield
3	Joint completion report on IDNP (2007)	Nagarjuna Sagar Right Command and Krishna Western Delta	–	1 1.0–1.2	60 75	Pipe drain Open drain	Pumped Geo-textile	Paddy crop yields have increased around 50 % in 2-year duration
4	Srinivasulu et al. (2005)	Konanki village in Prakasam district (21.63 ha)	1998–1999	0.9 and 1.1, 1	100 30, 60	Open Pipe Pipe	Gravity Geo-textile Nylon mesh	Late sowing kharif crop rice yield increase by 51 %
5	Ritzema et al. (2008)	Konanki and Uppugunduru (8 and 5 ha)	1999	0.90–1.75	45–150	Open drain Pipe drains	Pumped Geo-textile Nylon mesh	Water table has controlled and cropping intensity increased
Assam								
6	Bhattacharya et al. (1996)	Mornai	1996	–	–	Tile drain	–	Controlled water table and increased yield of tea leaves are observed
Gujarat								
7	Michael and Ojha (2013)	Saurashtra (22 km)	1972	1.2–1.8	–	Open drains	–	Land has reclaimed from salinity and waterlogging problems
8	Anonymous (1988)	Bidaj	1981	–	–	Open drains	–	Controlled salinity and waterlogging.
9	Ritzema (2009)	Segwa, Surat (5.6 ha, 2.6 ha, 5.7 ha and 4.6 ha)	1998–1999	0.9 1.0–1.2 1.2 1.2	45 30 45 60	Singular and Composite open pipe drains	Gravity	Water quality (EC reduced from 18.2 to 1.4 dS/m during 2001) has improved after one year and has used in irrigation
10		Sisodra, Surat (4.2 ha, 5.4 ha, 2 ha and 14 ha	2000	0.9 0.9 0.7 0.8	30 60 30 60	Singular and open pipe drains	Gravity	Cropping intensity has increased to 102 % after SSD installation
11	Kaledhonkar et al. (2009)	DabuoKheda (50 ha)	1987–1988	–	–	–	–	Water table has controlled, and this has given positive effect on crop production.
12		Muraj Kheda (52 ha)	1990–1991	–	–	–	–	Water table has controlled
Haryana								

Table 1 continued

S.no.	References	Study area description for SSD Installation		Drainage design parameters				Major conclusion
		Place and area (ha)	Year of installation	Drain depth (m)	Drain spacing (m)	Type of drain	Type of outlet and envelope material	
13	Jaiswal and Dhruvanarayana (1972)	Karnal, Haryana	1972	1.5	10–30	Tile and open drains	Gravity Pumped	Open drain has not been found suitable in alkaline soils
14	Kaledhonkar et al. (2009), Anonymous (1988)	Kailana khas (1.6 ha)	1980–1982	–	–	Open and Tile drains	Gravity	Groundwater table has been controlled
15	Rao et al. (1986), Kamra et al. (1991), Kaledonkar et al. (1998), Gupta (2002)	Sampla, Rohtak (1 ha, 10 ha, 27 ha, 13 ha)	1980–1981 1983–1984 1986–1987 1991–1992	1.5 1.5 1.75	20 50 25, 50, 75	Open drains Tile drains	Pumped and gravel for deep drain system Synthetic for shallow drain system	Wheat and barley crops yield was obtained 2.5–4.9 (t/ha) and 2.1–4.2(t/ha) at the zero production area Cropping intensity has been increased up to 200 % after drainage installation
16	Hirekhanet al. (2007), Pali et al. (2014)	Mudlana, Sonepat (50 ha)	1984–1985	1.75 and 1.6	50, 67, and 84	PVC corrugated pipes	Pumped	Salinity has been controlled
17	Kelleners et al. (2000)	CIRB, HAU Hisar (75 ha)	1990–1991	1.5	24	Tile drain	Pumped	Water table has controlled and crop yield increased
18	Kaledhonkar et al. (2009)	HLRDC farm, Hisar	1989–1990	–	–	Tile drain	Pumped	Groundwater table has controlled and increased crop yield
19		Ujana, Sonepat (32 ha)	1986–1987	–	–	Tile drain	Pumped	
20		Bhan Brahmana (105 ha)	1988–1990	–	–	Tile drain	Pumped	
21		Kolekhan, Jind (41 ha)	1989–1990	–	–	Tile drain	Pumped	
22		Ishapur Kher (58 ha)	1989–1990	1.0–1.2	–	Tile drain	Pumped	
23		Rori, Sirsa (1500 ha)	2004–2005	–	–	Poly-pipes	Pumped	
24	Joint Completion Report on IDNP (2007)	Gohana (1236 ha)	1979–1999	1.6	60–67	PVC corrugated pipes	Pumped Synthetic	Salinity has decreased 35 % from the year of installation to 2000
25	Lal (2011)	Jhajjar (Beri, 800 ha)	2004–2005	1.75	60–67	Poly-pipes	Pumped	Land has been reclaimed from salinity hazard
26		Dadri, Bibhani (800 ha)	2004–2005	–	–	Poly-pipes	Pumped	Agricultural land removes salinity at a desired level
27		Kalyant (1000 ha)	2004–2005	–	–	Poly-pipes	Pumped	Cropping intensity has increased
28		Sonipat (1500 ha)	2004–2005	–	–	Poly-pipes	Pumped	Water table has controlled
Karnataka								
29	Joint Completion Report on IDNP (2007)	Islampur, Karnataka (Upper Krishna Canal Command, 29.3 ha, 14.4 ha)	1998–1999 1998–1999-	1.0–1.1 1.0–1.1	50 50	PVC perforated corrugated Pipe Open Drain	Pumped Geo-textile	Cropping intensities have increased 27, 35 and 100 % in first three years
30				1.1–1.2 1.1–1.2	50 30	Pipe drain Pipe drain	Pumped	Cropping intensities have increased 114 and 156 % in first two years
31	Datta et al. (2004)	Mudhol Taluk Karnataka Bagal kot (19.35 ha)	1995–1996	–	1	–	Gravity	Water logged area has reclaimed

Table 1 continued

S.no.	References	Study area description for SSD Installation		Drainage design parameters				Major conclusion
		Place and area (ha)	Year of installation	Drain depth (m)	Drain spacing (m)	Type of drain	Type of outlet and envelope material	
32	Satyanarayana et al. (2002) Srinivasulu et al. (2014)	Uppugunduru	1999	0.85–1.1	50	Open	Pumped	Water table has controlled and cropping intensities increased to 170 % after drainage installation
		In Krishna Western Delta(7 and 5 ha)			30	Pipe	Geo-textile	
					45	Pipe	Nylon mesh	
					60	Pipe		
Kerala								
33	Mathew et al. (1993)	karumady	1993	–	–	Tile drain	–	Increased rice yield
Madhya Pradesh								
34	Yadav (1975), Anonymous (1988)	IARI, New Delhi	1975	1.0	32	Open drains	Gravity	Significant increase in sorghum and wheat yield
35	Yadav (1975), Anonymous (1988)	Indore, M.P.	1975	0.6–0.9	6.5–40	Tile drains	Gravity	Significant increase in crop yield and coarse sand filters are not suitable for alkaline vertisols
36	Sharma et al. (1995)	Barna, M.P.	1995	–	–	Open drain	–	Increased crop yield
37	Ramana et al. (2009)	CIAE Bhopal (M.P.)	–	0.60	2, 4 and 6	Tile drain and mole drain	Gravity	Water table has controlled
Maharashtra								
38	Gupta (2002), Kaledhonkar et al. (2009)	Manjri	1920	1.5	–	combination of open and tile drains	Gravity	Crop yield has increased
39	Anonymous (1988)	Manjri	1941	–	–	Tile drains and open drain	Gravity	Crop yield has increased
40		Parbhani	1982			Tile drain	Gravity	Increased net profit
41	Kamble et al. (2006)	Dhudgan Sangli (1100 ha)	–	2.5		Corrugated perforated PVC	Gravity	Water table has controlled
42	Kaledhonkar et al. (2009)	Kasabe Digraj Sangli (1065 ha)	2004–2006	2.5	10.75	Tile drains	Gravity	Water table has controlled
43	Kolekar et al. (2014)	College of Agril Engg and Tech., Parbhani	2006–2008	2, 4 and 6	–	Mole drain	–	The total yield of groundnut has highest in 4 m mole spacing, i.e., 69.20 %
Punjab								
44	Gupta (2002)Kaledhonkar et al. (2009)	Amritsar	1873	–	–	Stone, tile drains and 16-tube well (0.0425 m^3/sec)	Gravity	Malaria disease has mitigated and land free from waterlogging problem
45		Chakanwal	1920	0.9	33–66	Open drain	Gravity	This has played a vital role in crop production
46	Anonymous (1988)	Ludhiana	1967	–	–	Tile drains and open drain	Gravity	Maize yield increased

Table 1 continued

S.no.	References	Study area description for SSD Installation		Drainage design parameters				Major conclusion
		Place and area (ha)	Year of installation	Drain depth (m)	Drain spacing (m)	Type of drain	Type of outlet and envelope material	
47	Michael and Ojha (2013)	Ranbirpura, Patiala (800 h)	1959–1960, 1971	–	–	Open drain	–	This area has significantly free from salinity
48	Michael (1967), Michael and Ojha (2013)	Ludhiana	1966	1.2	–	Interceptor and combination of open and tile drains	Pumped	System was not much effective in rabi season
49	Shakya et al. (1995)	Golewala	1995	–	–	Multiple well one point system	Pumped	Permitted conjunctive use of groundwater
Rajasthan								
50	Lovas (1972), Anonymous (1988)	Digod	1972	0.6–1.0 1.0–1.3	2–2.5 8–12 15–20	Mole, plastic asbestos pipes	Gravity	Mole drains have found to be cheaper and less time-consuming
51	Datta et al. (2004)	Chambal command area RAJAD (1420 ha, 12580 ha)	1992–1993 1993–1997	1–1.2	30, 60	Tile and open drains	Gravity	Cropping intensity has increased
52	Anonymous (1995)	Kota	1995	–	–	Tile and open drains	Gravity	Project benefit cost ratio 2.5
53	Shekhawat (2007)	Lakhuwali (15 ha, 60 ha)	2000 2001	1.2	100, 150, 200	Pipe Drain	Pumped 300–350 Microns geo-textile nonwoven	Kharif (paddy, cotton,) and Rabi (wheat barley, mustard) crops has obtained just double yield and cropping intensity has increased 160 %
Uttar Pradesh								
54	Chauhan and Ram (1972)	Pant Nagar, U.P.	1972	1.0	15–35	Tile drains	Gravity	More water storage has taken p and less runoff from draining
West Bengal								
55	Maity et al. (1996)	Midnapur	1996	–	–	Open drain	–	Increase in observed crop yield
56	Bhattacharya (1999)	Dankuni Basin	1999	–	–	Raised bed pond	–	Salvaged waterlogged land for a

- In past, cement concrete pipes were very popular in alkaline and saline areas because these pipes have not given any complaints up to 20 years in alkaline areas and 8 years in saline zones after installation. But nowadays PVC pipes are more popular than others due to its portability and light in weight.

- Drainage depth is an issue of increase or decrease in cost of installation and also availability of machineries and labor. But it is depending on position of groundwater table, soil type and hydraulic characteristics. For agricultural drainage, recommended depth of the lateral is >1.2 m because agricultural activities may damage the laterals.

- Drain spacing is classified on the basis of the texture of the soil. The lateral spacing, size and grade (in %) for light-textured soils are ranging between 100 and 150 m, 100 mm and 0.10 %, respectively, 50 and 100 m, 125 mm and 0.075 % for medium-textured soil and 30 and 50 m, 150 mm and 0.05 % for heavy-textured soils.

- On the basis of climatic conditions, drainage coefficient has an optimum value which lies between 1 and 3 mm/day (i.e., arid region is a 1 mm/day, semiarid region is a 2 mm/day, and subhumid region is 3 mm/day).

- Continuous movement of water in the pipe and aquifer system collects sediment in the pipes which may affect the performance of the drainage system. For resolving the problem, the provision of filters and envelopes on the drain pipes has to be adopted. Filter/envelope material is used to filter that surrounds drain pipe, and these are commonly used along with drain pipes (geo-textile, polypropylene, coconut fiber, polystyrene and foam plastic). The traditional filter material is a combination of gravel and coarse sand.

- Normally the drain effluent is disposed in the canal, salt-making ponds, fishpond, or it can be reused in crop production. The several methods have been suggested by researchers for effluent reuse in irrigation such as blending and mixing.

2.3 Subsurface drainage pilot studies in Pakistan

As a consequence, waterlogging and salinity now are serious threats to irrigated agriculture of the 16.7 million hectare in the Indus Basin; about 2 million hectare is waterlogged, and 6 million hectare is salt-affected (Nijland et al. 2005).

Pakistan made an effort for salinity control by starting a series of salinity control and reclamation projects in 1959, and results show a very good agreement to reduce salinity level in the surface and soil profile salinity. The salinity in percentage has been compared for 17-year variation (between the initial salinity year 1960 and 1977–1979) after installation of drainage, and this is also indicated that the salt-free areas increased up to 20 % in surface and profile salinity from past decades (Bhutta 2007).

The subsurface drainage installations in Pakistan are still <1 % of the total cultivable commanded areas so it needs to be increased (Ghumman et al. 2012; Azhar et al. 2004). Bhutta (2007) has discussed salient findings of drainage research and its benefits in Pakistan. Pakistan has world's largest irrigation system, namely the Indus Basin Irrigation System that commands about 14.2 M ha canal irrigated area (Niazi 2008). Due to poor maintenance and neglecting behavior of academician of Pakistan, most of the parts of canal are still unlined and result in the basic cause of waterlogging problem, which directly covered major part of agricultural lands of Pakistan. Niazi (2008) and Ghumman et al. (2010) reported that Government of Pakistan has made effort to save major fertile land of agriculture by the installation of major subsurface drainage projects from past few decades. Various methodologies have been used to protect agricultural lands of Pakistan from the salinity, but most of the subsurface drainage project had not proved informative and beneficial to solve the problem (Smedema 1990; Sarwar 2000; Kahlown and Khan 2004). So many approaches like surface drainage, subsurface pipe drainage and tube wells had been installed for reclamations of fertile lands of Pakistan, but subsurface pipe drainage was proved more beneficial than other two methods (Azhar et al. 2010). Four decades ago, Pakistan government was more concerned about the protection of the productive agricultural lands, and they passed eight subsurface pipe drain projects, but Sarwar and Feddes (2000) and Azhar et al. (2004) have reported that the installed drainage projects had not proved beneficial as per the expectation because these projects have not been technically sound and designed in that particular decade. After that Pakistan's irrigation and drainage engineers and scientists continuously worked on the drainage design parameters such as drainage coefficient, drain spacing Naz et al.2009) and drain depth. Kahlown et al. (2007) and Azhar and Latif (2011)have been worked on three experimental sites NIA, Bughio and Nawazabad farms in Pakistan in order to find the drainage coefficient for the efficient performance of futuristic projects and problems. History of Pakistan's irrigation and drainage researches are mainly explained by existence of nine different subsurface drainage projects (Azhar et al.2005) as presented in Table 2.

Table 2. Major four pilot studies of Pakistan

S. no.	References	Site information	Main design parameters						Shortcomings
			Drain depth (m)	Drain spacing (m)	Design discharge (mm/day)	Design $D_{W.T.}$ (m)	Type of drainage	Outlet and sump	
1	Ghumman et al. (2012)	Khushab SCARP, Pakistan (42,000,000 ha)	1.68–1.98, 2.1	150–300	1.8	1.2	Pipe drainage	Pumped	In Khushab, the water salinity measured in terms of dS/m was reduced by 33 %, respectively, in 12 years
2		FDP (Fourth Drainage Project, Faisalabad), Pakistan (30,351 ha)	1.68–1.98	500–600	2.44	1.2	Pipe drainage	Pumped	In Faisalabad, the water salinity measured in terms of dS/m was reduced by 25 %, respectively, in 12 years
3	Azhar and Latif (2011), Ghumman et al. (2012)	Mardan SCARP Project (MSP) Pakistan (29,542 ha)	2.4	150–200	3	1.05	Pipe drainage	Pumped	Increase in rice, sugarcane, kharif fodder, wheat, sugar beet and tobacco yield was 46, 49, 50, 56,43 and 23 %, respectively
4		Mirpukhas Tile Drainage Project Pakistan (24,281 ha)	1.8–2.4	–	0.95		Pipe drainage	Pumped	Crop yield had increased by the use of saline drain water
5	Ritzema (2007)	East Khaipur Tile Drainage	1.95	–	2.5–3.5	1.0	Pipe drain	Pumped	Crop yield had increased
6	Sarwar and Mehammad (2008)	Swabi SCARP	1.8–2	100–150	2	1.0	Pipe Drain	Pumped	Yields of sugarcane, rice and wheat were increased by 16, 24 and 37 %, respectively
7	Asian Development Bank (1984)	Chashma Command Area Development (18160 ha)	2.10 1.8–2.7	–	1.2–2.6	1.4	Interceptor drain, PVC pipe drain	Pumped	Cropping intensities increased 105 to 135 % after installation of SSD
8	Ritzema (2007)	DC Khan SCARP	–	–	1.88	–	Pipe drain	Pumped	Crop yield has increased and water table is controlled
9	Ritzema (2007)	Fordwah Eastern Sadiqia (South)	2.1		1.5	1.2	Pipe drain	Pumped	Crop yield has increased

2.4 Pakistan drainage design guideline

In Pakistan, in drained areas where a deep water table is maintained, farmers sometimes complain about the increased need of irrigation water (Qureshi et al. 1997). A shallow water table, especially in the fine soils of the Indus plains, is capable of water delivery to the crops through capillary rise. In areas with an 'acceptable' groundwater quality, there is no need to maintain a deep water table. The following guideline arises from the studies of the different research works. On the bases of the pilot studies, following steps are recommended.

1 In Pakistan, drainage design depth of the subsurface drainage systems should be in between 1.8 and 2.4 m while collector depth >3 m.

2 SSD drain discharge has to be designed as 0.95–3.5 mm/day.

3 PVC pipes are generally popular in field drains with fixed dimensions, i.e., diameters 100–200 mm and length up to 800 m.

4 In collector drains, PVC and PE pipes are used with fixed dimensions of 200–380 mm in diameter and length >4 km.

5 Gravel filters are more popular in Pakistan subsurface drainage installation.

In Iraq, for example, more than 50 % of the lower Rafidain Plains faces a stern salinity and waterlogging problems (El-Hinnawi 1993). Similarly, salinity and waterlogging have been an inevitable problem in Egypt. These problems have existed during the pre- and post-Aswan High Dam periods

(Ritzema 2009; IPTRID Secretariat 2007). To overcome this twin problem, subsurface drainage projects were commissioned in 1942. About 55 % of agricultural land was reported as saline in Iran (FAO 1994). Waterlogging and drainage problems occur in the central and southern parts of the Saudi Arabia. In some projects, like Al-Hassa irrigation project, the agricultural drainage water is mixed with fresh groundwater and reused for irrigation. So, due to the poor quality of irrigation water, soil salinity problem is increasing (AQUASTAT 2008). Multi-level subsurface drainage has also proved beneficial (Hornbuckle et al. 2012). Hirekhan et al. (2007) have suggested that field observation must be taken before and after the installations of SSD in a semiarid climate area and have stressed on rigorous analysis before adopting a subsurface drainage technique. Chahar and Vadodaria (2010) and Chahar and Vadodaria (2012) investigated the optimal spacing in an array of fully penetrating ditches for subsurface drainage. They developed an explicit equation for computing the optimal spacing between the ditches. Eldeiry and Garcia (2010) have compared ordinary kriging, regression kriging and co-kriging techniques to estimate the soil salinity using various images. The best combinations have evaluated to estimate soil salinity with different crop types. Gammal El and Ali (2010) reported that subsurface drainage water has proved very beneficial for crop production in Egypt. Salinity and waterlogging problems affected countries are trying to reclaim their waste land, which are severely affected from these hazardous reasons presented in Table 3.

Table 3. Pilot studies on desalinization and waterlogging

S. no.	References	Location states and countries	Spacing(m) and depth (m)		Type of drains and type of outlet		Major conclusions
1	Kale 2011	Konya–C, umra Plain, Turkey	–	–	–	Pumped	Water table, soil salinity have been controlled, and wheat crop yield also increased
2.	Ozturk (2004)	Anatolia, Turkey	21.3	1.8–2.0	Tile drains	Gravity	Crop yield increased, and water table has been controlled
3.	El-Mowelhi et al. (1988)	Nile Delta Egypt	20, 75	1.5	Tile drains	Gravity	Crop yield increased

3. CONCLUSIONS

This paper briefly discussed the old to current state of art in subsurface drainage system, positive effects of drainage on crop production and also what the future hold for this technical approach. Pilot studies reviewed subsurface drainage investigations from 1969 to 2014 (India and abroad) with their outcomes. This study is to suggest the numeric value of drain depth which is kept greater than 1.2 m, i.e., >1.2–1.8 m and drain spacing depending on the soil texture classification, i.e., 100–150 m for light-textured soils, 50–100 m for medium-textured soils and 30–50 m for heavy-textured soils for Indian subsurface drainage installation in the problem area, and Pakistan subsurface drainage project studies show drain depth kept more than 1.8 m. With the help of this paper, researchers will be able to decide the drain depth, spacing, types of subsurface drainage and type of outlets.

REFERENCES

1. Anonymous (1988) Handbook for drainage of irrigated areas in India. Irrigation Management and Training Project: technical report no. 5. LBII/WAPCOS, Room 213, Ansal Chambers, Bhikaji Cama Place, New Delhi, pp 12.3 & 12.5

2. Anonymous (1995) Analysis of subsurface drainage design criteria. Rajasthan Agricultural Drainage Research Project, Chambal Command Area Development, Kota, Rajasthan, pp 7.3

3. AQUASTAT (2008). Irrigation in the Middle East Regionc in figures. Water Report No 34, FAO. Rome, Italy

4. Azhar AH, Latif M (2011) Subsurface drainage impact assessment on crop yield. J Anim Plant Sci 21(2):215–219

5. Azhar AH, Alam MM, Rafiq M (2004). Impact of subsurface drainage systems on land and water conditions. IWASRI Publication 261, WAPDA, Lahore

6. Azhar AH, Alam MM, Rafiq M (2005) Agricultural impact assessment of sub-surface drainage projects in Pakistan—crop yield analysis. Pak J Water Resour 9(1):1–8

7. Azhar AH, Bhutt MN, Latif M (2010) Reclamation irrigated agriculture through tile drainage at Fourth Drainage Project, Faisalabad. J Anim Plant Sci 20(3):211–216

8. Bhattacharya AK (1999) Drainage of agricultural lands. In: Singh GB, Sharma BR (eds) 50 Years of natural resource management research. Division of Natural Resource Management, ICAR, New Delhi, pp 347–362

9. Bhattacharya AK, Ray SB, Sinha AK, Ghosh G (1996) Drainage in tea plantation. AICRP on Agricultural Drainage, Water Technology Centre, IARI, New Delhi, vol 23, p 4

10. Bhutta MN (2007) Salient findings of drainage research and its benefits in Pakistan. Pak J Water Resour 11(1):31–38

11. Chahar BR, Vadodaria GP (2010) Optimal spacing in an array of fully penetrating ditches for subsurface. J Irrig Drain Eng ASCE 136(1):63–67

12. Chahar BR, Vadodaria GP (2012) Steady subsurface drainage of ponded surface by an array of parallel ditches. J Hydrol Eng 17:895–908

13. Channabasiah HSM (1972) Sub-surface drainage in salt affected medium to deep black soils in India. Lecture delivered at the Summer Institute on Saline and Alkali Soils and their Management at C.S.S.R.I., Karnal, India

14. Chauhan HS, Ram S (1972) Surface and sub-surface drainage hydrology. In: Proceedings symposium on waterlogging causes and measures for its prevention. CBIP. Publication, vol 118(1), pp 19–20

15. Datta KK, Tewari L, Joshi PK (2004) Impact of subsurface drainage on improvement of crop production and farm income in north-west India. Irrig Drain Syst 18:43–55

16. Devadattam DSK, Chandra RS (1995) Agricultural land drainage in coastal saline soils. AICRP on agricultural drainage, A.N.G. Ranga Agricultural University, Hyderabad, 40 p

17. Drainage Design Factors (1986) Irrigation and drainage. Paper no. 38. Food and Agriculture Organization of the United Nations, Rome, Italy

18. Drainage manual (1978) U.S. Department of the Interior, Bureau of Reclamation. U.S. Government Printing Office, Washington, DC

19. Eldeiry AA, Garcia LA (2010) Comparison of ordinary kriging, regression kriging, and cokriging techniques to estimate soil salinity using LANDSAT images. J Irrig Drain Eng ASCE 136(6):355–364

20. El-Hinnawi E (1993) Population, environment, and development in the Arab World. Paper presented to the Arab Population Conference, Amman, United Nations Economic and Social Commission for Western Asia (ESCWA)

21. El-Mowelhi N, El-Bershamgy A, Hoffman GJ, Chang AC (1988) Enhancement of crop yields from subsurface drains with various envelopes. Agric Water Manag 15(2):131–140

22. FAO (1994) Land degradation in south Asia: its severity, causes and effects upon the people. Food and Agriculture Organization of The United Nations, World Soil Resources Reports, Rome

23. Gammal EL HAA, Ali HMM (2010) Commissioning of abandoned drainage water reuse systems in Egypt: a case study of upgrading the Umoum project, Nile Delta. Irrig Drain 60(1)115–122

24. Gates TK, Grismer ME (1989) Irrigation and drainage strategies in a salinity affected region. J Irrig Drain Eng 115(2):87–99 **128(2)**

25. Ghassemi F, Jakeman AJ, Nix HA (1995) Salinization of land and water resources. University of New South Wales Press, Sydney

26. Ghumman AR, Ghazaw YM, Niazi MF, Hashmi HN (2010) Impact assessment of subsurface drainage on waterlogged and saline lands. J Environ Monit Assess 172:189–197

27. Ghumman AR, Ghazaw YM, Hashmi HN, Kamal MA, Niazi MF (2012) Environmental and socio-economic impacts of pipe drainage in Pakistan. Environ Monit Assess 184:1671–1681

28. Goel A, Tiwari P (2013) Review of computer based software tools for salinity management in agricultural lands. J Indian Water Resour Soc 33(4):25–32

29. Gupta SK (2002) A century of subsurface drainage research in India. Irrig Drain Syst 16:69–84

30. Gupta SK (2003) A historical perspective of subsurface drainage research in India. Asian Agric Hist 7(1):45–54

31. Hanumanthaiah CV, Satyanarayana TV (2003) Cost effectiveness of subsurface drainage technology under Indian conditions—a boon to farming community in the areas of saline soils, paper no 118. Presented at the 9th international drainage workshop, Utrecht, The Netherlands

32. Hefny MM, Mohamed E, Metwali R, Mohamed AI (2013) Assessment of genetic diversity of sorghum (Sorghum bicolor L. Moench) genotypes under saline irrigation water based on some selection indices. Australian J Crop Sci (AJCS) 7(12):1935–1945

33. Hirekhan M, Gupta SK, Mishra KL (2007) Application of Wasim to assess performance of a subsurface drainage system under semiarid monsoon climate. J Agric Water Manag 88:224–234

34. Hornbuckle JW, Christen EW, Faulkner RD (2012) Analytical solution for drainflows from bilevel multiple-drain subsurface drainage systems. J Irrig Drain Eng 138:642–650

35. IPTRID SECRETARIAT (2007) Egypt's experience in irrigation and drainage research uptake, final report. Electronic Publishing Policy and

Support Branch, Information Division, Food and Agriculture Organization, FAO, Viale delle Terme di Caracalla, 00153, Rome, Italy

36. Jaiswal SR, Dhruva Narayana VV (1972) Design of a sub-surface drainage system. In: Proceedings symposium on waterlogging causes and its prevention CBIP, vol 110(1), pp 91–100

37. Joint Completion Report on IDNP Result (2007) Recommendation on waterlogging and salinity control based on pilot area drainage research, p 90

38. Kahlown MA, Khan AD (2004) Tile drainage manual, Pakistan Council of Engineering in Water Resources (PCRWR), Islamabad, Pakistan, ISSB 969-8469-13-3, PCRWR Press, Islamabad

39. Kahlown MA, Marri MK, Azam M (2007) Design, construction and performance evaluation of small tile drainage systems in the Indus Basin. J Irrig Drain ICID 56(1):217–225

40. Kale S (2011) Field-evaluation of DRAINMOD-S for predicting soil and drainage water salinity under semiarid conditions in Turkey. Spanish J Agric Res 9(4):1142–1155

41. Kaledhonkar MJ, Gupta SK, Singh G.(2009) An overview of research, development, opportunities and challenges for subsurface drainage in India. In: 60th international executive council meeting & 5th Asian regional conference, New Delhi, India

42. Kaledhonkar MJ, Kamra SK, Sharma DP (1998) Drainage measures for control of waterlogging in semi-arid regions. In: Proceedings 9th National Symposium on Hydrology, Amritsar, 26-27th Nov. Amritsar, pp 97–104

43. Kamble BM, Chougule BA, Rathod SD, Rathod PK (2006) Improvement of salt affected-water logged soils in western Maharashtra through subsurface drainage (Ssd) system. Asian J Biol Sci 1(2):64–67

44. Kamra SK, Singh SR, Rao KVGK, van Genuchten MTh (1991) A semi-discrete model for water and solute movement in tile-drained soils 1. Governing equations and solution. Water Resour Res 27:2439–2447

45. Kelleners TJ, Kamra SK, Jhorar RK (2000) Modeling of drainage water salinity of pipe drains. J Hydrol 234:249–263

46. Kolekar OL, Patil SB, Rathod SD (2014) Effects of different mole drain spacings on the yield of summer groundnut. Int J Res Eng Technol 3(4):2321–7308

47. Kornecki TS, Fouss JL, Grigg BC, Southwick LM (2001) Drainage research to improve runoff water quality and soil trafficability. Land Water Mag 45(3):10–13

48. Lal M (2011) Impact of subsurface drainage on land quality, ground water quality and crops yields in Haryana. PhD, Thesis, National Institute of Technology Kurukshetra, Haryana, India

49. Lovas L (1972) Salinity control trials by sub-soil drainage at Digod. In: Proceedings All India symposium on soil salinity held at Kanpur, pp 156–160

50. Luthin JN (1978) Drainage Engineering. R. E. Krieger Publishing Co., Huntington

51. Maity SP, Ghosh RK, Das NC, Patra SK (1996) Drainage and Crop Productivity in Saline Waterlogged Soil. AICRP on agricultural drainage. Bidhan Chandra Krishi Vishwavidyalaya. Kalyani, West Bengal, vol. 55, p 11

52. Mandal AK, Sharma RC (2011) Delineation and characterization of waterlogged salt affected soils in IGNP using remote sensing and GIS. J Indian Soc Remote Sens 39(1):39–50

53. Mathew EK, Nair M, Raju TD (1993) Drainage and Rice Productivity. AICRP on Agricultural Drainage. Kerala Agricultural University, Karumady, p 27

54. Michael AM (1967) Drainage research scheme. Annual report. College of Agricultural Engineering Punjab. Agricultural University, Ludhiana. p 108

55. Michael AM, Ojha TP (2013) Principles of Agricultural Engineering. Jain Brothers (New Delhi) 2(4):383–395

56. Naz BS, Ale S, Bowling LC (2009) Detecting subsurface drainage systems and estimating drain spacing in intensively managed agricultural landscapes. Agric Water Manag 96(4):627–637

57. Niazi MFK (2008) Evaluating the impact of subsurface drainage on land and water. Ph.D. thesis, University of Engineering and technology, Taxila, Pakistan

58. Nijland HJ, Croon FW, Ritzema HP (2005) Subsurface drainage practices: guidelines for the implementation, operation and maintenance of subsurface pipe drainage systems. ILRI Publication 60, Wageningen, p 607

59. Ozturk A (2004) Multi-objective drainage requirement of Harran Plain. Pak J Biol Sci 7(2):150–154

60. Pali AK, Katre P, Khalkho D (2014) An unsteady subsurface drainage equation incorporating variability of soil drainage properties. Water Resour Manag 28:2639–2653

61. Patel RM, Prasher SO, Bonnell RB, Broughton RS (2002) Development of comprehensive soil salinity index. J Irrig Drain Eng 128(3):185–188

62. Qureshi AS, Iqbal M, Anwar NA, Aslam M, Ramzan MC (1997) Benefits of shallow drainage. In Proceedings of seminar on on-farm salinity, drainage, and reclamation. IWASRI Publication, p 179

63. Ramana RKV, Kishore R, Singh R (2009) Mole drainage to enhance soybean production in waterlogged Vertisols. J Agric Eng 46(4):54–58

64. Rao KVGK, Singh OP, Gupta RK, Kamra SK, Padey RS, KumbharePS, Abrol IP (1986) Drainage investigations for salinity control in Haryana. Bull. no. 10. CSSRI, Karnal, India. p 95

65. Ritzema H (2007) Performance assessment of subsurface drainage systems case studies from Egypt and Pakistan. Alterra-ILRI Wageningen University and Research Centre Wageningen The Netherlands, p 137

66. Ritzema PH (2009) Drain for gain, making water management worth its salt, subsurface drainage practices in irrigated agriculture in semi-arid and arid regions. Ph.D. Dissertation, Wageningen University and UNESCOIHE Institute for Water Education, Wageningen, The Netherlands

67. Ritzema HP, Satyanarayana TS, Raman S, Boonstra J (2008) Subsurface drainage to combat waterlogging and salinity in irrigated lands in India: lessons learned in farmers' fields. Agric Water Manag 95:179–189

68. Sarwar A (2000) A transient model approach to improve on-farm irrigation and drainage in semi-arid zones. Ph.D. Dissertation, Wageningen University & Research Centre, The Netherlands

69. Sarwar A, Feddes RA (2000) Evaluation drainage design parameters for the fourth drainage project, Pakistan by using SWAP model: Part ii-modeling result. Irrig Drain Syst 14:281–299

70. Sarwar T, Muhammad N (2008) Effect of enhanced water allowance on the performance of tile drainage system at Shabaz Ghari pilot project area Swabi SCARP. Sarhad J. Agric 24(1):59–68

71. Satyanarayana TV, Lakshmi GV, Srinivasulu A, Hanumanthaiah CV, Ratnam M, Hemakumar HV (2002) Drainage and water management for salinity control in canal commands. A comprehensive report of research achievements of Bapatla Network Center, Indo-Dutch Network Operational Research Project, Acharya N.G. Ranga Agricultural University, Bapatla, India

72. Sethi M, Bundela DS, Sethi M, Lal K, Kamra SK (2010) Remote sensing and geographic information system for appraisal of salt-affected soils in India. J Environ Qual 39(1):5–15

73. Shakya SK, Gupta, PK Kumar D(1995) Innovative drainage techniques. AICRP on drainage. Punjab Agricultural University, Ludhiana, Punjab. p 34

74. Sharma RK, Yadav RD, Mourya BM, Kushwaha SS, Sharma AK (1995) Research on agricultural drainage in black soils of Barna command. AICRP on agricultural drainage. Jawaharlal Nehru Krishi Vishwavidyalaya, Powerkheda, Madhya Pradesh, India. p 35

75. Shekhawat RS (2007) Economic analysis of sub-surface drainage under Indira Gandhi Nahar Priyojna command area—A case study. Agric Econ Res Rev 20:361–374

76. Smedema LK(1990). Comparative study of the East Khairpur, Mardan and drainage-IV pipe drainage project. Project: PAK/83/022. Technical report no. 90/01, IWASRI Pub. No. 8, Lahore, Pakistan

77. Srinivasulu A, Satyanarayana TV, Kumar HVH (2005) Subsurface drainage in a pilot area in Nagarjuna Sagar right canal command, India. Irrig Drain Syst 19:61–70

78. Srinivasulu M, Rao AR, Babu MR, Babu GR, Kumar HVH (2014) Performance of subsurface drainage system installed at Appikatla region of Andhra Pradesh. Int J Argric Eng 7(1):180–185

79. Thatte CD, Kulkarni SA (2000) Status and strategy for prevention and reclamation of waterlogged and salt-affected lands from irrigated areas of India. 8th ICID international drainage workshop, New Delhi

80. Tiwari P (2011) Simulation of salt and water dynamics in subsurface drained land using hydrus. M.Tech. Dissertation, College of Agricultural Engineering, JNKVV Jabalpur, Madhya Pradesh India

81. Yadav JSP (1975) Annual Report. All India coordinated scheme for research on water management and salinity (unpublished)

82. Zimmer D, Bermond M, Melliand ML, Tato-Serrano S (2002) Global drainage status and needs, private communication, report will be published later by Cemagref/ICID

CHAPTER 2

Thermal effect of a borehole thermal energy store on the subsurface

Philipp Mielke1*, Dan Bauer2, Sebastian Homuth1, Annette E Götz1 and Ingo Sass1

1Department of Geothermal Science and Technology, Institute of Applied Geosciences, Darmstadt University of Technology, Schnittspahnstrasse 9, Darmstadt 64287, Germany. 2Institute of Thermodynamics and Thermal Engineering, University of Stuttgart, Pfaffenwaldring 6, Stuttgart 70550, Germany.

ABSTRACT

Background

The thermal effect on the subsurface of a large borehole thermal energy store (BTES) has been investigated by coupling measured rock properties with an enhanced FEFLOW simulation.

Methods

The finite element model has been validated against measured data from a 2-year operation period. The thermal changes in the subsurface have been predicted by simulation for a 30-year operation period. The model is based on three 80-m core sections drilled in Triassic carbonates, which have been analyzed in detail with respect to lithology, facies, and thermal and hydraulic parameters.

Results

The model shows thermal effects of the BTES on the subsurface at a distance of approximately 350 m after 10 years and a distance of approximately 850 m after 30 years of operation. At a distance of 100 m, the temperature of the subsurface rises by 2 K after 30 years.

Conclusions

The simulation describes the real BTES in an accurate manner and is suited for predicting the thermal changes in the subsurface for long-term operational durations.

KEYWORDS

Borehole thermal energy store; Borehole heat exchanger system; Heat and mass transport model; Geothermal parameters; Triassic carbonates

1. BACKGROUND

The present work intends to predict the thermal effect of a borehole thermal energy store (BTES) on the subsurface in the case of flowing groundwater by simulation with the software FEFLOW. Predicting thermal effects on the subsurface or grouting using finite element modeling has been done by other authors before (Wołoszyn and Gołaś [2013]; Rees and He [2013]; Wagner et al. [2013]; Zhang et al. [2011]), but the properties of the surrounding geological formation itself are generally simplified or focus on the grouting material only. Predicting thermal effects on the subsurface is essential for construction approvals by authorities and should not exceed a threshold value of 6 K ([VDI-Richtlinien 2011]). In this study, a detailed examination of the geology has been included into the FEFLOW model. The model has been further enhanced by integrating a newly developed add-on, which significantly reduces computation times and allows the coupling of TRANSYS with FEFLOW (Bauer et al. [2011]). The resulting model has then been validated against measured data spanning a 2-year period of operation of the BTES in Crailsheim.

1.1 Borehole thermal energy store in Crailsheim

A BTES is an underground heat exchanger system used to store large quantities of excess heat. A common application is the storage of heat from a combined heat and power plant or the capture of solar heat in summer for later use in winter. BTES consists of multiple arranged and interconnected borehole heat exchangers (BHEs) installed with a spacing of 2 to 5 m (VDI-Richtlinien [2001]). The thermal interaction of the individual borehole heat exchangers is essential for an efficient storage process.

The investigated BTES was constructed in 2008 in Crailsheim, Germany (Figure 1) (Riegger [2008]) and forms a major part of a solar district heating system (Bauer et al. [2008]). The BTES consists of 80 double U-tube BHEs, drilled to a depth of 55 m and separated at a distance of 3 m to each center. The well pad occupies a circular area of 30-m diameter. Each BHE is installed in a 0.13-m borehole and is cemented in place with thermally enhanced grout. The BTES is covered with 0.4 m of a thermal insulation material and 1.6 m of elevated soil from a nearby earthwork. The BTES in Crailsheim operates with pure water as working fluid. The BHEs are interconnected by pairs, with one BHE being situated more inside of the BHE grid and one more outside (Figure 2). The water flow direction is from the inner BHE to the outer during heat injection and the other way round during heat extraction.

The BTES was monitored in detail during a 2-year operation period. The flow rate within the pipes from and to the heat storage, as well as the in- and outflow temperatures, is logged continuously during the operation phase of the system. From this data, the loading and the unloading heat amount can be calculated. Furthermore, 24 temperature sensors have been installed at 12 different depths on measuring lances M1 and M31 (see Figure 2). Both measuring lances record temperatures of the solid underground up to a depth of 80 m.

In addition to monitoring the flow rates and fluid temperatures, groundwater temperatures and levels are measured in several groundwater monitoring wells. GWM2, GWM3, and GWM4 are 80 m below ground surface (aquifer in the Upper Muschelkalk), and in GWM2a, GWM3a, and GWM4a, wells are 20 m below ground surface (aquitards in the Lower Keuper) (see Figure 3).

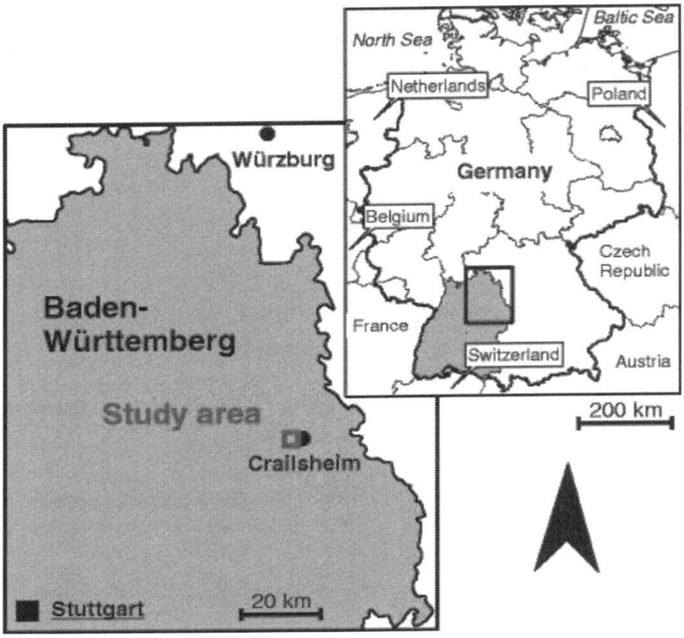

Figure 1. Map showing the study area near Crailsheim.

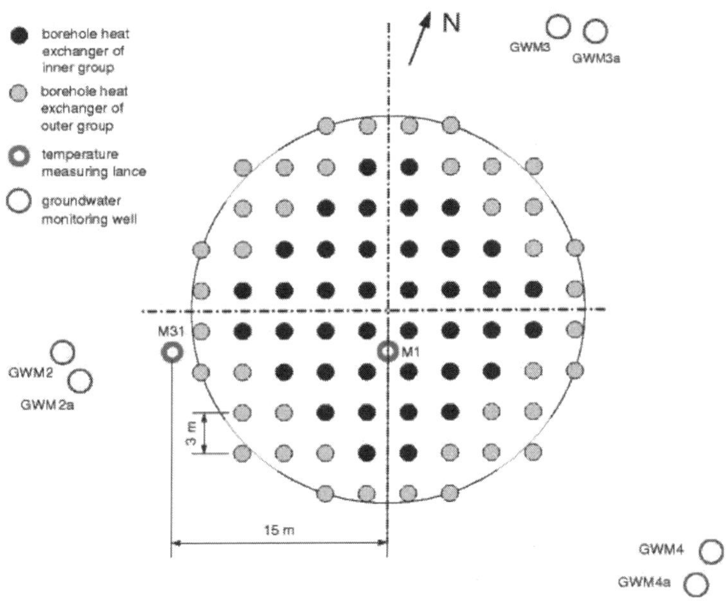

Figure 2. Layout of Crailsheim BTES with position of groundwater monitoring wells and temperature measuring lances.

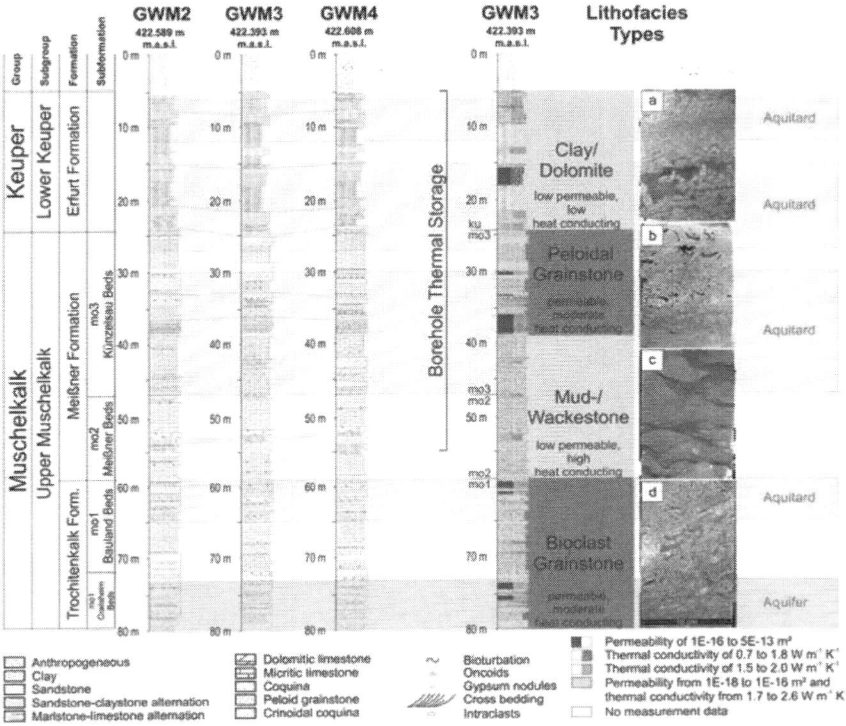

Figure 3. BTES Crailsheim well section - stratigraphy, facies, and hydrogeology parameters. (a) Clay/dolomite. **(b)** Peloidal grainstone. **(c)**Mud/wackestone. **(d)** Bioclast grainstone.

1.2 Geological setting

The study area is part of the Western Germanic Basin. During Triassic times, the Germanic Basin was situated in a peripheral position towards the Tethys spreading center and also affected by syndepositional tectonism. Generally, two main systems governed the basin's tectonic evolution. The first, mainly in the eastern and southern parts of the basin, was controlled by reactivated Variscan structures; the second developed in the NW part of the basin related to the North Sea rifting belt (Feist-Burkhardt et al. [2008]). During Middle Triassic times, the Variscan structures were reactivated and thereby influenced the sedimentary processes and tectonic subsidence within the study area. Today's fracture system resulted from the stress field that has developed since the Late Cretaceous. Main fracture directions are NNE-SSW and NE-SW oriented. These two main directions were also measured in the quarries Heldenmühle and Neidenfels near Crailsheim showing a dominant NE-SW orientation.

The study site is located in the Hohenloher Ebene which composes the largest Muschelkalk karst area in Germany as part of the South German Scarplands. The Hohenlohe area is dominated by sedimentary rocks of Triassic age. The stratigraphic formations are (from the oldest to the youngest): the Middle Triassic Muschelkalk subdivided into Lower Muschelkalk (Wellenkalk), Middle Muschelkalk, and Upper Muschelkalk (Hauptmuschelkalk), and the Upper Triassic Keuper which is subdivided into Lower Keuper (Lettenkeuper), Middle Keuper (Gipskeuper), and Upper Keuper. Bedding is nearly horizontal with a dip of about 3° SE.

The Muschelkalk in the Hohenloher Ebene reaches a total thickness of 250 m with Lower, Middle, and Upper Muschelkalk series of about 80-m each. The Upper Muschelkalk is overlain by a 10- to 30-m-thick series of sandstones and claystones of the Lower Keuper.

1.3 Hydrogeology

Hydrogeologically, drill logs and flow meter measurements performed in all groundwater monitoring wells suggest multiple aquitards separated by layers of clay in the topmost 50 m of the groundwater monitoring wells. Fracture-controlled aquifers in the drilled zone exist in depth deeper than 50 m. Groundwater level measurement data indicates a flow direction of NE-SW for the fracture-dominated aquifer. This direction is supported by the general NE-SW strike of the Upper Muschelkalk. The flow direction of the aquitards can only be inferred on a regional scale to be SSW-NNE.

No regional groundwater flow rates were available for this area. Flow meter measurements in the wells could not detect any significant flow rates, neither through fractures nor through matrix. However, studies by the Bavarian Environment Agency ([2007]) in southwest Germany suggest hydraulic conductivities of 1.0×10^{-5} to 1.0×10^{-4} m s^{-1} for the Middle and Upper Muschelkalk and 1.0×10^{-6} to 1.0×10^{-4} m s^{-1} for the Lower Keuper. Pumping and recharge tests conducted at the six monitoring wells returned transmissivities in a range of 1.9×10^{-7} m^2 s^{-1}to 5.1×10^{-5} m^2 s^{-1}. Storage coefficients (S_S) returned from hydraulic tests in all groundwater monitoring wells imply unconfined groundwater conditions (S_S 8.0×10^{-3} m^{-1} to 2.0×10^{-2} m^{-1}) for the aquifer in the Upper Muschelkalk and confined groundwater conditions (S_S 6.0×10^{-5} m^{-1} to 3.0×10^{-4} m^{-1}) for the aquitards in the Lower Keuper.

1.4 Well logs

The geological formations penetrated by the 55-m-deep BTES belong to the Upper Muschelkalk and the Lower and Middle Keuper. The Middle Muschelkalk was reached by an adjacent exploration well at about 100-m depth. Groundwater monitoring wells GWM2, GWM3, and GWM4 have been cored over the whole length (80 m each); thus, the lithology of all drill cores can be logged in detail (Figure 4). Stratigraphic correlation of all three groundwater monitoring wells is very good, whereas most marker horizons commonly differ in less than 2 m in intersected depths. Additionally, the petrography of GWM3 has been examined in detail using thin sections and acetate peels of representative stratigraphic intervals. Stratigraphic correlation of all three wells was possible since they are located within an area of about 150 m^2 in horizontally layered formations.

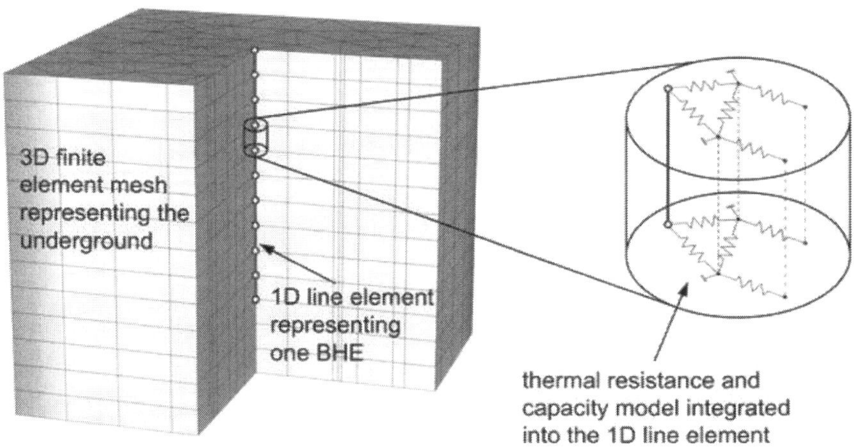

Figure 4. Borehole heat exchanger modeled as 1D line element in a 3D finite element mesh.

GWM3 starts with coquina beds of the Crailsheim Beds (Trochitenkalk Formation, mo1) in 80-m depth, extending up to about 72-m depth below ground surface. The Crailsheim Beds are characterized by a limestone-marl alternation with peloidal packstones. The following Bauland Beds range up to about 59 m and consist of up to about 63 m of clayey limestone with intercalated microcrystalline packstones. The topmost 4 m of the Bauland Beds comprises bioclastic packstones, grainstones, and rudstones.

The following Meißner Beds (mo2) form the lowest subunit of the Meißner Formation ranging up to about 47 m. This unit primary consists of packstones, rudstones, and coquina. Clay horizon β, a regional marker horizon, is located at about 53-m depth. The Meißner Beds are separated from the overlaying Künzelsau Beds (mo3) by the clay horizon γ, a second marker horizon in this area. Künzelsau Beds consist of micritic limestones with intercalated microcrystalline packstones and gypsum nodules. A third marker horizon, clay horizon δ, is located at about 44-m depth. Fine-grained coquina beds in 41-m depth are followed by bioclastic and peloidal grainstones up to about 35-m depth. The topmost 10 m of the Meißner Formation are formed by limestone with intercalated dolomitic marls, micritic limestone, and coquina. The boundary between the Upper Muschelkalk and the Lower Keuper is located at about 24-m depth. The Lower Keuper starts with dolomitic marl, sandstone, limestone, and intercalated claystone. At about 15-m depth, massive sandstones are present, followed by layers of marl and dolomite. The topmost 4 m of the well is formed by anthropogeneous clay.

1.5 Modeling

In general, the geological and hydrogeological settings at the site are very suitable for validating the recently developed FEFLOW add-on in conjunction with a BTES (Bauer et al. [2011]). Geological and hydrogeological settings such as the low, but still measurable, groundwater flow through the BTES and the aquifer below the BTES with its higher flow rate enable the investigation of the thermal effect of a BTES on the subsurface and the verification of the developed simulation software.

The developed simulation software is an add-on to the finite element flow and transport modeling tool FEFLOW (Zheng [2007]; Diersch et al. [2011a, b]). The add-on is capable of taking groundwater flow into account for large borehole heat exchanger fields. Borehole heat exchangers are modeled in FEFLOW as one-dimensional line elements in a three-dimensional finite element model of the underground. Heat and mass transport processes inside the BHE are modeled based on thermal resistance and capacity models (Bauer et al. [2011]). Discretizing the surrounding ground with a relatively coarse mesh (compared to a three-dimensional discretization of a BHE) means that the number of finite elements can radically be reduced; this saves computation time during simulation. BTES can be modeled and simulated in an efficient and variable manner with this technique. Figure 4 shows a one-dimensional line element representing a borehole heat exchanger in the three-dimensional finite element modeling of the surrounding ground.

2. METHODS

The thermal conductivity and thermal diffusivity, permeability, porosity, and density have been measured on 76 representative samples of drill cores from GWM3. Conventional methods for determining permeability commonly return an integrative result that refers to the whole sample. A gas pressure permeameter permits a numerous number of pointwise measurements, which can be assigned to the individual points of a rock sample.

2.1 Laboratory measurements

Permeability measurements were performed with a combined columnar and minipermeameter after Jaritz ([1999]) and Filomena et al. ([2013]). The minipermeameter is a compressed air-driven gas permeameter. It allows a fast and precise measurement of the permeability at the surface of a rock sample in comparison to other permeability-measuring techniques. The sample's surface can be plane or curved (e.g., a drill core), but has to be smooth. The measurement is fast and nondestructive. Multiple measurements have been done on each sample (scattered over all suitable surfaces), and the readings have been averaged to get one value per sample.

Furthermore, a columnar permeameter is used to validate the results of the minipermeameter. The columnar permeameter requires test slices that are sawed off from selected drill cores. It is necessary to consider moisture and temperature of the ambient air when measuring permeability. For this reason, the measuring accuracy is significantly improved by using ovendry samples. Similarly, earth-moist samples cannot be measured. A detailed study on the technique and accuracy of gas permeameters, including the device used for this study, can be obtained from Filomena et al. ([2013]).

Porosity and density have been measured using a commercial gas-driven pycnometer combined with a displacement-measuring technique that utilizes a fine-grained powder as quasi-fluid for the determination of the sample volume. The method allows fast but highly accurate measurements (with 0.03% accuracy and 0.02% reproducibility) of high quantities of samples, but only returns the gas-effective porosity.

Thermal conductivity and thermal diffusivity were measured on ovendry samples applying the optical scanning method after Popov et al. ([1999]). The method allows relatively fast measurements of samples showing differing quality with an accuracy of 3% (Bär et al. [2011]).

2.2 Model setup

The heat and mass transport model was created using the numerical simulation modeling software for subsurface flow and transport processes FEFLOW 6.0. The model dimension extends over 1.5 km in the downstream direction and 500 m in the upstream direction along the aquifer. The width of the model was set to 1 km, with the BTES being placed in the middle. The depth of the model was set to 150 m with the BTES ranging from 0 to 55 m in depth.

Model layers and their thickness have been chosen individually to fit the geological realities. However, as some sections have an intensely interbedded structure with layers often changing every few centimeters (e.g., the Bauland Beds, Meißner Beds, and Künzelsau Beds), they could not be modeled in detail. Consequently, multiple single geological layers had to be merged into reasonable model layers. The maximum discretization level for the vertical element heights was set to 1 m and the minimum to 10 m. Generally, layers in the vicinity of the BTES (i.e., the topmost 75 m) have a thickness of 1 to 4 m while layers far below the BTES have a thickness of 5 to 10 m. The upper part (about 5 m) of the boreholes has a different diameter and is filled with a different grouting material (Riegger [2008]). This was not modeled as only a minor effect on the thermal behavior of the store is expected. The covering of the BTES, consisting of a thermal insulation material (0.4 m) and elevated soil from a nearby earthwork (1.6 m), was modeled by two additional continuous layers that taper off to 0.01 m each outside the BTES. The predefined mesh size changes with distance towards the BTES. Element sizes around each individual BHE are finest with horizontal triangle areas of 0.02 to 0.08 m^2, increasing to 0.1 to 0.4 m^2 in the interspace between each BHE and increasing further to 0.5 to 2 m^2 in an oval-shaped area that encloses the BTES and points in the downstream direction. The model has its coarsest elements (200 to 500 m^2) in the remaining area. The total model consists of 57 layers and about 1,445,000 finite elements (Figure 5).

Groundwater flow directions were defined by setting suitable hydraulic heads as a boundary condition of the first kind at the model borders at the depth of the aquitards and the aquifer. The hydraulic gradient of the aquitards was set to 0.01, while the aquifer has a value of 0.0045. These values were given by the Regional Authorities for Geology, Raw Materials and Mining as typical values for the site.

Based on the available measuring results, hydraulic conductivities were assigned to the individual model layers. Measurements were averaged when multiple measurements exist within the range of an individual model layer. For layers deeper than 80 m, data from unpublished internal studies were used.

After setting up the hydraulic parameters, the initial hydraulic conditions for the model were simulated in steady flow mode (Table 1).

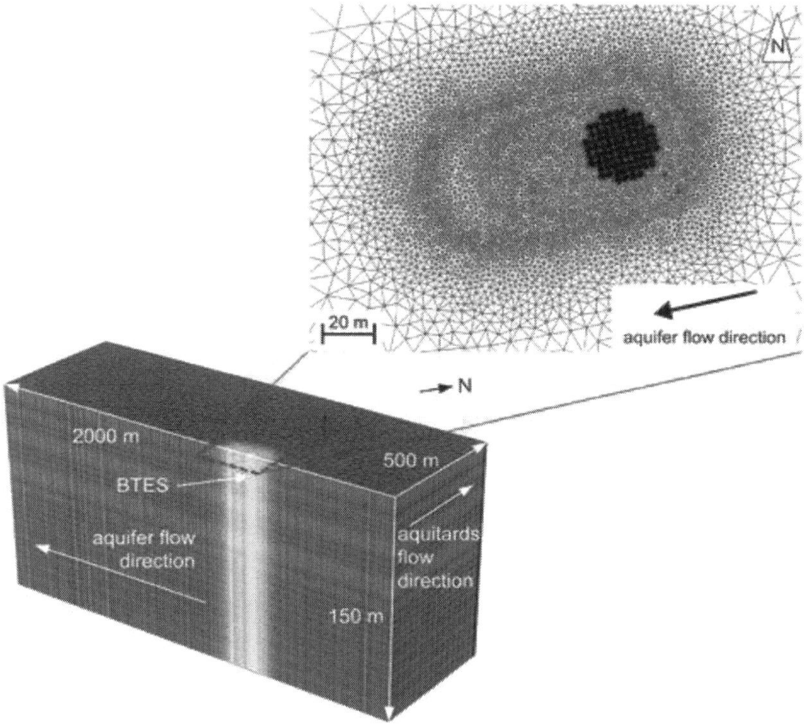

Figure 5. Model meshing and layer dimensions of the 3D finite element mesh. Vertical cut through the middle of the thermal storage along the aquifer flow direction; vertical exaggeration, ×5.

Subsequently, the unsteady heat transport simulation was then set up by assigning porosity, thermal conductivity, and heat capacity to each layer of the flow model. Measurement data of the thermal conductivity and heat capacity of the solids were measured on dry samples and thus have to be converted to fully saturated conditions. In FEFLOW, this can be achieved by inputting the thermal properties of the fluid (i.e., the groundwater), which were set to $4.19 \times 10^{+6}$ J m^{-2} K^{-1} for the volumetric heat capacity and 0.6 W m^{-1} K^{-1} for the thermal conductivity.

Table 1. Layer dimensions and input data of the FEFLOW model

Depth (m.b.g.s)	Layer number	Layer height (m)	Hydraulic conductivity ($m\ s^{-1}$)	Thermal conductivity ($W\ m^{-1}\ K^{-1}$)	Volumetric heat capacity ($J\ m^{-3}\ K^{-1}$)	Porosity (%)	Stratigraphic unit
−2	1	1.6	6.3×10^{-9}	2.1	2.2	0.01	Elevated soil
−0.4	2	0.4	6.3×10^{-9}	0.8[a]	0.13[a]	1.0×10^{-6}[a]	Thermal insulation
0	3	1	6.3×10^{-9}	1	3	0.02	Anthropogeneous clay
1	4	1	6.3×10^{-9}	1	3	0.02	
2	5	1	6.3×10^{-9}	1	3	0.02	
3	6	1	6.3×10^{-9}	1	3	0.02	
4	7	1	6.3×10^{-9}	1	3	0.02	Lower Keuper
5	8	1	6.3×10^{-9}	1	3	0.05	
6	9	1	1.0×10^{-6}	1	3	0.05	
7	10	1	1.0×10^{-6}	1	3	0.05	
9	11	2	1.0×10^{-6}	1.5	3	0.05	
13	12	4	7.9×10^{-11}	1.5	3	0.01	
15	13	2	1.0×10^{-6}	1.5	3	0.01	
16	14	1	1.0×10^{-6}	1.5	3	0.01	
17	15	1	1.0×10^{-6}	1.5	3	0.01	
19	16	2	1.0×10^{-6}	2.1	3	0.01	
23	17	4	1.0×10^{-6}	2.1	3	0.02	
27	18	4	6.6×10^{-11}	2.1	3	0.02	Upper Muschelkalk Künzelsau Beds (mo3)
29	19	2	6.6×10^{-11}	2.1	3	0.03	
30	20	1	1.0×10^{-6}	1.9	2.8	0.03	
30	20	1	1.0×10^{-6}	1.9	2.8	0.03	
31	21	1	1.0×10^{-6}	2	2.8	0.03	
32	22	1	1.0×10^{-6}	2	2.8	0.03	
33	23	1	1.0×10^{-6}	1.8	2.6	0.03	
34	24	1	1.0×10^{-6}	2	2.6	0.03	
35	25	1	1.0×10^{-6}	2	2.6	0.03	
36	26	1	1.0×10^{-6}	1.9	2.7	0.03	
37	27	1	1.0×10^{-6}	1.9	2	0.03	
39	28	2	1.0×10^{-6}	2.1	1.6	0.01	
43	29	4	1.0×10^{-6}	2.1	1.6	0.01	
48	30	5	4.0×10^{-11}	2.1	1.7	0.02	Upper Muschlkalk Meißner Beds (mo:
52	31	4	4.0×10^{-11}	2	1.5	0.02	
54	32	2	6.2×10^{-11}	2	1.5	0.01	
55	33	1	6.2×10^{-11}	2	1.5	0.01	
56	34	1	1.8×10^{-11}	2.3	1.4	0.01	
57	35	1	1.8×10^{-11}	2.3	1.4	0.01	
58	36	1	1.8×10^{-11}	2.3	1.4	0.01	
59	37	1	1.0×10^{-5}	2	1.6	0.02	Upper Muschelkalk Bauland Beds (mo1
60	38	1	1.0×10^{-5}	2	1.6	0.02	
61	39	1	1.0×10^{-5}	2.3	1.4	0.02	
62	40	1	1.0×10^{-5}	2.3	1.4	0.02	
63	41	1	1.0×10^{-5}	2.3	1.4	0.02	

63	41	1	1.0×10^{-5}	2.3	1.4	0.02	
64	42	1	1.0×10^{-5}	2.3	1.4	0.02	
66	43	2	2.9×10^{-11}	2.3	1.4	0.02	
70	44	4	2.9×10^{-11}	2.3	1.4	0.02	
72	45	2	1.0×10^{-4}	2	2	0.02	Upper Muschelkalk Crailsheim Beds (mo1)
73	46	1	1.0×10^{-4}	1.7	1.5	0.02	
75	47	2	1.0×10^{-4}	2	1.6	0.02	
80	48	5	1.0×10^{-4}	2	1.6	0.02	
85	49	5	1.0×10^{-4}	2	1.5	0.02	
90	50	5	1.0×10^{-4}	2	1.5	0.02	
95	51	5	7.7×10^{-10}[b]	2[b]	1.5[b]	0.02[b]	Middle Muschelkalk[c]
100	52	5	7.7×10^{-10}[b]	2[b]	1.5[b]	0.02[b]	
110	53	10	7.7×10^{-10}[b]	2[b]	1.5[b]	0.02[b]	
120	54	10	7.7×10^{-10}[b]	2[b]	1.5[b]	0.02[b]	
130	55	10	7.7×10^{-10}[b]	2[b]	1.5[b]	0.02[b]	
140	56	10	7.7×10^{-10}[b]	2[b]	1.5[b]	0.02[b]	
150	57	10	7.7×10^{-10}[b]	2[b]	1.5[b]	0.02[b]	

m.b.g.s, meter below ground surface. [a]On top of the BTES only; for the remaining area, values are equal to layer 1; [b]from unpublished internal data; [c]no sample material, stratigraphy has been adopted from Hagdorn and Simon (1988).

Porosity measurements using a gas pycnometer return the gas-effective porosity and thus have to be converted to the water-effective porosity first. As no water-based measurements where possible due to decomposition and/or swelling of the samples, known published porosity data from similar rocks (Götz and Lenhardt [2011]) were used as a reference. By decreasing all porosity measuring results by a factor of 5, the resulting porosity magnitude accords quite well with the published values, while the porosity varieties of the measured samples are preserved.

The initial temperature distribution of the model has been calculated by using a fixed surface temperature of 10.5°C (Blocon [2014]) and a heat inflow of 0.065 W m^{-2} (Pollack et al. [1993]) at the base of the model. The resulting temperature gradient is about 3.5 K per 100-m depth.

The 80 BHEs have been integrated as a boundary condition of the fourth kind, which is capable of simulating injection and extraction at nodal points. Grout and pipe properties have been set according to data provided by the manufacturers. After setting up all required BHE parameters, all of the BHEs were interconnected pairwise using FEFLOW's internal BHE Integrator.

At first, the transient simulation was conducted over 2 years in order to validate FEFLOW against measured data from a 2-year period of operation. The BTES was only charged during the first 2 years. Then the measured

BHE thermal load was set as a boundary condition to the model (see Figure 6). Then the computed and measured temperatures at measuring lances M1 and M31, as well as groundwater temperatures at GWM2a to GWM4a and GWM2 to GWM4, were compared.

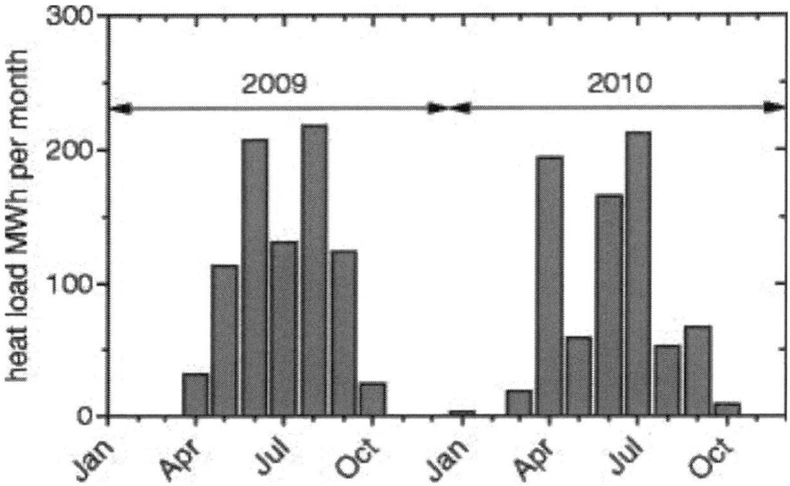

Figure 6. Measured BHE thermal load set as a boundary condition to the model for validation purpose.

Finally, the thermal changes in the subsurface were simulated over a time span of 30 years. The semiannually alternating operation of the BTES was simulated according to the planning with inflow temperatures of 75°C in the summer months (April to September) and 20°C in the winter months (October to March). Simulation duration using a common workstation (Quadcore with 2.66-GHz clock speed and 24-gigabyte internal memory) took about 12 h.

3. RESULTS AND DISCUSSION

Figure 7a,b depicts the comparison between measured and computed underground temperatures at measuring lances M1 (middle of the BTES) and M31 (margin of the BTES) of the 2-year validation run. Figure 7c,d depicts the comparison between measured and computed groundwater temperatures at groundwater monitoring wells GWM2a to GWM4a (20-m depth) and GWM2 to GWM4 (80-m depth).

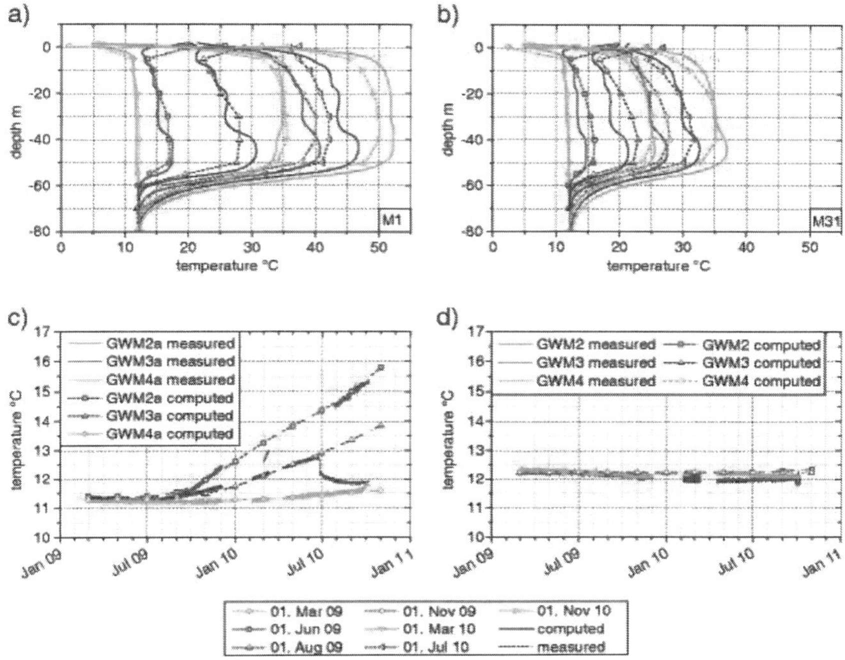

Figure 7. Measured and computed underground and groundwater temperatures. Measured and computed underground temperatures at measuring lances M1 **(a)** and M31 **(b)** and measured and computed groundwater temperatures at wells GWM2/a, GWM3/a, and GWM4/a at 20-m depth **(c)** and 80-m depth **(d)**.

It appears as though both, the temperatures of the solid ground and the groundwater, can be properly computed with the FEFLOW model. Minor differences in results should be accepted because the measured BHE thermal load set as a boundary condition to the model is slightly too high. This is because it includes the unknown thermal losses of the 300-m-long pipes to and from the BTES causing underground temperatures to be computed higher than normal at the end of the second year. Furthermore, it is not possible to install the BHE and the measuring lances absolutely vertical. Hence, an uncertainty of measurement due to inexact sensor positioning must be considered. The measured groundwater temperatures at GWM2a and GWM3a show an unexpected course in March 2010 and the end of June 2010, respectively. This may be explained by the introduction of surface water which led to an erroneous reading.

Altogether, it can be stated that the simulation model describes the real BTES in an accurate manner. It is suited for predicting the thermal changes in the subsurface for an operational duration of 30 years.

Figure 8 shows the computed temperature profiles in the subsurface in vertical cross sections parallel to the aquifer flow direction after 3, 10, and 30 years of operation. The results of the simulation show a clear impact of the BTES on the aquifer, while the aquitards are only slightly affected. Apparently, the transmissivity of the aquitards is too low to significantly impact the temperature profile.

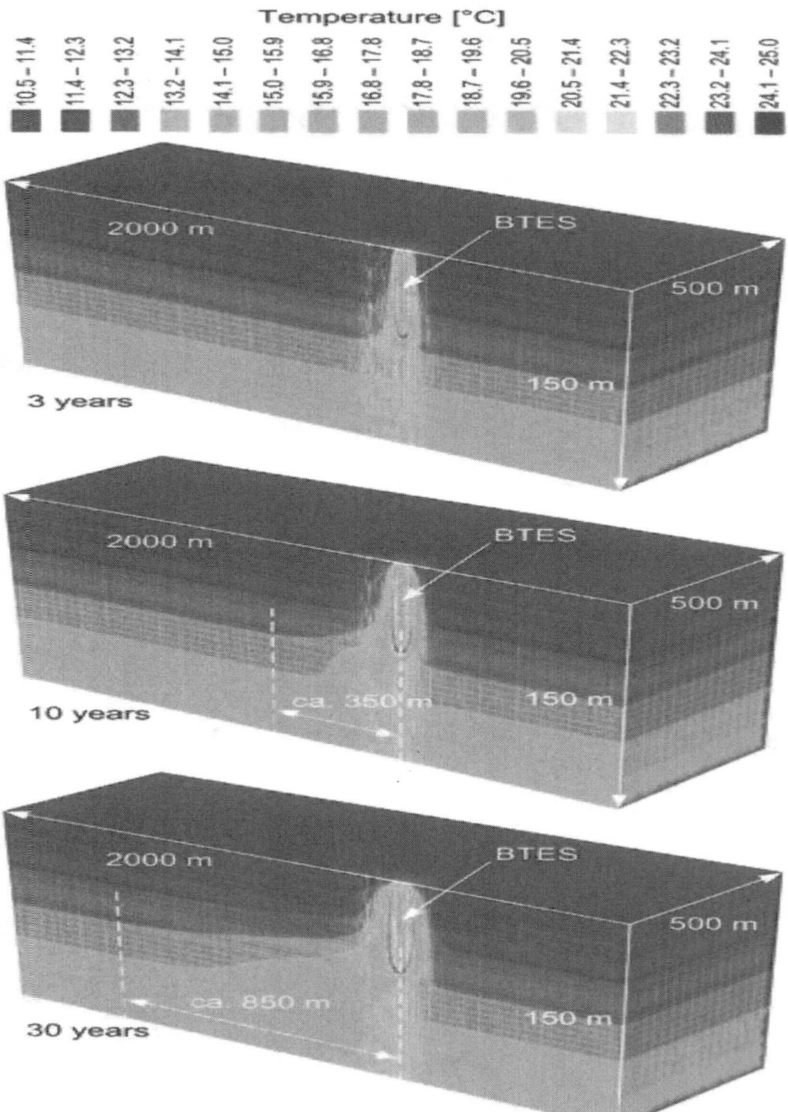

Figure 8. Computed temperature profiles in the subsurface in vertical cross sections parallel to the aquifer flow direction. After 3, 10, and 30 years of operation; vertical exaggeration, ×5.

After 3 years of operation, the advective-caused heat transport in the main flow direction is almost invisible. However, after 10 years, an obvious temperature propagation up from 350-m distance to the BTES is visible in the aquifer. The propagation increases up to 850 m after 30 years of operation. Nevertheless, the temperature change in this distance is below 1 K. In 100-m distance to the BTES, in the aquifer flow direction, the computed temperature rise at the most affected depth (71 to 76 m) is 2.1 K.

The temperature profile inside the BTES is only slightly affected by advective effects (Figure 9). This implies a low impact of the groundwater flow on the heat storage efficiency. Furthermore, results show a distinctly higher temperature spreading in the deeper parts of the BTES between a completely charged and discharged condition caused by the higher thermal conductivities of the Bauland Beds. By volume, this allows increased heat storage, albeit the heat loss increases, too.

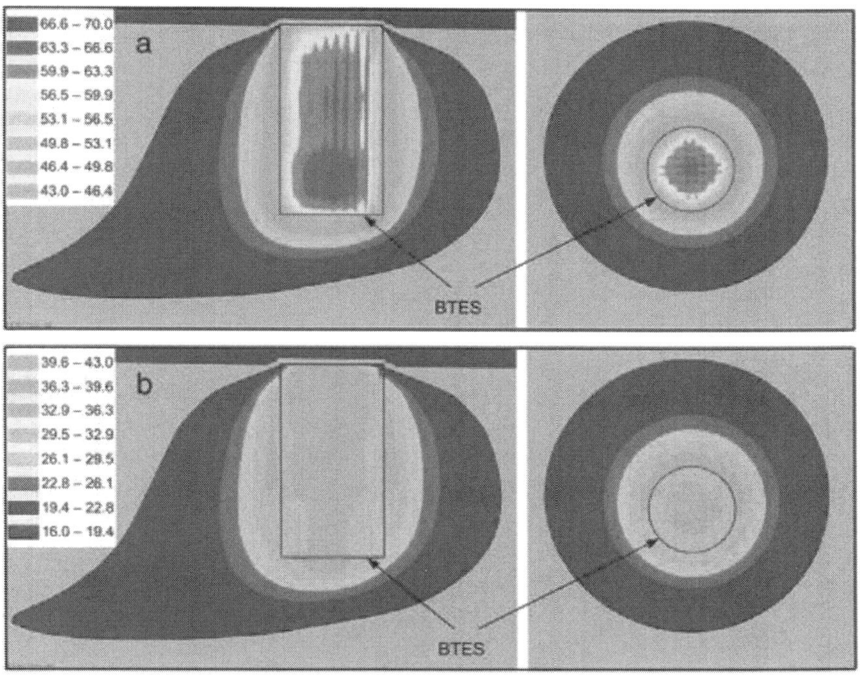

Figure 9. Computed temperature profiles of the BTES in the 30th year of operation. Completely charged **(a)** and completely discharged**(b)**; left: vertical cross section parallel to the aquifer flow direction; right: horizontal cross section at approximately 40-m depth; temperature scale in degrees Celsius.

4. CONCLUSIONS

Three-dimensional modeling and simulation of a BTES located predominantly in Middle Triassic carbonates with multiple aquitards at the upper part of the BTES and one aquifer below the BTES were carried out in order to determine the thermal effect of the BTES on the subsurface. As the BTES was modeled in detail with real technical and measured geological and hydrogeological data, the resulting model is of high quality. This was confirmed by validating the computed transient temperature changes of the solid underground and the groundwater against measured data from a 2-year period of operation.

Possible thermal changes in the surrounding subsurface have been predicted by simulation for an operation duration of 30 years. The model shows thermal effects of the BTES on the subsurface at a distance of approximately 350 m after 10 years and a distance of approximately 850 m after 30 years of operation. At a distance of 100 m, the temperature of the subsurface rises by 2 K.

All considered, the examined simulation program, FEFLOW, is a powerful tool for projecting the thermal effects of seasonal underground heat storage systems on the subsurface in a time-efficient manner. Thus, FEFLOW is of importance for authorities and investors for granting and planning similar systems in future projects.

ACKNOWLEDGEMENTS

This work is funded by the German Federal Ministry for Environment, Nature Conservation and Nuclear Safety. The authors gratefully acknowledge this support. Raw data for this study orginates from the unpublished diploma thesis "Faziesabhängigkeit geothermischer Kennwerte von Kalkstein am Beispiel des Oberen Muschelkalk (Crailsheim, Baden-Württemberg)" by Hanna Wicke (2009).

COMPETING INTERESTS

The authors declare that they have no competing interests.

AUTHORS' CONTRIBUTIONS

PM carried out the data analysis, modeling and wrote the manuscript. DB carried out the result analysis and participated in drafting the manuscript. SH participated in the design and coordination of the study and participated in writing the manuscript. AEG participated in the stratigraphic study. IS participated in design and coordination of the study and participated in drafting the manuscript. All authors read and approved the final manuscript.

REFERENCES

1. *Blocon (2014) EED 3.16. Earth Energy Designer. Blocon, Lund http://www.buildingphysics.com. Accessed 15 June 2011*

2. *Bär K, Arndt D, Fritsche J-G, Götz AE, Kracht M, Hoppe A, Sass I:* **3D modelling of the deep geothermal potential of the Federal State of Hesse (Germany) - input data and identification of potential.** *ZDGG 2011,* **162:** *371–388.*

3. *Bauer D, Heidemann W, Marx R, Nußbicker-Lux J, Ochs F, Panthalookaran V, Raab S: Solar unterstützte Nahwärme und Langzeit-Wärmespeicher (solar assisted district heating and seasonal thermal energy stores), Forschungsbericht zum BMU-Vorhaben 0329607J (Juni 2005 bis Juli 2008). University of Stuttgart, Stuttgart; 2008.*

4. *Bauer D, Heidemann W, Müller-Steinhagen H, Diersch H-JG:* **Thermal resistance and capacity models for borehole heat exchangers.***Int Journal of Energy Research 2011,* **35:** *312–320. Bavarian Environment Agency (2007) Hydrogeologische Übersicht Bayerns, Teilraum Muschelkalkplatten. . Accessed 11 Dec 2013, [*http://www.lfu.bayern.de/geologie/hydrogeologie_daten/hydrogeologische_raumgliederung/teilraum/doc/teilraum_muschelkalkplatten.pdf*]*

5. *Diersch H-JG, Bauer D, Heidemann W, Rühaak W, Schätzl P:* **Finite element modeling of borehole heat exchanger systems - part 1: fundamentals.** *Comput Geosci 2011,* **37:** *1122–1135.*

6. *Diersch H-JG, Bauer D, Heidemann W, Rühaak W, Schätzl P:* **Finite element modeling of borehole heat exchanger systems - part 2: numerical simulation.** *Comput Geosci 2011,* **37:** *1136–1147.*

7. *Feist-Burkhardt S, Götz AE, Szulc J, Borkhataria R, Geluk M, Haas J, Hornung J, Jordan P, Kempf O, Michalík J, Nawrocki J, Reinhardt L, Ricken W, Röhling H-G, Rüffer T, Török Á, Zühlke*

R: **Triassic.** *In Geology of Central Europe, volume 2 - Mesozoic and Cenozoic. Edited by: McCann T. Geological Society, London; 2008.*

8. *Filomena CM, Hornung J, Stollhofen H:* **Assessing accuracy of gas-driven permeability measurements: a comparative study of diverse Hassler-cell and probe permeameter devices.** *Solid Earth Discuss 2013,* **5:** *1163–1190.*

9. *Götz AE, Lenhardt N:* **The Anisian carbonate ramp system of Central Europe (Peri-Tethys Basin): sequences and reservoir characteristics.** *Acta Geol Pol 2011,* **61:** *59–70.*

10. *Hagdorn H, Simon T: Geologie und Landschaft des Hohenloher Landes. Jan Thorbecke, Sigmaringen; 1988.*

11. *Jaritz R: Quantifizierung der Heterogenität einer Sandsteinmatrix am Beispiel des Stubensandsteins (Mittlerer Keuper, Württemberg). Dissertation, Universität Tübingen. 1999.*

12. *Pollack HN, Hurter SJ, Johnson JR:* **Earth's interior: analysis of the global data set.** *Rev Geophys 1993,* **31:** *267–280.*

13. *Popov YA, Pribnow DFC, Sass JH, Williams CF, Burkhardt H:* **Characterization of rock thermal conductivity by high-resolution optical scanning.** *Geothermic 1999,* **28:** *253–276.*

14. *Rees SJ, He M:* **A three-dimensional numerical model of borehole heat exchanger heat transfer and fluid flow.** *Geothermics 2013,***46:** *1–13.*

15. *Riegger M:* **Saisonaler Erdsonden-Wärmespeicher in Crailsheim.** *bbr 2008,* **9:** *24–32.*

16. *VDI-Richtlinien (2001) Thermische Nutzung des Untergrundes - Unterirdische thermische Energiespeicher (thermal use of the underground - underground thermal energy storage). VDI 4640, Berlin*

17. *Wagner V, Blum P, Kübert M, Bayer P:* **Analytical approach to groundwater-influenced thermal response tests of grouted borehole heat exchangers.** *Geothermics 2013,* **46:** *22–31.*

18. *Wołoszyn J, Gołaś A:* **Modelling of a borehole heat exchanger using a finite element with multiple degrees of freedom.** *Geothermics 2013,* **47:** *13–26.*

19. *Zhang Y, Pan L, Pruess K, Finsterle S:* **A time-convolution approach for modeling heat exchange between a wellbore and surrounding formation.** *Geothermics 2011,* **40:** *261–266.*

20. *Zheng C:* **FEFLOW: a finite-element ground water flow and transport modeling tool.** *Ground Water 2007,* **45:** *525–528.*

CHAPTER 3

Detection of Surface and Subsurface Cracks in Metallic and Non-Metallic Materials Using a Complementary Split-Ring Resonator

Ali Albishi and Omar M. Ramahi *

Electrical and Computer Engineering Department, University of Waterloo, 200 University Avenue West, Waterloo, ON N2L 3G1, Canada

ABSTRACT

Available microwave techniques for crack detection have some challenges, such as design complexity and working at a high frequency. These challenges make the sensing apparatus design complex and relatively very expensive. This paper presents a simple method for surface and subsurface crack detection in metallic and non-metallic materials based on complementary split-ring resonators (CSRRs). A CSRR sensor can be patterned on the ground plane of a microstrip line and fabricated using printed circuit board technology. Compared to available microwave techniques for sub-millimeter crack detection, the methods presented here show distinct advantages, such as high spatial resolution, high sensitivity and design simplicity. The response of the CSRR as a sensor for crack detection is studied and analysed numerically. Experimental validations are also presented.

KEYWORDS

crack detection; complementary split-ring resonator (CSRR); microstrip line technology; near-field detection; metamaterials

1. INTRODUCTION

The responses of materials to electrodynamic fields depend on their electrical properties, such as permittivity, permeability and conductivity. Characterising such properties for the purpose of detecting changes in material is important in many areas, such as the food industry, bio-sensing applications and surface and subsurface crack detections, to name but a few [1–7]. Material defects, such as sub-millimeter cracks in metallic and non-metallic materials, affect the near-field distribution of the electric and magnetic fields. Such a change in the field distribution (magnitude and phase) is a consequence of change in the medium effective constitutive parameters. These changes can be used as signatures for crack detection purposes, which is the underlying physics behind the detection of cracks in metallic and non-metallic media in general.

Inspection for surface and subsurface cracks in metallic and non-metallic materials is critical for quality assessment and maintenance. Microwave techniques have been proposed and utilized for crack detection. In [8], an open-ended waveguide operating at around 20 GHz was used as a microwave technique for crack detection in metallic materials, where a 840 μm wide surface crack was detected by direct correlation with the change in the reflection coefficient. While capable of detecting sub-millimeter cracks, the waveguide gives rise to a small perturbation in the near field, leading to small changes in the reflection coefficient; thus, the demand for higher accuracy was needed [9]. In [10–12], methods were introduced to increase the sensitivity of open-waveguide sensors. In [9], the detection of micro-scale cracks measuring 200 μm in width in steel plates were studied using near-field microwave dual-behavior resonator (DBR) filters, where a frequency shift in the resonance frequency indicated the presence of a crack. In the DBR technique, two filter probes were used with operating frequencies of 13.577 GHz and 8.008 GHz. The sensor exhibited resonance frequency shifts between 39 and 51 MHz. However, the changes in the resonance frequencies were small, indicating low sensitivity. In addition, the minimum transmission and reflection coefficient (indicative of the presence of a crack) was spread over a relatively wide band, which made finding the resonance frequency difficult, especially if the sensor were to be integrated in a stand-alone detection system rather than a high precision spectrum analyzer.

Electromagnetic fields in the microwave regime can penetrate and interact with dielectric material, provided that the permittivity of the media is not excessively high. The result of the interaction can be used in many applications, such as detecting sub-millimeter cracks in dielectric material. Several microwaves techniques for crack detection in non-metallic material have been reported. In [13], an open-ended waveguide antenna was used to detect flaws (holes) ranging in diameters from 0.025 and 0.59 mm in ceramic materials (Si_4N_4) used in fabricating gas turbines. Since the technique in [13] is based on the far field, which is based on reflection and refraction, the perturbation in the size of the defects or anomalies in the material under test needs to be proportional to the wavelength, thus requiring operation in the W band (75 GHz to 110 GHz). Such operation frequency makes the measurement setup expensive and complex. To overcome this problem, microwave near-field techniques were introduced to detect cracks in non-metallic materials. Near-field techniques are dependent on the evanescent type of fields having effective wavelengths much higher than the wavelength in free space, which makes detection of particles having sub-wavelength features possible. In [12], an open-ended waveguide, similar to the one reported in [8], was used to detect cracks in cement-based materials where the basic detection principle is the complex interaction between the near field at the end of the waveguide and the material under test. To the best knowledge of the authors and based on the available literature, the available detection techniques for crack detection in metallic and non-metallic materials have at least one of the following challenges: (1) working at a relatively high frequency; (2) requiring a complex measurement setup; (3) a relatively high manufacturing cost; (4) low sensitivity; and (5) low resolution.

In [7], a method for sub-millimeter surface cracks in metallic material based on the near-field microscopy technique using a complementary split-ring resonator (CSRR) was introduced for the first time and was validated experimentally. The technique used a CSRR as a sensor, which was able to detect sub-millimeter surface cracks with a resolution of λ600 and to resolve two adjacent cracks separated by 1 mm. A recent paper by Taehwa *et al.* [14] used the same CSRR reported in [7] for crack detection in metallic material, but with a substrate integrated waveguide (SIW) for excitation.

In this paper, we study CSRR sensors for crack detection in metallic and non-metallic materials. We introduce design formulae that can be used by engineers and scientists to easily design CSRR sensors and predict their resonance frequencies. We then investigate inexpensive techniques to increase the sensitivity of CSRR sensors and to decrease the operating frequency. We focus on the surface current distribution as the underlying physical mechanism behind the CSRR sensor. We consider the effect of the geometrical features on the CSRR resonance frequency with the aim of achieving an overall small form factor. Finally, we show that the CSRR sensor is effective, not only for sensing cracks in metallic surfaces, but also non-metallic material, such as fiberglass, plastics, carbon fibers, amongst other dielectric materials that are increasingly used in aircraft structures and high performance applications [15].

2. COMPLEMENTARY SPLIT-RING RESONATOR AS A NEAR-FIELD SENSOR

Quasi-static resonators are designed such that the resonators' structures are electrically small compared with the electromagnetic field wavelength of the excitation source. The resonance phenomenon in quasi-static resonators is based on the effective tightly-distributed capacitance and inductance of the structure, unlike resonators of dimensions comparable to the wavelength, where the resonance phenomenon is based on phase propagation. Several works provided circuit models for CSRRs [16–18]. Generally, the resonance frequency can be formulated as.

$$f_0 = \frac{1}{2\pi\sqrt{L_T C_T}} \tag{1}$$

where L_T and C_T are the total distributed inductance and capacitance in the structure, respectively. There have been many designs used to realize quasi-static resonators, where loops create current circulations that directly contribute to the effective inductance and gaps affect the net capacitances. Examples include split-ring resonators (SRRs) [19], spiral resonators [20] and Hilbert curves [21]. Like SRRs, CSRRs are quasi-static resonators that were originally conceived of based on Babinet's principle. The CSRRs can be designed by creating the complement of the SRR, whereby copper is replaced by air and *vice versa*. Strictly speaking, the CSRR etched on a substrate is semi-complementary to the SRR due to the presence of the dielectric substrate essential for the construction of the sensor.

At the resonance frequency of a resonator, the electric and magnetic energy densities are enhanced significantly at certain locations in the close proximity of the resonator [19]. Any disturbance of the electromagnetic fields around the resonator causes the resonance frequency to shift. The shift in the resonance frequency of any resonator and the electrical properties of a material under test (MUT), such as the permittivity and permeability, can be related to each other according to [22]:

$$\frac{\triangle f_r}{f_r} = \frac{\int_v (\triangle \epsilon \mathbf{E}_1 \cdot \mathbf{E}_0 + \triangle \mu \mathbf{H}_1 \cdot \mathbf{H}_0) dv}{\int_v (\epsilon_0 |E_0|^2 + \mu_0 |H_0|^2) dv} \quad (2)$$

where Δf_r is the shift in the resonance frequency, f_r, $\Delta\epsilon$ and $\Delta\mu$ are the change in the permittivity and permeability and υ is the perturbed volume. $\mathbf{E}_0$ and $\mathbf{H}_0$ are the field distributions without the perturbation and $\mathbf{E}_1$ and $\mathbf{H}_1$ are the field distributions with the perturbation. If the perturbation is small and the field distribution is assumed unchanged, thenEquation (2) reduces to:

$$\frac{\triangle f_r}{f_r} = \frac{\int_v (\triangle \epsilon |E_0|^2 + \triangle \mu |H_0|^2) dv}{\int_v (\epsilon_0 |E_0|^2 + \mu_0 |H_0|^2) dv}. \quad (3)$$

The perturbation approach is used in resonant cavity methods and near-field sensors to extract the constitutive properties of materials (permittivity and permeability) [23].

When the electrodynamic near field of a resonator is disturbed due to the presence of either new material or deformation (such as topological anomalies) in the original material, the resonance frequency exhibits a shift that can be used as a signature for detection or for characterizing the new environment in general. In this paper, we used CSRR excited by a microstrip line as a near-field sensor. Figure 1 shows a CSRR sensor that is used to detect defects in metallic surfaces, such as cracks. The CSRR sensor is a planar structure excited by a microstrip line, which produces an electric field perpendicular to the surface of the sensor. The CSRR sensor has a conducting center island surrounded by a conducting plane that connects the center island through two narrow bridges, as shown in Figure 2.

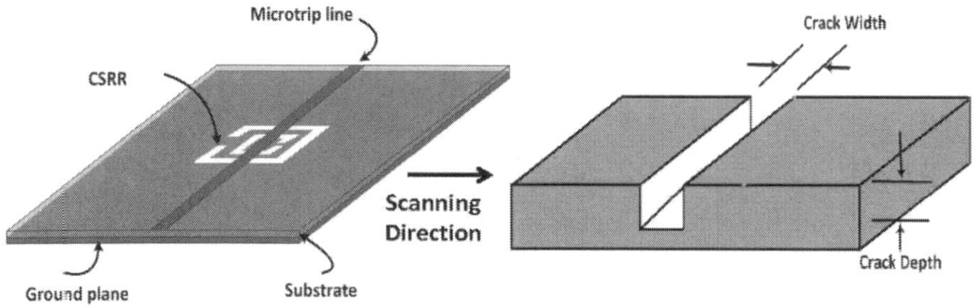

Figure 1. A schematic of a CSRR excited by a microstrip line used as a near-field sensor to detect defects in metallic surfaces.

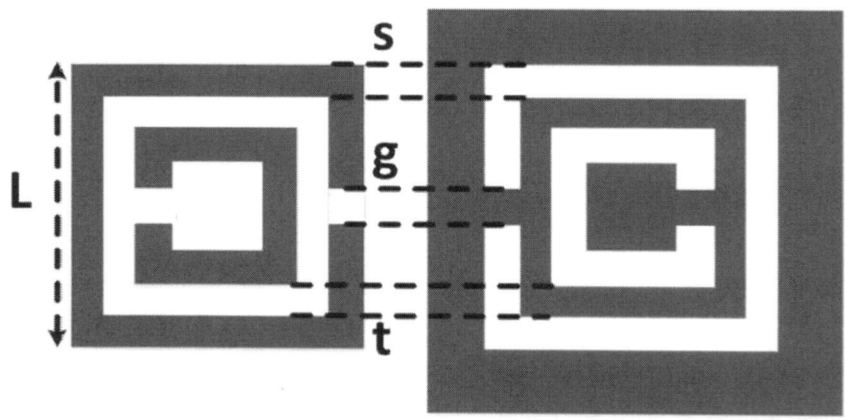

Figure 2. The geometry of a CSRR and SRR, where $L = 3$ mm, $g = 0.2$ mm, $s = 0.2$ mm and $t = 0.2$ mm.

The primary objective when designing a sensor is to obtain a high measurement sensitivity, reduce measurement setup complexity and use widely available relatively inexpensive microwave components. Since the near field of any electromagnetic source occupies a space directly proportional to the size and frequency of the radiator, to effectively sense small changes, such as sub-millimeter cracks in materials in the near field of the sensor, the overall size of the CSRR must be sufficiently small so as not to be much larger than the features to be detected. In the examples we considered here, the gap separations and trace widths of the sensor were chosen to be 0.2 mm as the smallest feature in this design. Generally, there is flexibility in designing a CSRR sensor at a wide range of frequencies associated with a target crack size.

3. CSRR SENSOR DESIGN APPROACH

Unlike conventional wire antennas, where the resonant frequency is directly proportional to the length of the radiator, the resonant frequency of electrically-small resonators, such as CSRR and SRR resonators, is typically found through measurement or simulation [7,16,18,24,25]. In general, however, the larger the overall size of the electrically-small resonator, the lower its resonant frequency Here, we consider correlating the topological dimensions of a CSRR element to its resonant frequency for the purpose of obtaining useful design guidelines. Because the sensor is a two-port microstrip line exciting a CSRR, a vector network analyzer (VNA) is used to measure the reflection and transmission parameters. In order to match the internal impedance of the VNA in our measurement setup, the microstrip line was designed to have a characteristic impedance of 50 Ω (the microstrip line can be designed to match the internal impedance of other measurement setup devices). To achieve a 50 Ω characteristic impedance for the microstrip line using Rogers RO4350 substrate with a thickness of 0.75 mm, a permittivity of 3.66 and a loss tangent of 0.0031, the width of the microstrip line was 1.7 mm.

For the purpose of developing CSRR sensor design guidelines, the CSRR length was varied from 2 to 4 mm. Two different values of 0.1 mm and 0.2 mm were considered for the gap separations and trace widths of the structure. The resonance frequencies *versus* the size of CSRRs (considered to be its length) are plotted in Figure 3. An exponential fitting function was used to find the resonance frequency of a CSRR as a function of its length, expressed as:

$$f_r = e^{a+bL+cL^2} \tag{4}$$

where Lis the length of the resonator and a, b and c are the fitting function constants. For gap separations and trace widths of 0.1 mm, the constants were $a = 4.041$, $b = -1.085$ and $c = 0.104$. In the case of the gap separations and trace widths of 0.2 mm, the constants were $a = 4.34$, b = -1.093 and $c = 0.098$.

Figure 3 shows that the resonance frequencies decrease exponentially by increasing the overall lengths of the resonator. Of course, lowering of the resonant frequency is expected due to the increase in the effective inductance of the SCSR; however, the exponential relationship could not have been intuitively predicted.

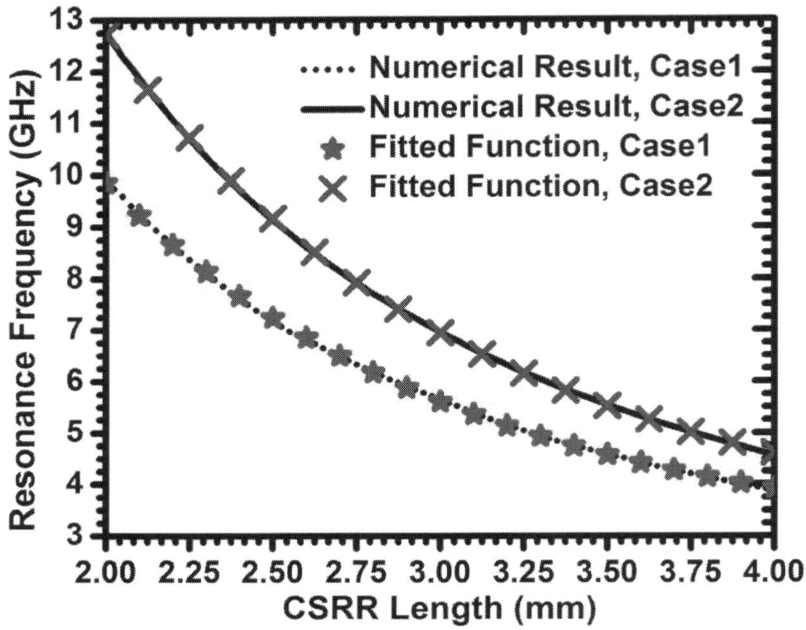

Figure 3. Resonance frequencies *versus* CSRR length obtained form full-wave numerical simulation and from the fitting exponential function for gap separations and trace widths of 0.1 mm and 0.2 mm labeled as Case 1 and Case 2, respectively.

A CSRR sensor was designed to resonate at around 6.95 GHz with resonator dimensions of $L = 3$ mm and $s = g = t = 0.2$ mm (see Figure 2). For the purpose of crack detection, designing a sensor to operate at a specific resonance frequency is unnecessary as long as the resonance frequency and dimensions of the sensor are chosen to detect cracks with dimensions falling within a specific range. Figure 4 shows a photo of the designed CSRR with dimensions of $L = 0.27$ mm and $g = s = t = 0.16$ mm.

Figure 5 shows a comparison between the simulation and measurements for a sensor with $L = 0.27$ mm and $g = s = t = 0.16$ mm. The feeding microstrip line had a width of 1.7 mm on Rogers RO4350 substrate with a thickness of 0.75 mm, permittivity of 3.7 and a loss tangent of 0.003. The agreement between the full-wave simulation and measurements is observed to be very strong, especially at the frequency of minimum transmission. This gives strong confidence in the capability of numerical simulation to determine the response of different sensors and different crack topologies.

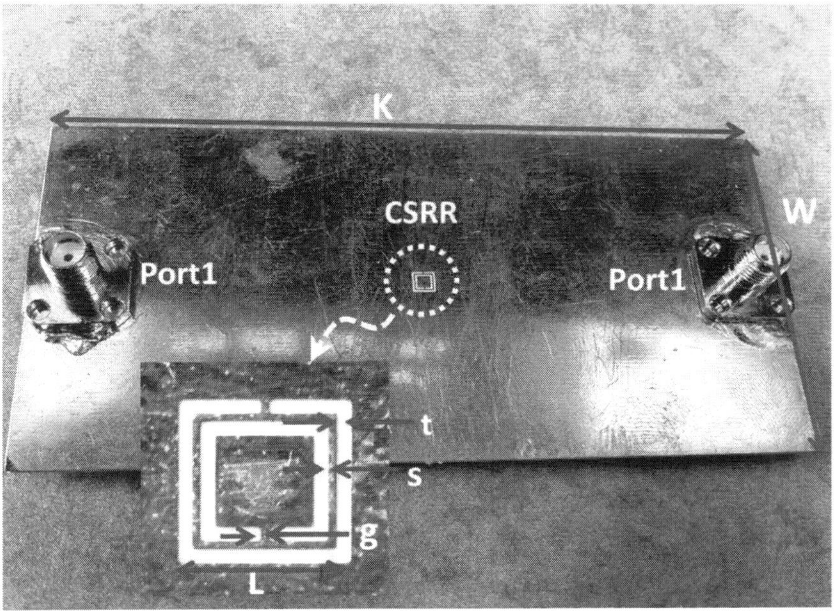

Figure 4. The CSRR printed on the ground plane of a microstrip line with its fabricated dimensions, L = 0.27 mm, $g = s = t$ = 0.16 mm, K = 100 mm and W = 50 mm.

4. SENSING MECHANISM IN CSRR SENSORS

The sensitivity of the CSRR sensor is determined by how much the near field of the resonator can be disturbed by defects in materials causing the resonance frequency to shift. In addition, the size of detects is essential for determining the length of the CSRR, which, in turn, corresponds to the resonance frequency of the sensor. In this section, we investigate factors affecting the sensitivity of the CSRR sensor by careful analysis of the surface current distribution on the resonator. We considered two CSRRs of gap separations and trace widths of 0:2 mm with different lengths of 3 mm and 6 mm. Figure 6shows the current surface distribution density on the narrow bridges, which was found to be the highest over the entire sensor CSRR surface. The maximum surface current density yields the highest magnetic field from any part of the sensor, and thus, the most sensitive part of the CSRR sensor is the narrow bridge linking the inner island to the outer plane conductor. Figure 7 shows the magnetic field lines around the sensing element.

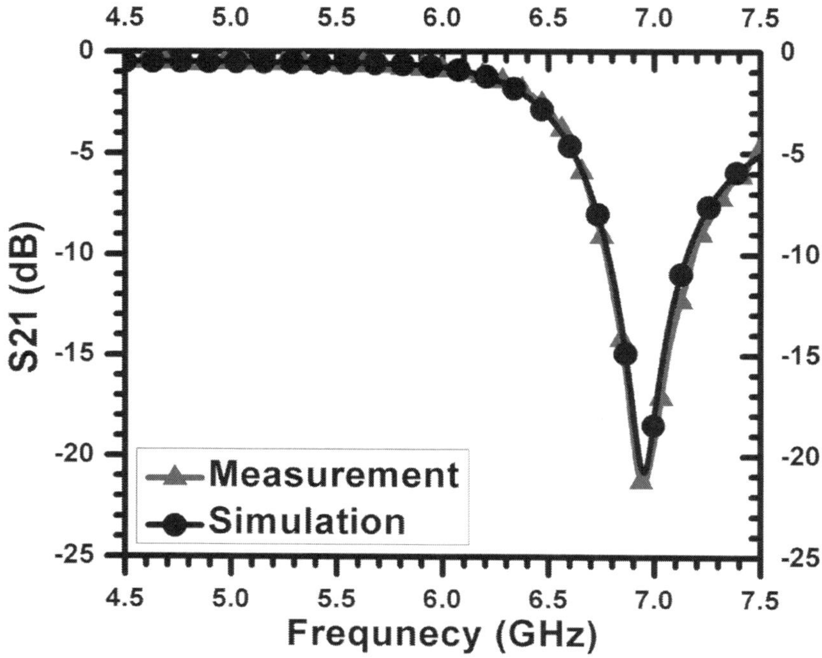

Figure 5. Frequency response of the sensor obtained from full-wave simulation using HFSSand measurements for a 3-mm sensor.

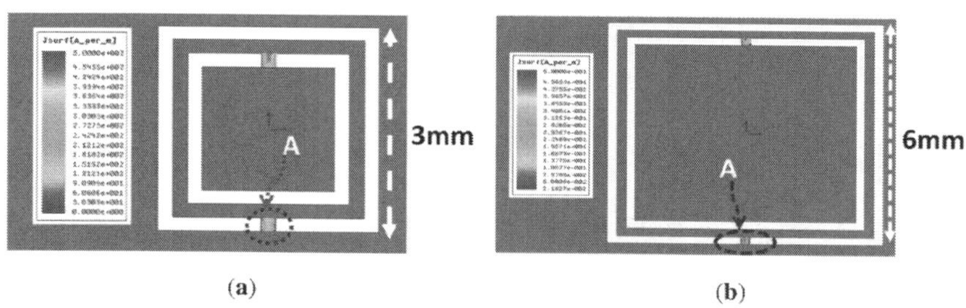

Figure 6. Current distribution density on CSRRs at the resonance frequency. The bridge labeled A shows the area of high current density. (**a**) A CSRR of a length of 3 mm. (**b**) A CSRR of a length of 6 mm. The surface current on the 3-mm CSRR is one order of magnitude higher than that of the 6-mm CSRR.

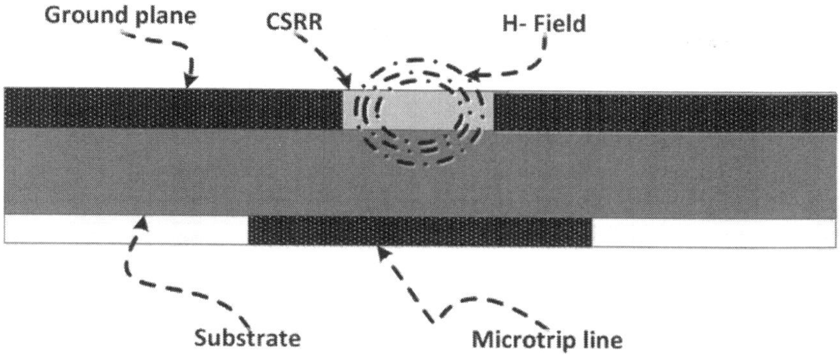

Figure 7. Magnetic field lines around the bridge line labeled A in Figure 11.

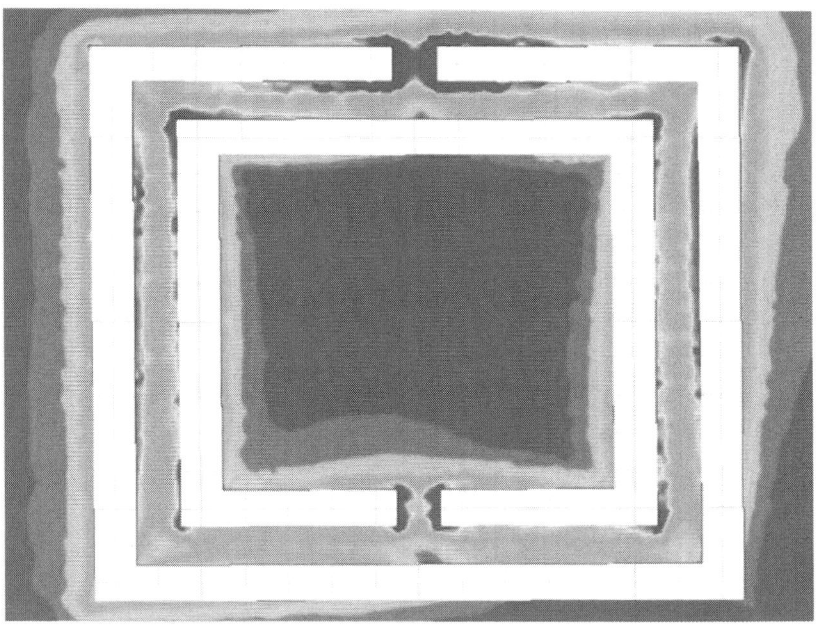

Figure 11. Simulated current surface distribution on a 3-mm CSRR at the resonance frequency.

The resonance frequency of the CSRR sensor (plotted in Figure 3) represents the sensor in the absence of the metallic surface to be detected. When the sensor is placed on the solid metallic surface, the resonance frequency shifts. The resonance frequency of the sensor in the presence of the solid crack-free surface is used as the reference resonance frequency. Any shift in the resonance frequency due to the presence of a crack is then compared to the reference frequency. The crack considered in [7] extended across the entire width of the aluminum block under test. Here, we use numerical simulation to test the effectiveness of the CSRR sensor to detect cracks of finite length, which represents a scenario more practical than that of very long cracks. Figure 8 shows the resonance frequency shift experienced by the 3-mm CSRR sensor for different crack lengths (a crack width of 0.2 mm and a depth of 2 mm). For a crack with a very short length, Figure 8 shows a very small shift in the resonance frequency, as would be expected. As the crack length increases, the shift in the resonance frequency increases, until the crack length becomes approximately equal to the length of the sensor (in this example, a 3-mm CSRR sensor was used). Increasing the crack length further did not lead to any additional increase in the resonance frequency shift. In Figure 9, we show the effect of varying the crack depth on the resonance frequency shift for a 0.2 mm-wide and 15 mm-long crack.

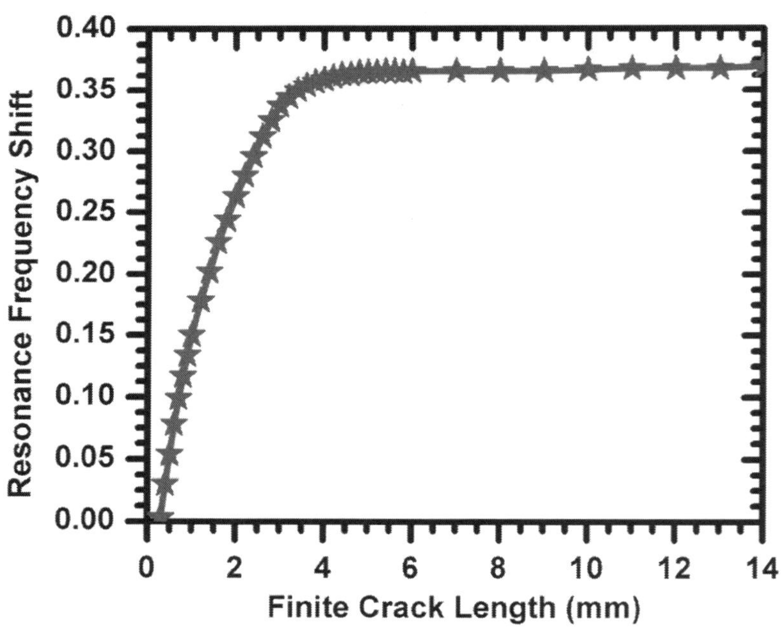

Figure 8. Resonance frequency shift as a function of the crack length. The crack is 0.2 mm wide and 2 mm deep.

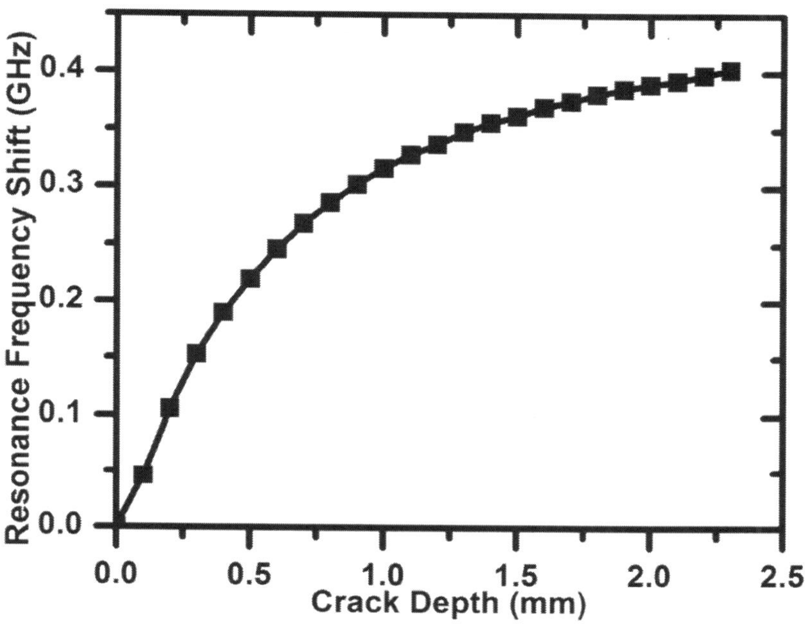

Figure 9. Resonance frequency shift as a function of the crack depth. The crack is 0.2 mm wide and 10 mm in length.

5. INCREASING THE SENSITIVITY OF CSRR SENSORS

Defects in metallic surfaces, such as sub-millimeter cracks, introduce capacitance that is parallel to the effective capacitance of the sensor. The crack-induced capacitance leads to a downward shift in the resonance frequency in accordance with Equation (1). Therefore, the sensitivity of the sensor can be enhanced by increasing the capacitances of the crack by filling the crack with liquid dielectric material. For this purpose, a crack was filled with silicon oil of a dielectric constant of 2.70. The test sample was fabricated using an aluminum plate with a crack width of 200 μm and a depth of 2 mm. Figure 10 shows the measured frequency of minimum transmission coefficient for the reference case (solid aluminum surface without a crack) and for the cases of the crack with and without silicon oil. The filling of the crack with silicon increased the sensitivity of the sensor by 150 MHz (a shift of more than 435 MHz was registered with respect to the case of solid aluminum without the crack). It is expected that liquids with higher dielectric constants can increase the sensitivity further. In practical crack detection scenarios, silicon oil or other liquids can be applied to the surface of the metallic material to be detected prior to the sensing procedure.

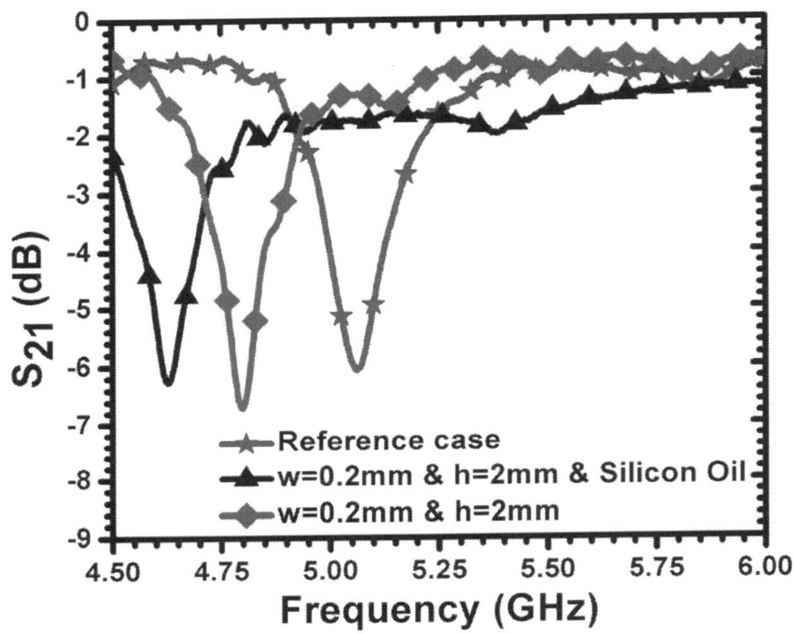

Figure 10. Minimum transmission of the sensor for a crack width of 200 μm and a depth of 2 mm for the reference case (without a crack) and for the cases of a crack with and without silicon oil.

6. CSRR MINIATURIZATION USING LUMPED AND DISTRIBUTED ELEMENTS

Near-field sensors operating at lower frequency are relatively larger than sensors operating at higher frequencies. The sensitivity of relatively large sensors to detect small defects in materials, such as sub-millimeter cracks, becomes lower in comparison to smaller sensors. Sensors with a small form factor are more advantageous than larger sensors, since they are easier and less expensive to integrate in practical stand-alone sensing systems. Therefore, miniaturization of sensors, while maintaining a relatively low resonant frequency, has clear advantages.

One immediate choice for miniaturization of CSRR sensors is the introduction of lumped inductance in the path of the dominant surface current. By understanding the surface current distribution on the surface of the CSRR, a lumped inductor can be placed at the location of the highest current. Figure 11 shows the current distribution at the resonance frequency of a CSRR of a length of 3 mm. Clearly, the current surface distribution is highly concentrated at the two narrow bridges, indicating that the magnetic field of the sensor is highest around the two bridges, as discussed above. Replacing the narrow bridges by a lumped inductor adds additional inductance, which results in shifting the resonance frequency downward.

To test the effectiveness of the added lumped inductor on the sensitivity of the CSRR sensor, a CSRR of a length of 6 mm was designed using full-wave numerical simulation and tested experimentally. Then, the narrow bridges were replaced by two 20 nH inductors, as shown in Figure 12. Figure 13 shows the minimum transmission coefficient of the CSRR sensor with the lumped inductors. A strong agreement between the measurement and simulation is observed. Figure 14 shows the experimental result of the CSRR before and after replacing the narrow bridges. We observe almost a reduction of the resonance frequency by a factor of five. Using numerical simulation, the miniaturization, however, was not found to yield sufficient sensitivity for detecting the types of cracks considered above (the resonance frequency exhibited a very small shift, not shown here for brevity). Therefore, while the introduction of lumped elements indeed achieved miniaturization, the sensitivity of the sensor decreased due primarily to the high localization or confinement of the magnetic field within the added inductors.

Instead of using lumped inductors, inductance can be added to the CSRR sensor by employing a different design involving meander lines, as shown in Figure 15. Figure 15 depicts the current distribution at the resonance frequency of the CSRR sensor. Figure 16 shows the transmission coefficient of the CSRR and meandering CSRR. The meandering increased the total inductance, such that the resonance frequency shifted downward by approximately 50% with respect to the original CSRR design response. Figure 17 shows that for a long crack with a width of 100 μm and a depth of 2 mm, the meander CSRR sensor resulted in a frequency shift of more than 70 MHz. Therefore, we conclude that increasing the total inductance of a CSRR for miniaturizing purposes using either lumped or distributed elements is possible; however, the sensitivity of the CSRR is decreased for crack detection purposes, but the sensor has the advantage of achieving resonance at a lower frequency.

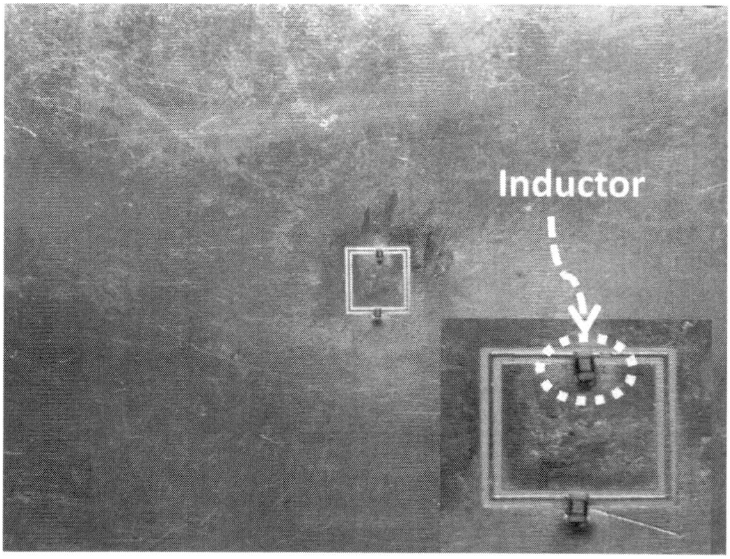

Figure 12. A CSRR after replacing the two narrow conducting bridges with lumped inductors. The red color indicates the highest current magnitude.

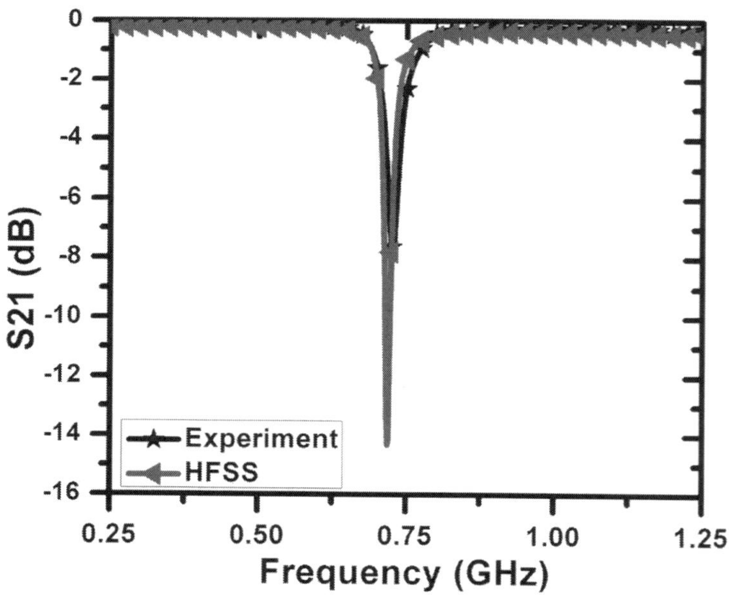

Figure 13. Minimum transmission coefficient of the sensor obtained from HFSS and measurement of 6-mm CSRR with respect to the case without the crack after adding the two inductors of a value of 20 nH.

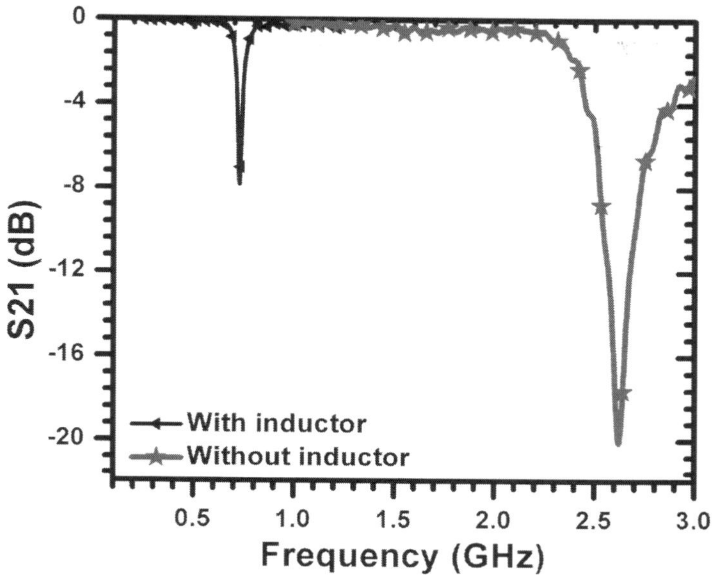

Figure 14. Minimum transmission coefficient of the sensor obtained from measurement with and without inductors of 6-mm CSRR.

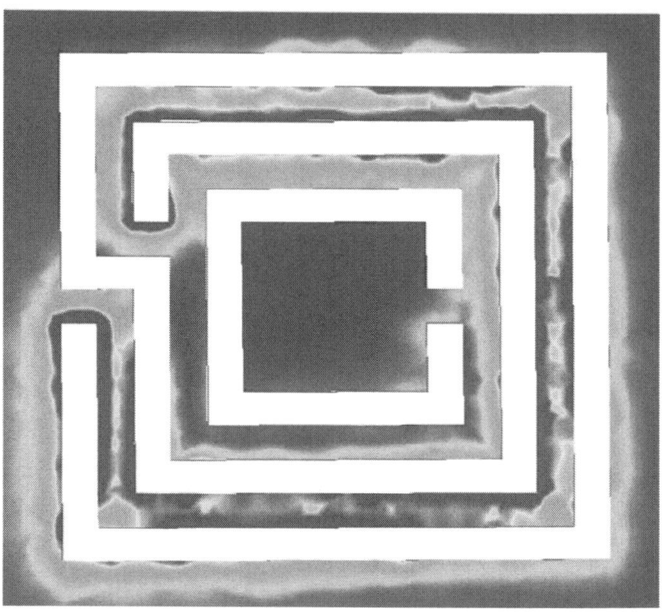

Figure 15. Simulated current distribution on a new CSRR at a resonance frequency of around 3.781 GHz. The red color indicates the highest current magnitude.

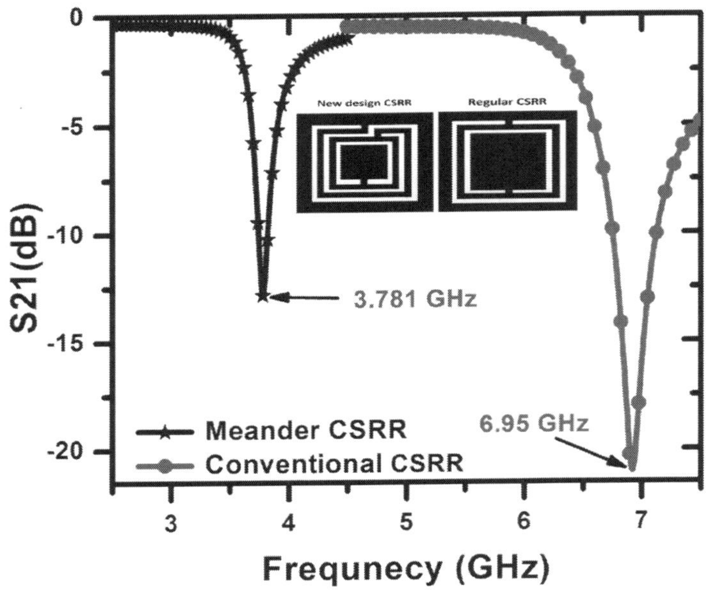

Figure 16. Minimum transmission coefficient of the two sensor obtained from HFSS at the resonance frequency of the 3-mm CSRR.

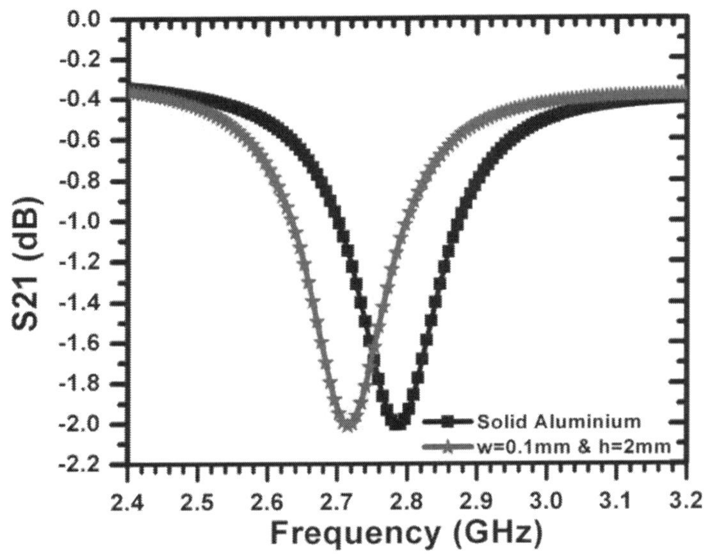

Figure 17. Minimum transmission of the new CSRR sensor for a crack width (w) of 100μm, and the crack depth (h) is 2 mm compared with solid aluminium.

7. CRACK DETECTION IN NON-METALLIC MATERIALS

Detection of anomalies, such as cracks in dielectric material, is based on a change in the electric field due to the presence of the anomaly. Therefore, the sensing element for the near-field detection of anomalies, such as cracks in dielectric surfaces, needs to have a dominant electric field. By investigating the total electric field on the surface of the CSRR sensor, we observed a dominant electric field along one side of the CSRR. Figure 18 shows the maximum electric field distribution on a 3-mm CSRR at the resonance frequency. The sensing element for the purpose of detecting dielectric anomalies, therefore, can be considered as the side depicted by the letter C in Figure 18.

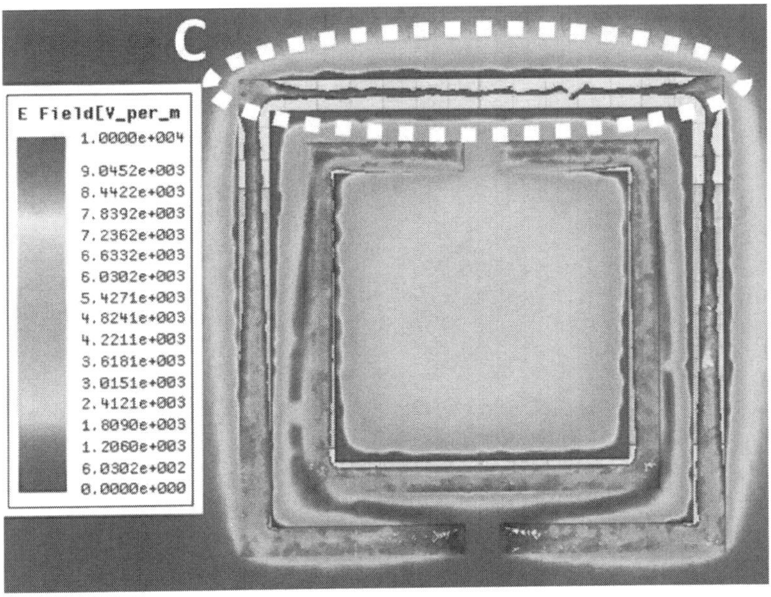

Figure 18. Total electric field distribution on a CSRR at the resonance frequency of a 3-mm CSRR. The red color indicates the highest current magnitude.

For validation purpose, a CSRR of a length of 3 mm is used to detect sum-millimeter cracks in dielectric materials. Fiberglass and ceramic materials with dielectric constants of 2.3 and 9.8, respectively, were used. Figure 19 shows the frequency of minimum transmission coefficient obtained experimentally when the sensor passes over a fiberglass material with different crack widths of 150 μm and 200 μm (depths of 1.5 mm). A maximum shift of more than 312 MHz was observed with respect to the reference case. For the ceramic material considered, Figure 20 shows a maximum shift of more than 477 MHz for the two crack topologies considered. These results indicate that the sensitivity increases with the dielectric constant of the material under test. Based on Figures 19 and 20, we observe that the cracks result in an upward shift in the resonance frequency (in contrast to the downward shift observed in the case of detecting cracks in metallic surfaces), which is consistent with the prediction of Equation (1), due to the decreased capacitance experienced by the sensor.

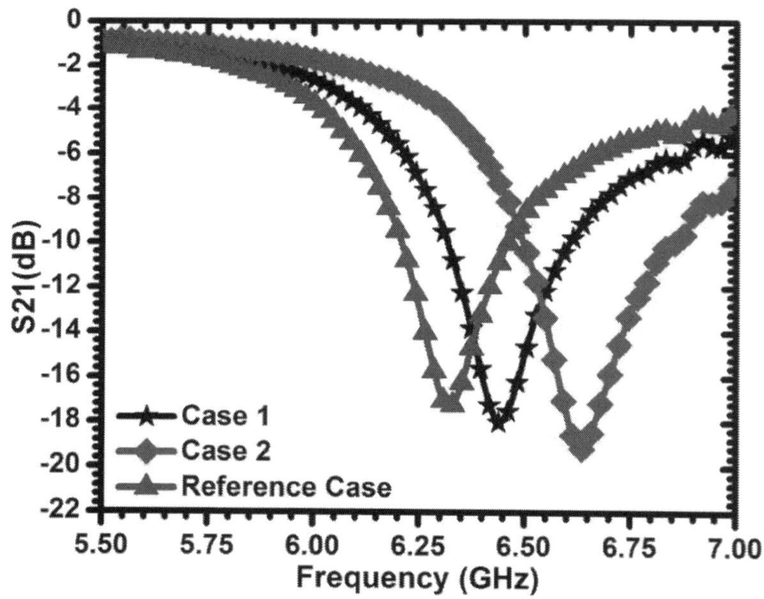

Figure 19. Minimum transmission of the sensor for a crack width (w) of 150 μm (Case 1) and 200 μm (Case 2) compared with solid dielectric material (fiberglass) with a dielectric constant of 2.3 (reference case). The crack depth (h) is 1.5 mm.

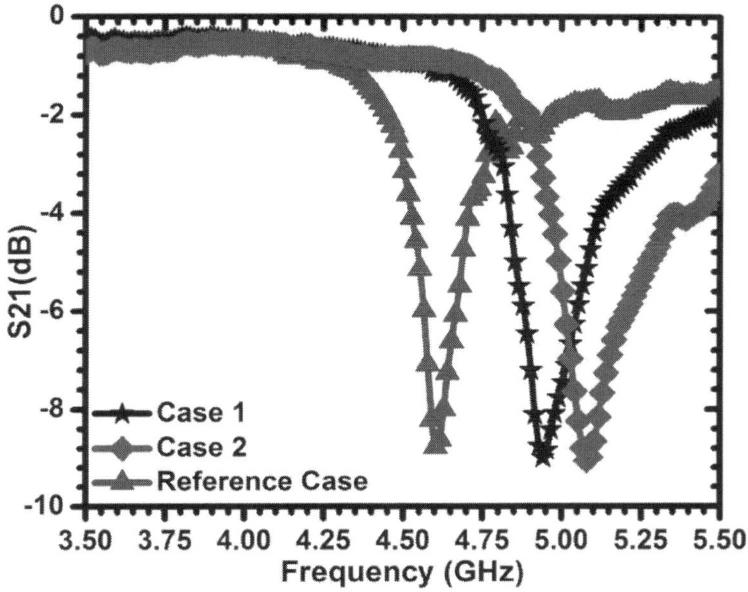

Figure 20. Minimum transmission of the sensor for a crack width (w) of 150 μm (Case 1) and 200 μm (Case 2) compared with a solid dielectric material, ceramic (ceramic material), with a dielectric constant of 9.8 (reference case). The crack depth (h) is 1.5 mm.

8. CONCLUSIONS

This work presented a comprehensive analysis of the capabilities of the CSRR sensor to detect sub-millimeter cracks in metallic and non-metallic surfaces. Design formulae were derived based on full-wave numerical simulation to provide the designer with guidelines that can eliminate expensive full-wave numerical simulations. Analysis of the surface currents on the CSRR revealed important clues as to the active sensing element that is most sensitive to changes in the surface under test. Miniaturization of the CSRR sensor was studied using two approaches using lumped elements and meandering. The lumped element approach achieved miniaturization, but did not yield any appreciable sensitivity, while the meandering achieved a reduction in the form factor of the sensor by 50% while achieving sufficient sensitivity for the detection of sub-millimeter cracks. Based on the physics of the CSRR sensing mechanism, a dielectric liquid, such as silicon oil, was introduced to enhance the sensitivity of the sensor by approximately 150 MHz for the case considered. Finally, the CSRR sensor was also found to be highly effective for detecting sub-millimeter cracks in dielectric surfaces.

ACKNOWLEDGMENTS

This work was supported by King Saud University and the National Science and Engineering Research Council of Canada (NSERC). We would like to acknowledge CMC Microsystems for the provision of products and services that facilitated this research, including CAD tools and test support.

AUTHOR CONTRIBUTIONS

Ali Albishi and Omar M. Ramahi conceived and designed the experiment; Ali Albishi performed the experiments; Ali Albishi and Omar M. Ramahi analyzed the data; Ali Albishi and Omar M. Ramahi wrote the paper.

CONFLICTS OF INTEREST

The authors declare no conflict of interest.

REFERENCES

1. Nelson, S.O.; Guo, W.-C.; Trabelsi, S.; Kays, S.J. Dielectric spectroscopy of watermelons for quality sensing. *Meas. Sci. Technol.* **2007**, *18*, 1887.

2. Nelson, S.; Trabelsi, S. Dielectric spectroscopy measurements on fruit, meat, and grain. *Trans. ASABE* **2008**, *51*, 1829–1834.

3. Guo, W.-C.; Nelson, S.O.; Trabelsi, S.; Kays, S.J. 10-1800-MHz dielectric properties of fresh apples during storage. *J. Food Eng.* **2007**, *83*, 562–569.

4. Nelson, S.O.; Trabelsi, S. Influence of water content on rf and microwave dielectric behavior of foods. *J. Microw. Power Electromagn. Energy* **2009**, *43*, 13–23.

5. Dalmay, C.; Pothier, A.; Blondy, P.; Lalloue, F.; Jauberteau, M.-O. Label free biosensors for human cell characterization using radio and microwave frequencies. Proceedings of the 2008 IEEE MTT-S International Microwave Symposium Digest, IEEE, Atlanta, GA , USA, 15–20 June 2008; pp. 911–914.

6. Kim, J.; Babajanyan, A.; Hovsepyan, A.; Lee, K.; Friedman, B. Microwave dielectric resonator biosensor for aqueous glucose solution. *Rev. Sci. Instrum.* **2008**, *79*, 086107.

7. Albishi, A.; Boybay, M.S.; Ramahi, O.M. Complementary split-ring resonator for crack detection in metallic surfaces.*Microw. Wirel. Compon. Lett. IEEE* **2012**, *22*, 330–332.

8. Yeh, C.-Y.; Zoughi, R. A novel microwave method for detection of long surface cracks in metals. *IEEE Trans. Instrum. Meas.* **1994**, *43*, 719–725.

9. Kerouedan, J.; Quéffélec, P.; Talbot, P.; Quendo, C.; Blasi, S.; Brun, A. Detection of Micro-Cracks on Metal Surfaces Using Near-Field Microwave Dual-Behavior Resonator Filters. Available online: http://stacks.iop.org/0957-0233/19/i=10/a=105701 (access on 1 June 2014).

10. Zoughi, R.; Ganchev, S.; Huber, C. Measurement parameter optimization for surface crack detection in metals using an open-ended waveguide probe. Proceedings of the Instrumentation and Measurement Technology Conference, 1996. IMTC-96. Conference Proceedings. 'Quality Measurements: The Indispensable Bridge between Theory and Reality', IEEE, Brussels, Belgium, 4–6 June 1996; Volume 2, pp. 1391–1394.

11. Mazlumi, F.; Sadeghi, S.; Moini, R. Using open-ended rectangular waveguide probe for detection and sizing of fatigue cracks in metals. *Electron. Lett.* **2005**, *41*, 334–335.

12. Nadakuduti, J.; Chen, G.; Zoughi, R. Semiempirical electromagnetic modeling of crack detection and sizing in cement-based materials using near-field microwave methods. *IEEE Trans. Instrum. Meas.* **2006**, *55*, 588–597.

13. Bahr, A.J. Nondestructive microwave evaluation of ceramics. *IEEE Trans. Microw. Theory Tech.* **1978**, *26*, 676–683.

14. Yun, T.; Lim, S. High-Q and miniaturized complementary split ring resonator-loaded substrate integrated waveguide microwave sensor for crack detection in metallic materials. *Sens. Actuators A Phys.* **2014**, *214*, 25–30.

15. Ramahi, O.M.; Albishi, A. Complementary Split Ring Resonators for Detection of Cracks in Metallic and Non-Metallic Materials. U.S. Patent No. 62,030,400, 2014.

16. Baena, J.; Bonache, J.; Martin, F.; Sillero, R.; Falcone, F.; Lopetegi, T.; Laso, M.A.G.; Garcia-Garcia, J.; Gil, I.; Portillo, M.; *et al.* Equivalent-circuit models for split-ring resonators and complementary split-ring resonators coupled to planar transmission lines. *IEEE Trans. Microw. Theory Tech.* **2005**, *53*, 1451–1461.

17. Bonache, J.; Siso, G.; Gil, M.; Iniesta, A.; Garcia-Rincon, J.; Martin, F. Application of composite right/left handed (CRLH) transmission lines based on complementary split ring resonators (CSRRs) tothe design of dual-band microwave components. *Microw. Wirel. Compon. Lett. IEEE* **2008**, *18*, 524–526.

18. Wu, B.; Li, B.; Su, T.; Liang, C.-H. Equivalent-circuit analysis and lowpass filter design of split-ring resonator dgs. *J. Electromagn. Waves Appl.* **2006**, *20*, 1943–1953.

19. Pendry, J.; Holden, A.; Robbins, D.; Stewart, W. Magnetism from conductors and enhanced nonlinear phenomena.**1999**, *47*, 2075–2084.

20. Baena, J.D.; Marqués, R.; Medina, F.; Martel, J. Artificial Magnetic Metamaterial Design by Using Spiral Resonators. Available online: http://link.aps.org/doi/10.1103/PhysRevB.69.014402 (accesse d on 1 June 2014).

21. Yousefi, L.; Ramahi, O.M. New artificial magnetic materials based on fractal hilbert curves. Proceedings of the 2007 International Workshop on Antenna Technology: Small and Smart Antennas Metamaterials and Applications, Cambridge, UK, 21–23 March 2007; pp. 237–240.

22. Pozar, D.M. *Microwave Engineering*, 3rd ed.; Wiley: Danvers, MA, USA, 2005.

23. Gao, C.; Xiang, X.D. Quantitative microwave near-field microscopy of dielectric properties. *Rev. Sci. Instrum.* **1998**, *69*, 3846–3851.

24. Boybay, M.; Ramahi, O.M. Material characterization using complementary split-ring resonators. *IEEE Trans. Instrum. Meas.* **2012**, *61*, 3039–3046.

25. Aydin, K.; Cakmak, A.O.; Sahin, L.; Li, Z.; Bilotti, F.; Vegni, L.; Ozbay, E. Split-Ring-Resonator-Coupled Enhanced Transmission through a Single Subwavelength Aperture. Available online: http://link.aps.org/doi/10.1103/PhysRevLett.102.013904 (acce ssed on 1 June 2014).

CHAPTER 4

A Mobile Acoustic Subsurface Sensing (MASS) System for Rapid Roadway Assessment

Yifeng Lu [1,*], Yi Zhang [1], Yinghong Cao [1], J. Gregory McDaniel [2] and Ming L. Wang [1]

[1] Department of Civil and Environmental Engineering, Northeastern University, 360 Huntington Avenue, Boston, MA 02115, USA

[2] Department of Mechanical Engineering, Boston University, 110 Cummington Street, Boston, MA 02215, USA

ABSTRACT

Surface waves are commonly used for vibration-based nondestructive testing for infrastructure. Spectral Analysis of Surface Waves (SASW) has been used to detect subsurface properties for geologic inspections. Recently, efforts were made to scale down these subsurface detection approaches to see how they perform on small-scale structures such as concrete slabs and pavements. Additional efforts have been made to replace the traditional surface-mounted transducers with non-contact acoustic transducers. Though some success has been achieved, most of these new approaches are inefficient because they require point-to-point measurements or off-line signal analysis. This article introduces a Mobile Acoustic Subsurface Sensing system as MASS, which is an improved surface wave based implementation for measuring the subsurface profile of roadways. The compact MASS system is a 3-wheeled cart outfitted with an electromagnetic impact source, distance register, non-contact acoustic sensors and data acquisition/processing equipment. The key advantage of the MASS system is the capability to collect measurements continuously at walking speed in an automatic way. The fast scan and real-time analysis advantages are based upon the non-contact acoustic sensing and fast air-coupled surface wave analysis program. This integration of hardware and software makes the MASS system an efficient mobile prototype for the field test.

KEYWORDS

mobile; acoustic; subsurface; MASS; radiating surface wave

1. INTRODUCTION

The highway pavement and bridge deck health situation is critical for our society. It affects the safety of millions of people daily and requires financial investments that benefit from detailed knowledge. Pavement and bridge deck health diagnosis has become very important as the traffic efficiency and personal security should be protected. The conventional pavement and bridge deck inspection technologies are designed to diagnose surface defects such as different sizes of cracks, potholes, and roughness, since it is quite easy to visually inspect the surface. However, the deterioration of pavement initiates not only from surface cracks and potholes due to tire scratching, but also from the de-bonding of subsurface layers, or stripping due to material aging and corrosion. Defects caused by subsurface damage often develop before the surface defects are visible.

For the inspections of the subsurface of pavement and bridge roadway, many methods have been developed and are being utilized for fieldwork testing. Impact Echo (IE), Impulse Response (IR), Ground Penetrating Radar (GPR), Chain-Drag, and Spectral Analysis of Surface Waves (SASW) are typical technologies. IE is limited to identifying the de-bonding and thickness of a shallow top layer [1]. IR could test the overall dynamic stiffness/mobility of the entire pavement structure [2]. GPR is best for locating the metal materials, such as reinforcement rebar [3]. Chain-drag could find the de-bonded or corroded areas by hearing the resulting hollow sound with expert experience [4]. The SASW and its related methods are frequently used as their capability of estimating the thickness and stiffness of subsurface layers [5–7].

Since it was first proposed in the 1980s [5], the SASW method has been widely applied in geology field tests to estimate the underground soil profile without coring or opening the ground [8–10]. Figure 1 shows representative experimental layout for implementing the method. It utilizes the dispersion features of the surface wave that propagates horizontally in the soil when it is subject to an impact load. The dispersion curve represents the relationship between the wave speed, wave length and frequency. It is measured by phase difference between two receivers with known distance. Once the dispersion curve is obtained from the test data, the layer profile as shear velocities can be estimated by inversion procedures. Extensive research has been conducted to improve the accuracy and efficiency of the method. For the inversion procedure, analytical forward modeling for dispersion curve is needed. One typical method is the stiffness matrix method [11], which predicts the analytical dispersion curve used as the matching object for an inversion iteration. In recent years, the SASW method was extended to investigate shallow layered structures such as pavement systems [12,13] and concrete structures [14]. Some similar methods were successfully developed based on the same principle of SASW. For example, the Multichannel Analysis of Surface Wave method uses multiple sensors to record the complete wave field and resolves the different wave modes [15].

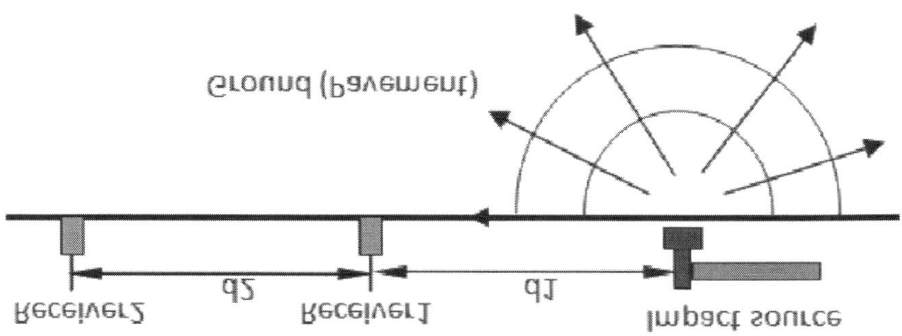

Figure 1. Test configuration of SASW Method.

Despite these past achievements of SASW and related methods, the methods are labor intensive in two ways. Firstly, the method requires the installation and repositioning of ground-mounted sensors to measure the ground vibration, such as accelerometers and geophones. Secondly, the inversion algorithms often require human adjustments based on experience. Another critical drawback is the disruption to traffic flow due to the requirement of attaching sensors to the ground. Recently, attempts were made to improve the testing efficiency by using noncontact sensors, such as microphones, to replace the traditional contact accelerometers [16], in which the surface wave measured by the microphones is used to obtain the dispersion feature. A mobile testing concept was also investigated which could perform the fast scanning on concrete plates successfully [17].

This paper presents an integrated mobile acoustic sensing system which was developed to estimate the thickness and elastic modulus for pavement layers at walking speed. An electrically powered hammer is used to produce an impact force to the ground. A microphone array is used to collect acoustic signals radiated from the surface wave. Multi-channel DAQ is assembled to collect and process data with a laptop computer. An encoder-based register locates the distance along the scanning path and triggers the hammer impact with the desired scan resolution. A key feature of this prototype is its real-time display of subsurface properties. To achieve this, the inversion procedure in air-coupled SASW is conducted with a fast inversion algorithm. This algorithm connects the dispersion curve with layered shear velocity profile directly by the *in-situ*particle displacement distribution [18]. The benefit of the algorithm is the avoidance of the need for expertise knowledge to initialize the forward profile model as well as trial and error iteration. More details of the algorithm are introduced by [18]. The prototype was tested at several fields, including the asphalt pavement and bridge deck. The test results demonstrate the concept and capability of the MASS prototype.

2. PHYSICS BEHIND MASS SYSTEM

2.1. Air-Coupled SASW

Consider a layered pavement structure subjected to a vertical point impact at the surface, as shown in Figure 2(a). Besides the propagation of body waves, including P-waves and S-waves, the surface wave (Rayleigh wave) propagates horizontally near the free surface with dominant energy. Above the surface, the ground vibration due to the surface wave acts as an acoustic source to radiate an acoustic wave in the air. This acoustic radiation (radiating surface wave) from the ground surface wave brings the same information about the dispersion properties with frequency/wavelength. A visualization snapshot by FEM is presented in Figure 2(b) [19]. It indicates the dominant radiating surface wave arrives earlier than the direct sound wave caused by impact. According to the Snell's laws, the Rayleigh angle θ is determined by:

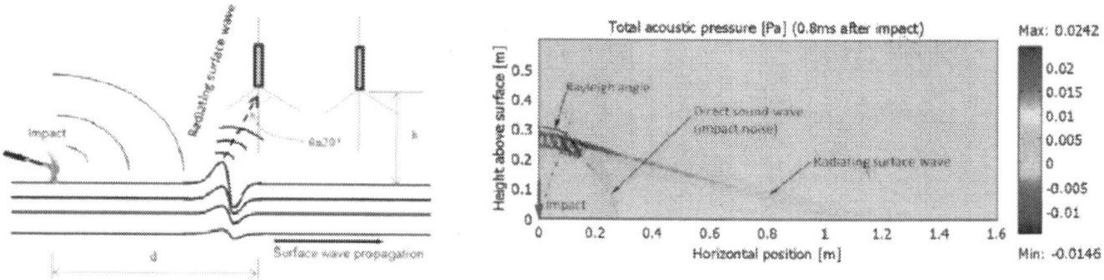

(a) (b)

Figure 2. (**a**) Propagation of radiating surface wave. (**b**) Visualization of radiating surface wave by FEM.

$$\sin(\theta) = \frac{C_a}{C_R} \tag{1}$$

where, C_a and C_R are the speed of sound in the air and the surface wave speed in the pavement, respectively. The speed $C_a$is approximately 340 m/s, and C_R is approximately at the level of 1,000 m/s for many road constructions. These speeds lead to a radiation Rayleigh angle θ of approximately 20°. The radiating surface wave can be detected with directional microphones, which usually have an effective conical angle of around 100°. Aligning the angle of the directional microphones to match the angle of the radiating surface wave helps improve the signal to noise ratio of the measured acoustic data.

By using acoustic radiation to sense the surface wave, the SASW test can be performed with microphones instead of accelerometers. Figure 3 shows a schematic configuration of the air-coupled SASW (similarly to the MASW) test. Two or more microphones are placed at a small distance above the ground and connected to the data acquisition device and computer. When a vertical impact is applied, the measured acoustic signal contains radiation from the surface wave (radiating surface wave), as well as the direct acoustic wave from the vibration in the vicinity of impact. The radiating surface wave brings the dispersion feature to indicate both the layer thickness as H_i and stiffness as $Stiff_i$ in Figure 3. When the microphones are near the surface and the shear velocity of the pavement is much larger than the sound speed, the direct acoustic wave from the impact arrives to the microphone later than the radiating surface wave. The difference of their arrival time can be calculated as:

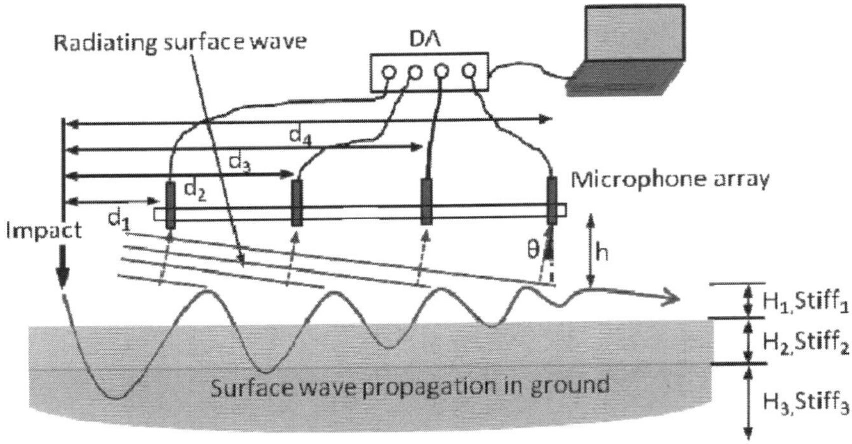

Figure 3. Air-coupled SASW configuration (like MASW with more channels).

$$\Delta t = \frac{\sqrt{d_i^2 + h^2}}{C_a} - \frac{d_i - h\tan\theta}{C_R} - \frac{h}{C_a\cos\theta} \tag{2}$$

For a typical configuration with $d_1 = 0.5$ m and $h = 0.05$ m, the time delay is approximately 1 ms. Therefore, the radiating surface wave can be extracted from the intact acoustic signal by applying an appropriate time window.

Figure 4 shows an example of the surface wave measured by accelerometer (a) and the radiating surface wave measured by microphones (b). According to the comparison, it can be seen the first part of the microphone data arrives nearly at the same time and shows the similar feature as the accelerometer data. This part of data can be identified as the radiating surface wave. The second part of the microphone data is regarded as the direct sound wave from the hammer knocking because no signal is present in the acceleration signal counterpart at the same time.

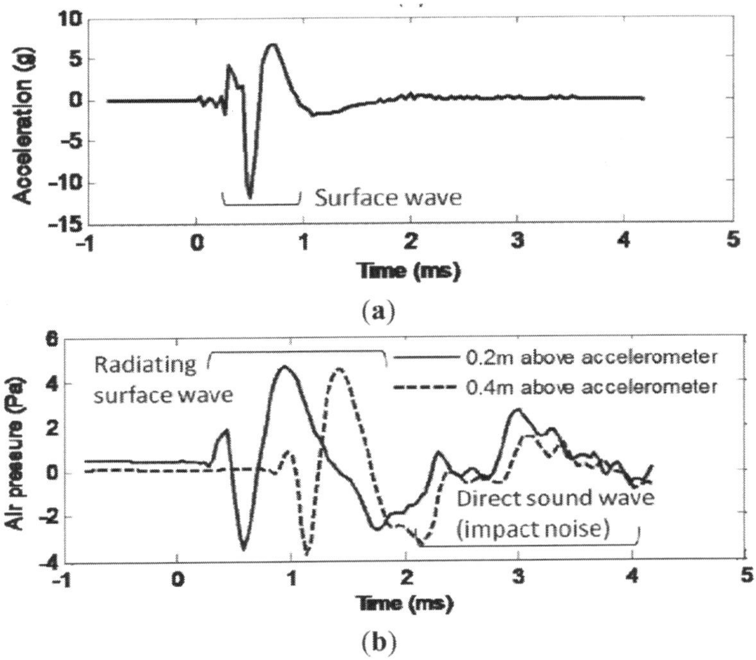

Figure 4. Comparison of microphone signal with accelerometer signal. (**a**) Measured surface acceleration. (**b**) Measured acoustic above surface signal.

Currently, some efforts have driven surface wave-based methods from traditional ground-contact sensing into non-contact sensing, either through an air-coupled transducer or an optical laser transducer. Earlier efforts like [20] sought to detect subsurface defects on bridge decks with acoustics. Recently, the authors of [16] described the air-coupling sensing theory and application. In their work, it was demonstrated that the acoustic radiation from the surface wave (radiating surface wave) could be acquired and processed similarly to the acceleration or velocity of the structure response. The replacement of ground-contact sensors with non-contact microphones would significantly improve the testing efficiency. However, there still exist difficulties to realize this efficient concept. One thing is the de-noising when the acoustic methods are used in the actual fieldwork subject to environmental noise. For this point, both the time windowing in the software and the de-noising in the hardware are applied in this paper. The time windowing in software was realized in an autonomous manner by using radiating surface wave arrival time estimation. The implementation of this automatic pre-processing to extract the radiating surface wave is introduced in the subsequent section.

2.2. Radiating Surface Wave Signal Extraction

Since the MASS system is designed with an orientation as an autonomous signal processing prototype, a pre-processing program implementing the automatic windowing for radiating surface wave is embedded in the software platform. The data acquisition is triggered by the impact acceleration through a professional shock accelerometer mounted coaxially on electromagnetic hammer core. The accelerometer monitors the hammer impact and triggers the data acquisition as the starting point in the time history. The time stamp where the first apparent peak or valley (max absolute amplitude) appears in the recorded acoustic data is identified as the arrival time of the radiating surface wave. A time-window with a Hanning shape is applied to truncate the signal of the radiating surface wave. The window center is aligned with the time stamp of the radiating surface wave, while the window size as arrival time difference from the radiating surface wave to the direct acoustic wave (Equation (2)). This windowing is realized automatically.

Additionally, several criteria are used in the pre-processing program to ensure acquired signal quality, especially in a field work environment. Firstly, the impact quality itself is verified through its time historical shape as acceleration, and only sharp impacts with broadband excitation would be accepted. Secondly, the magnitude squared coherence between sensing channels are also verified within 100 Hz to 10,000 Hz (Equation (3)):

$$C_{xy}(f) = \frac{\left|P_{xy}(f)\right|^2}{P_{xx}(f)P_{yy}(f)} \tag{3}$$

Signals with low coherence in this range imply poor signal quality and would be neglected. Recorded signals that failed to pass the pro-processing criteria would be discarded for further processing. A message window would display to indicate more trials being needed. Only the acquired signals that satisfy the criteria are used as raw data for further processing. The raw data with poor quality due to random environment noise such as low frequency ground vibration would not be used for analysis.

2.3. De-Noising Hardware Design

In order to improve the signal to noise ratio of acoustic data for the application in fieldwork, a sound barrier enclosure is designed to block the direct sound wave (impact noise) as well as the environmental noise. Figure 5 shows the design and a hardware realization. A steel cylinder is selected as the skeletal structure. A sound reflecting material with denser skin material wraps over the external surface of cylinder skeleton to reduce the penetration of outside noise. The acoustic noise would be mainly reflected on the denser skin surface first when it touches the skin. Then the noise penetrating the first reflection surface could be reflected by the steel cylinder wall again and attenuates in the form layer between the steel cylinder wall and the denser skin. A pyramid array pad is fitted in the internal surface of the skeleton and also the bottom circular surface to absorb reflections inside the chamber (Figure 5(b)). The microphone is mounted in the center of the cylinder with 1 cm above the ringed end plane as red circle marked in Figure 5(b).

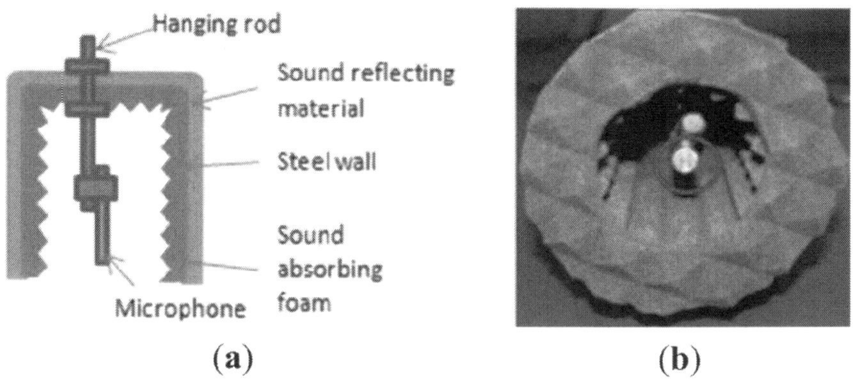

Figure 5. De-noise sound enclosure for microphone. (**a**) The schematic design of sound enclosure. (**b**) Built-up sound enclosure.

Such a design would block the ambient noise and impact noise whose incident directions are almost perpendicular to the enclosure cylindrical surface. In contrast, the radiating surface wave from the ground vibration would easily reach the microphone (Figure 2(a)).

Figure 6 indicates the effect of this sound enclosure. The microphone is 0.2 m away from the impact source. The comparison is realized with two impacts, one without enclosure and one with enclosure. Figure 6(a) shows that without enclosure, the noise level is so high that the radiating surface wave can hardly be separated from the environment noise in recorded data. After the enclosure is applied to the microphone (Figure 6(b)), the radiating surface wave can be easily identified and extracted from the time history signal. Moreover, the enclosure blocks other ambient noise arriving before the radiating surface wave. As could be observed in Figure 6(b), the recorded acoustic signal before the radiating surface wave is satisfactorily close to zero. This indicates a high signal to noise ratio.

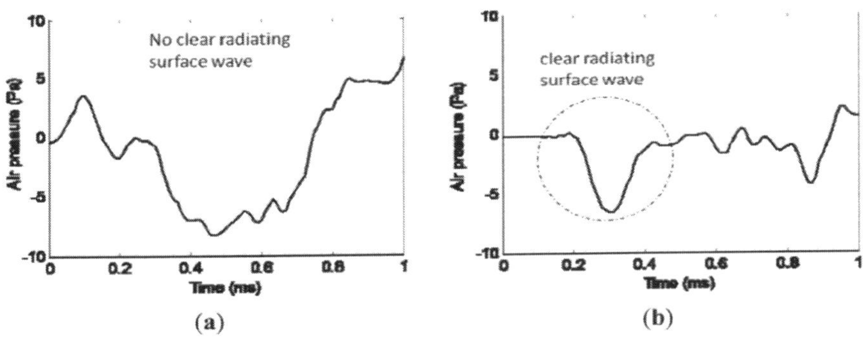

Figure 6. Effects of the sound enclosure for microphone in noisy environment. (**a**) Without enclosure. (**b**) With enclosure.

After the efforts as noted earlier to extract the radiating surface wave signal, the signal to noise ratio in the field work environment is studied using the Pingree Bridge Deck test (Section 5.3). As Figure 7(a) shows, the extracted radiating surface wave signal during impact is compared with the ambient noise during impact intervals. The average dB level for radiating surface wave is 96 dB, while the ambient noise contributes 75 dB. The signal to noise ratio is also indicated inFigure 7(b) within 10 kHz. The SNR is around 20 dB, except at the very low frequency end. This could be caused by the low frequency vibration of the bridge structure. However, this low frequency data would not affect the profile analysis for shallow depths within 1 m, since the corresponding penetration depth respect to wavelength is large.

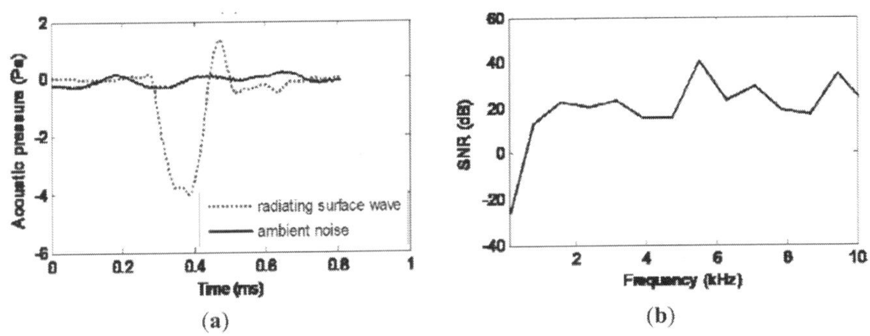

Figure 7. Ambient noise. (**a**) Comparison of extracted radiating surface wave and ambient noise. (**b**) Signal to noise ratio.

3. MASS HARDWARE PLATFORM INTEGRATION

The Mobile Acoustic Subsurface Sensing (MASS) system is constructed on a 3-wheeled cart. The cart platform is designed with an emphasis on stability during movement and impacting through vibration damping components. The MASS and its subsystems are presented in Figure 8. The first critical subsystem is the electronic hammer subsystem indicated as (a) inFigure 8. The hammer is based on two coaxial solenoids which lift and shoot the impact core. The impact properties could be controlled by tuning the power time on each solenoid with a time relay delay unit (528-TDRSOXP-24V). Due to the requirement of broad frequency band impact, the ideal impact would be close to a Dirac delta function. In order to monitor the impact quality and trigger the data acquisition, a professional shock accelerometer (PCB 350B03) is mounted on the impact core to watch the impact acceleration in real-time. Due to the variation in the roadway properties, the impact core is designed with replaceable tips with different stiffness.

The distance encoder/register subsystem (Figure 8(c)) is assembled with encoder mounted on the rear wheel and a PLC control unit (Figure 8(e)). The sensing spatial resolution is adjustable by changing the parameter setting at the PLC unit.

The MASS DAQ consists of a signal conditioner, a DAQ mother board, a Single Board Computer (SBC), DC power converter, and solid state hard drive. In the system, a SCM5B48 unit is selected as the signal conditioner. The PC104P-24DSI12 is selected as a 12-channel DAQ board which features low noise, 24-bit resolution, low phase distortion and multi-board synchronization. An embedded single board computer

(SBC) called *Mamba* from Versalogic (Tualatin, OR, USA) with a high-performance Intel Core 2 Duo processor, computes at up to 2.26 GHz. The Operation System is QNX. The system is realized as multiple-channel and high sampling rate.

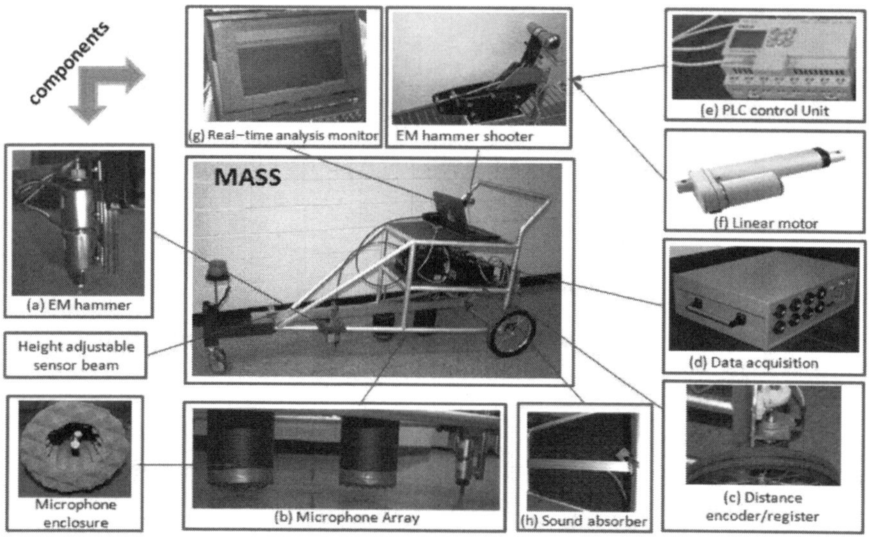

Figure 8. Integration of hardware of MASS.

The microphone array is mounted as shown Figure 8(b), the array is isolated from the cart with professional dampers to isolate the vibration from the cart frame, especially while the impact is acting. The clearance from microphones to ground is adjustable so that the system is adaptable to different ground surface profiles.

4. SOFTWARE FRAMEWORK

The MASS software framework consists of three modules: (1) the data acquisition program under QNX operating system which sets up the sampling configuration; (2) a fast air-coupled SASW analysis program [18] under the MATLAB framework in the laptop; and (3) the real-time data communication between the DAQ and laptop.

One benefit of this design is that the convenient MATLAB programming environment is robust enough for the analysis program updating for the prototype development. Consequently, the architecture of hardware and software integration is shown is Scheme 1.

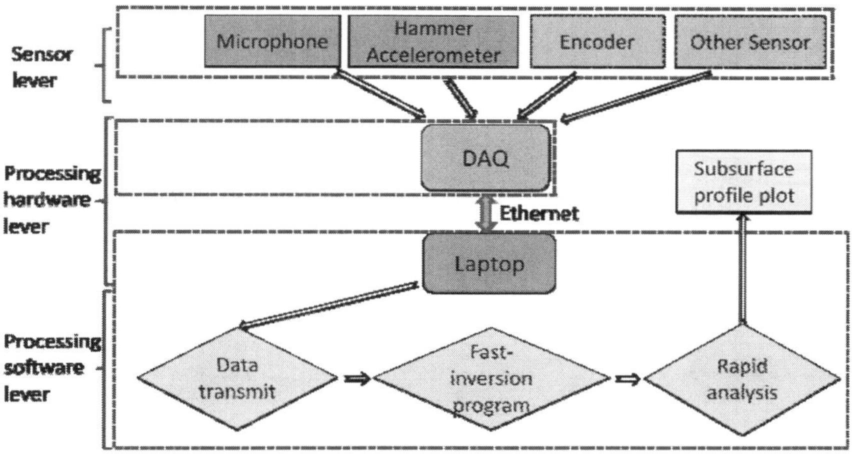

Scheme 1. Hardware/software integration architecture.

5. IMPLEMENTATION AND DISCUSSIONS

5.1. Asphalt Pavement

The MASS system was tested on an asphalt pavement located near the Northeastern University campus. The testing path was along a white chalk line as shown in Figure 9(a). The testing configuration is shown in Figure 9(b), where the microphone enclosure is 1 cm above the ground while the near (channel A) and father (channel B) microphones are 20 cm and 60 cm apart from the impact source, respectively. The sampling rate is set as 200 kHz for all sensing channels such as the microphones and hammer accelerometer. The entire system could be operated by only one person at walking speed after set-up.

A typical testing point result is shown in Figure 10. After the pre-processing in software as mentioned in Section 2.2 and with the de-noising sound enclosure as mentioned in Section 2.3, the radiating surface wave is extracted. Figure 10(a)indicates that the extracted signals are clean and smooth. The inversion result to indicate the stiffness profile is shown inFigure 10(b). Without prior information about the paving material, a uniform mass density of 1,800 kg/m^3 and a Poisson ratio of 0.3 are assumed in the calculation of the elastic modulus as general civil material properties. According to the profile, the first asphalt layer has an estimated depth of 0.1 m (3.94 inches) and estimated elastic modulus of 6.5 GPa (942.7 ksi), which falls into the range of a regular asphalt concrete material from 4.4 GPa (638.2 ksi) to 6.8 GPa (986.3 ksi). Also the thickness of the top layer matches with the typical roadway paving design. Underneath the top layer, the stiffness apparently decreases, as the modulus approaches 0.218 GPa (31.6 ksi) around 0.4 m (15.7 inches) depth. This is estimated to be compacted soil material.

(a) (b)

Figure 9. Remotion test on asphalt pavement. (**a**) Mobile sensing with MASS. (**b**) Test configuration.

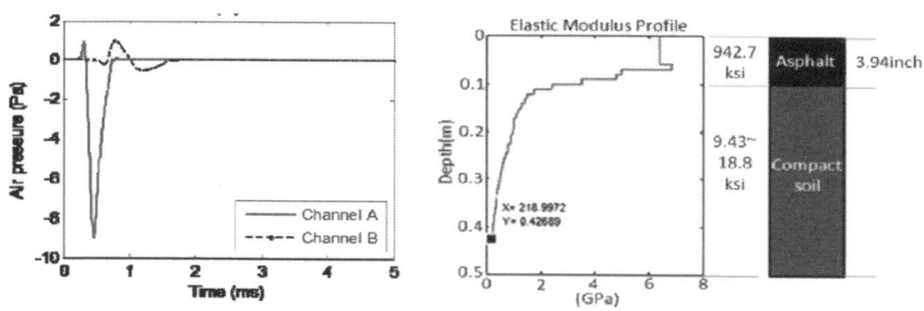

(a) (b)

Figure 10. Testing on asphalt pavement. (**a**) Extracted radiating surface wave. (**b**) Estimated elastic modulus profile.

5.2. Concrete Slab

The MASS system was also tested on a concrete slab in the Laboratory for Structural Testing of Resilient and Sustainable Systems (STReSS Lab) at George J. Kostas Research Institute for Homeland Security at Northeastern University. The same test configurations as above are employed. Figure 11(a) records the test scenario, where the MASS cart is working in the central area of the slab in order to avoid the reflections of the stress waves due to the limited slab size. The cross-section of the slab is photographed as Figure 11(b), with a symbolic diagram underside. The slab is embedded with two rows of metal rebar located 8.9 cm (3.5 inches) below the surface for the top row while the bottom row is close to the bottom surface. The thickness of the slab is 20.32 cm (8 inches).

(a) (b)

Figure 11. Concrete slab test configuration. (**a**) Test set-up. (**b**) Slab cross-section profile.

The testing results are presented in Figure 12. The extracted radiating surface wave is presented in Figure 12(a), where the signals are clean enough for post-processing after pre-processing as mentioned in Sections 2.2 and 2.3. Comparing with the radiating surface wave extracted from the asphalt pavement shown in Figure 10(a), the signals coming from the stiff concrete slab present fluctuations with higher frequency. When the stiffness of the top layer is much larger than that of the sub-grades underneath, the surface wave approaches the Rayleigh-Lamb wave. Without using multiple-channel sensing to resolve the higher modes, the fast air-coupled SASW procedure with only two channels still offers an acceptable estimation of the profile as seen in Figure 12(b). The estimated total thickness of around 20.32 cm (0.67 ft) coincides accurately with the actual thickness (Figure 11(b)). The average elastic modulus of the concrete slab is estimated as 11.9 GPa (1,720 ksi). This modulus estimation yields the normal concrete slab modulus range and is a little on the low side. The minimum stiffness of 2.1 GPa (300 ksi) roughly deeper than the slab thickness could be caused by the wood beam foundation (Figure 11(a)). Due to the large stiffness contrast between concrete slab and the wood beam foundation, the deeper profile of the lab ground would not be identified from Figure 12(b). This test demonstrates the capability of the MASS system for concrete slab sensing in a floating style through acoustics.

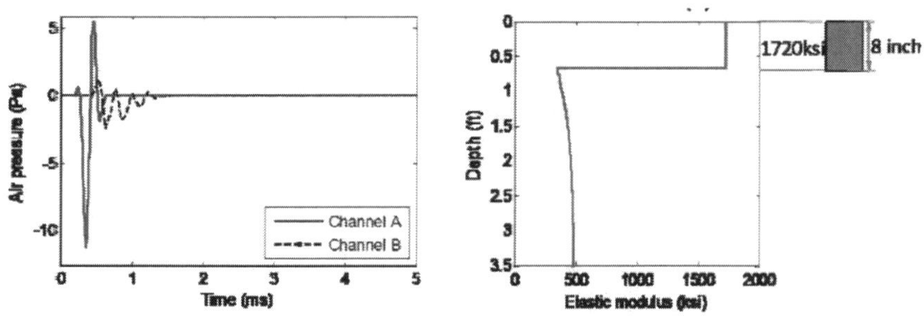

(a) (b)

Figure 12. Concrete slab test results. (**a**) Extracted radiating surface wave. (**b**) Estimated elastic modulus profile.

5.3. Concrete Bridge Deck

Beside the capability for sensing the pavement structure, MASS system has the potential for bridge deck structure sensing. The fast air-coupled surface wave analysis program could be used for the bridge deck test with a Lamb wave inversion when the half-space foundation is deleted. The MASS system tests were performed on a bridge deck on the Pingree Bridge (Pingree Farm Rd, Rowley, MA, USA). The bridge consists of single layer concrete deck with as designed 9 inch thickness and spans around 240 ft above the Interstate 95 highway (Figure 13(a)).

The test is implemented in the real traffic environment with traffic noise. The MASS system configuration is maintained the same as in the preceding tests. The scanning is executed along a path (the red dotted line with arrow in Figure 13(a)) where no support beam exists underneath. The continuous scanning test results are presented in Figure 14. In total,m 23 test points are scanned with 1.22 m (4 ft) spacing along the scanning path. Figure 14(a) shows the filtered acoustic signals in the form of the extracted radiating surface wave. It could be observed that the signal features are close to those of the concrete slab as in the previous case. Higher frequency signal components are apparent. The successive profiles acquired through the mobile scanning at a walking speed are summarized in Figure 14(c). The results are consistent along the same path and the estimated thickness and close to the designed thickness at 0.23 m (0.75 ft) (red star line in Figure 14(c)). The averaged modulus of the deck is 17.7 GPa (2580 ksi). The profile at points 20th, 21st and 22nd give an abnormally low modulus around 2.7 GPa (400 ksi). This abnormity is caused by passing over the slab joint crack as shown in Figure 14(b). The slab joint crack can be identified sensitively. This fieldwork test verifies the potential for bridge deck testing using the MASS system, which is designed originally for pavement structures, when some modifications are made to the program.

(a) (b)

Figure 13. Pingree bridge deck test (Rowley, MA, USA). (**a**) Scanning path on bridge deck. (**b**) MASS operation on bridge deck.

6. CONCLUSIONS AND FUTURE WORK

The transportation community has seen the great importance of technologies that can provide fast inspections of highways and bridge deck health assessments without interfering with or stopping traffic. In this paper, a mobile acoustic testing system is developed to estimate the subsurface profiles for both the thickness and stiffness of layers. The air-coupled non-contact sensing strategy is adopted in the MASS development. A fast air-coupled surface wave analysis program [18] is embedded to enhance the signal processing efficiency. The integration of the hardware utilizes the distance register/trigger together with an electronic hammer as impact source to supply repeatable and stable excitation. The de-noising sound enclosure and vibration isolation design improve the feasibility of acoustic sensing in actual fieldwork. The prototype hardware and software combination is demonstrated successfully. The system has been tested for asphalt pavement, a concrete slab and a concrete bridge deck. The reasonable matching of estimated results with exact subsurface profiles demonstrates the strength of the

MASS system concept. Despite these achievements, more efforts are still underway. A better impact source with adjustable force and highly robust contact shapes for the test targets are under construction. Additionally, more efficient de-noising devices are also under research to supply cleaner and higher signal to noise ratios for acoustic sensing under heavy traffic noise conditions. Also, more field tests are planned to enhance MASS system's practical capability.

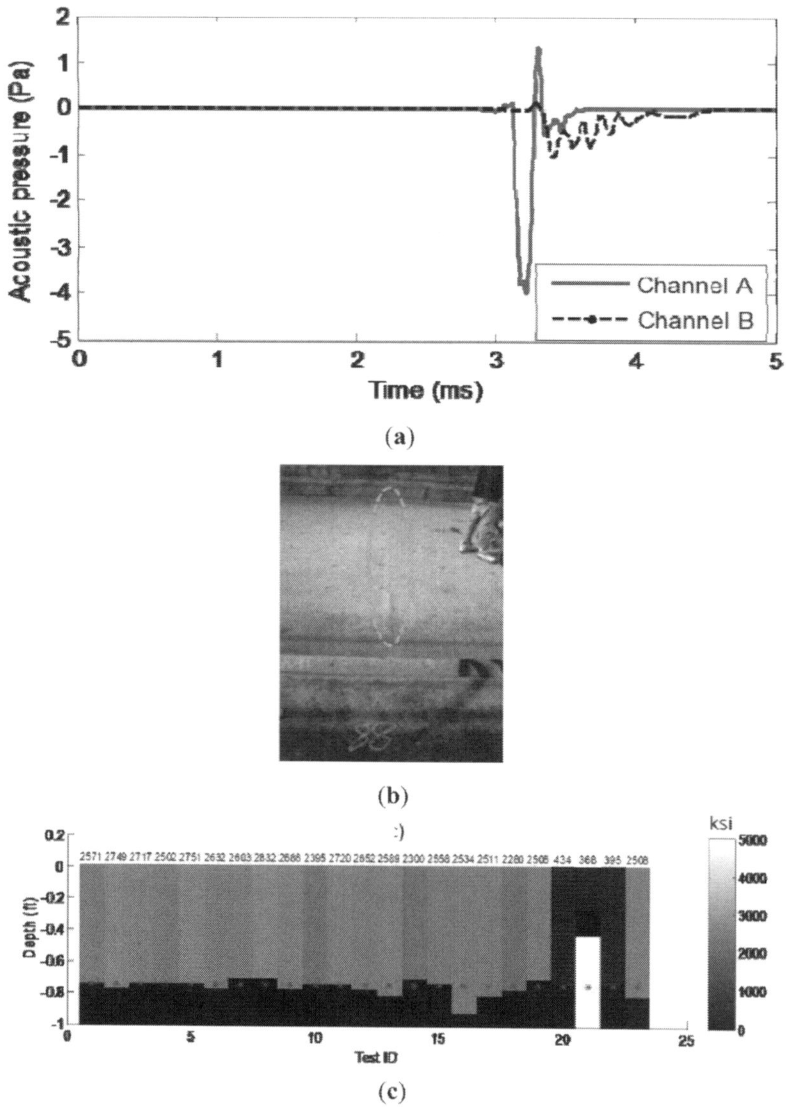

Figure 14. Pingree bridge deck test results. (**a**) Extracted radiating surface wave. (**b**) Joint crack. **(c)** Continuous scanning results.

ACKNOWLEDGMENTS

This research was sponsored by the USA Department of Commerce, National Institute of Standards and Technology, Technology Innovation Program, with Cooperative Agreement Number 70NANB9H9012. The authors gratefully acknowledge this support.

CONFLICT OF INTEREST

The authors declare no conflict of interest.

REFERENCES

1. Sansalone, M.; Streett, W.B. *Impact-Echo: Non-Destructive Evaluation of Concrete and Masonry*; Bullbrier Press: Jersey Shore, PA, USA, 1997.

2. Davis, A.G. The non-destructive impulse response test in North America: 1985–2001. *NDT E Int.* **2003**, *36*, 185–193.

3. Bungey, J.H. Sub-surface radar testing of concrete: A review. *Constr. Build. Mater.* **2004**, *18*, 1–8.

4. *Finding Delaminations in Concrete Bridge Decks*. Available online: http://www.acoustics.org/press/146th/Costley.htm(accessed on 20 January 2013).

5. Stokoe, K.H., II; Nazarian, S. Effectiveness of Ground Improvement from Spectral Analysis of Surface Waves. Proceedings of the European Conference on Soil Mechanics and Foundation Engineering, Helsinki, Finland, 23–26 May 1983; pp. 91–94.

6. Nazarian, S.; Stokoe, K.H., II. *In situ* Determination of Elastic Moduli of Pavement Systems by Spectral-Analysis-of-Surface-Waves Method: Practical Aspects. Available online: http://library. ctr. utexas.edu/pdf2/368-1F.pdf (accessed on 15 January 2013).

7. Krstulovic-Opara, N.; Woods, R.D.; Al-Shayea, N. Nondestructive testing of concrete structures using the rayleigh wave dispersion method. *ACI Mater. J.* **1996**, *93*, 75–86.

8. Nazarian, S.; Yuan, D.; Baker, M.R. Rapid Determination of Pavement Moduli with Spectral-Analysis-of-Surface-Waves Method. Available online: http://library.ctr.utexas.edu/pdf1/1243-1.pdf (accessed on 15 January 2013).

9. Joh, S.H. Advances in the Data Interpretation Technique for Spectral Analysis of Surface Wave (SASW) Measurements. Ph. D. Thesis, University of Texas at Austin, Austin, TX, USA, 1996.

10. Bay, J.; Gilbert, J.; Park, K.; Sasankul, I. Shear Wave Velocity Profiling Using Spectral Analysis of Surface Waves (SASW) at Williams Street Park and Coyote Creek Borehole. Available online: http://pubs.usgs.gov/of/2005/1169/chapters/of2005-1169_part2_ 05_ Bay.pdf (accessed on 15 January 2013).

11. Kausel, E.; Roesset, J.M. Stiffness matrices for layered soils. *Bull. Seismol. Soc. Am.* **1981**, *71*, 1743–1761.

12. Olson, L.D.; Miller, P. Comparison of surface wave tests for pavement system thicknesses/moduli. *Geotech. Sp. Public.* **2009**, *189*, 174–179.

13. Ismail, M.A.; Samsudin, A.R.; Rafek, A.G.; Nayan, K.A.M. *In situ* determination of layer thickness and elastic moduli of asphalt pavement system by Spectral Analysis of Surface Wave (SASW) method. *Geotech. Sp. Public.* **2009**, *192*, 70–76.

14. Cho, Y.S. NDT response of spectral analysis of surface wave method to multi-layer thin high-strength concrete structures. *Ultrasonics* **2002**, *40*, 227–230.

15. Park, C.B.; Miller, R.D.; Xia, J. Multichannel analysis of surface waves. *Geophysics* **1999**, *64*, 800–808.

16. Zhu, J.; Popovics, J.S. *Non-Contact NDT of Concrete Structures Using Air Coupled Sensors*; NSEL Report Series, Report No. NSEL-010; University of Illinois at Urbana-Champaign: Urbana, IL, USA, 2008.

17. Ryden, N.; Lowe, M.J.S.; Cawley, P. Non-Contact Surface Wave Testing of Pavements Using a Rolling Microphone Array. Proceedings of the NDTCE, Nantes, France, 30 June–3 July 2009.

18. Cao, Y.; Lu, Y.; Zhang, Y.; McDaniel, J.G.; Wang, M.L. A fast inversion analysis algorithm for the spectral analysis of Surface Wave (sasw) Method. *Proc. SPIE* **2011**.

19. Lu, Y.; Cao, Y.; McDaniel, J.G.; Wang, M.L. Finite element analysis of surface wave radiation for pavement debonding.*Proc. SPIE* **2010**, *7649*, 76490W:1–76490W:10.

20. Moore, W.M.; Swift, G.; Milberger, L.J. An instrument for detecting delamination in concrete bridge decks. *Highway Res. Record* **1973**, *451*, 44–52.

CHAPTER 5

The Search for Sustainable Subsurface Habitats on Mars, and the Sampling of Impact Ejecta

Magnus Ivarsson [1,*] and Paula Lindgren [2]

[1] Department of Palaeozoology, Swedish Museum of Natural History, Svante Arrheniusväg 9, P.O. Box 50007, SE-104 05 Stockholm, Sweden

[2] Department of Geographical and Earth Sciences, University of Glasgow, Gregory Building, Lilybank Gardens, Glasgow, G12 8QQ, UK

ABSTRACT

On Earth, the deep subsurface biosphere of both the oceanic and the continental crust is well known for surviving harsh conditions and environments characterized by high temperatures, high pressures, extreme pHs, and the absence of sunlight. The microorganisms of the terrestrial deep biosphere have an excellent capacity for adapting to changing geochemistry, as the alteration of the crust proceeds and the conditions of their habitats slowly change. Despite an almost complete isolation from surface conditions and the surface biosphere, the deep biosphere of the crustal rocks has endured over geologic time. This indicates that the deep biosphere is a self-sufficient system, independent of the global events that occur at the surface, such as impacts, glaciations, sea level fluctuations, and climate changes. With our sustainable terrestrial subsurface biosphere in mind, the subsurface on Mars has often been suggested as the most plausible place to search for fossil Martian life, or even present Martian life. Since the Martian surface is more or less sterile, subsurface settings are the only place on Mars where life could have been sustained over geologic time. To detect a deep biosphere in the Martian basement, drilling is a requirement. However, near future Mars sample return missions are limited by the mission's payload, which excludes heavy drilling equipment and restrict the missions to only dig the topmost meter of the Martian soil. Therefore, the sampling and analysis of Martian impact ejecta has been suggested as a way of accessing the deeper Martian subsurface without using heavy drilling equipment. Impact cratering is a natural geological process capable of excavating and exposing large amounts of rock material from great depths up to the surface. Several studies of terrestrial impact deposits show the preservation of pre-impact biosignatures, such as fossilized organisms and chemical biological markers. Therefore, if the Martian subsurface contains a record of life, it is reasonable to assume that biosignatures derived from the Martian subsurface could also be preserved in the Martian impact ejecta.

KEYWORDS

subsurface biosphere; hydrothermal systems; impact ejecta; life on Mars

1. INTRODUCTION

Life is a strong force on our planet, and it has endured for billions of years despite global catastrophies, such as meteorite impacts, global glaciations, mass extinctions, and climate changes. From the fossil record, we know that microbial life has been present on Earth since at least 3.5 Giga annum (Ga) [1,2,3] and perhaps as early as 3.8 Ga [4]. Eukaryotes have been present since ~1.8 Ga [5,6], and the atmosphere has been oxygenated due to photosynthesis since ~2.4 Ga [7]. Life is intimately involved with most of the processes that shape the surface of our planet, like weathering and the distribution of vegetation on the continents, as well as the influence on the chemical composition of the atmosphere. Earth's appearance from space is highly characterized by the presence of life, and from an alien perspective, it would not be difficult to detect life on our planet. Spectral signatures that indicate habitability, and that would be detectable from the other side of our galaxy, are, for example, atmospheric spectra of O_2, O_3, CH_4, and N_2O gases, and spectra from surface reflections indicating absorptions by pigments [8]. From such a remote perspective, Earth is the only planetary body in our solar system that harbors life. Nevertheless, it is possible that other planets harbored life earlier in their history, but that this life later went extinct. It could also be a possibility that life still exists on other planets, but that it has been forced to abandon the surface, or never inhabited the surface due to environmental reasons, and now only exists in isolated subsurface havens, where it is not detectable from Earth [9].

The last two decades have involved a change of view regarding the distribution of life on Earth, which may have consequences for our understanding of habitability on other terrestrial bodies. The knowledge of a deep subsurface biosphere that thrives at depths of as much as ~5 km in continental settings [10], as well as in the sub-seafloor crust [11], is constantly growing. This biosphere is not directly exposed to sun radiation and exists at conditions that previously were thought of as hostile for life. The subsurface communities are composed of extremophiles that can survive at high and low temperatures, high pressures, and extreme pHs. This life is more or less independent of the sun's radiation as an energy source (even though some oxidants used by the microorganisms may be due to photosynthesis at the surface). Instead, the microorganisms gain energy from inorganic sources, like minerals or volcanic gases, and extract carbon from inorganic carbon sources, like carbon dioxide.

The fossil record tells us that microbial life has existed in deep oceanic crust that is associated with hydrothermal activity throughout life's history on Earth. Some of the earliest fossils on Earth are found in pillow lavas [12] and hydrothermal deposits [13]. In fact, deep subsurface settings probably served as protected havens for life on a young Earth, which was constantly subject to heavy bombardment [14].

The knowledge of life's capability to inhabit the terrestrial subsurface also displays new niches of where to search for life on other terrestrial planets, like Mars [15]. Hopefully, a world does not necessarily need to be lifeless just because its surface appears to be sterile and inhospitable. Life can survive at depths for long periods of time, and thus be out of reach for our detection. The current conditions on the Martian surface are extremely hostile to life. The subsurface, on the other hand, provides a more hospitable environment and stimulates speculation on possible life on Mars. Future Mars Sample Return (MSR) missions will be targeted to areas with great potential to detect present, or perhaps, more likely, fossil life. Potential subsurface settings, where water can be liquid and life can survive would be at depths of several km and out of reach for future MSRs. However, the current goal of the MSR missions is to reach areas unaffected by radiation and oxidation, and where organic molecules indicative of life or fossilized microorganisms might have been preserved. To reach areas unaffected by the radiation and oxidation at the Martian surface, the missions need to drill between 2 and 5 metres [16,17,18]. This need exceeds their capabilities. The payload of European Space

Agency (ESA)'s planned 2018 Exomars rover is restricted to a drill with the capacity of 2 m [19]; that is why alternative ways to reach the subsurface are discussed. Areas with recent hydrothermal outflow due to impacts or volcanic activity have been suggested for sampling [20]. Another way to reach the Martian subsurface without drilling is by sampling from impact ejecta [9]. Examination of impact ejecta might be the best alternative way of reaching the fossil record of the subsurface. Impact ejecta are composed of material excavated from great depths below the surface. Impact cratering is a high energy event, but several studies show that fossil biomarkers can survive a hypervelocity impact [21,22]. Therefore, the study of impact ejecta on the Martian surface in future missions would enable a search for a fossil record of a sustainable Martian deep biosphere without using heavy drilling equipment.

In this paper, we review the current knowledge of sustainable subsurface ecosystems on Earth and their analogous nature to possible Martian counterparts. We also give suggestions of how to search for a putative Martian subsurface biosphere, and how to access the subsurface in future missions, through the sampling of Martian impact ejecta.

1.1 The Subsurface as a Protected Haven for Life

Subsurface settings, usually associated with hydrothermal activity, have been suggested by several authors to serve as protected havens for life on planetary bodies with less favorable surface conditions for life [9,14,15,23]. There are mainly two reasons for this. The first reason is that subsurface environments can supply a deep biosphere with the important prerequisites for sustaining life: water and accessible nutrients. This has led several authors to suggest subsurface or subseafloor hydrothermal settings as the most plausible environment where life originated on the early Earth [24,25]. However, we will not go into detail regarding biogeochemical conditions for the origin of life.

The second reason is that the subsurface environments can protect life against hostile surface conditions like damaging UV-radiation or hypervelocity impacts. The early Earth, for example, at the time when life is thought to have originated and evolved (~3.9–3.8 Ga), was experiencing heavy meteorite bombardments that probably sterilized the surface of the planet at several occasions [26]. Life could have existed on the surface between the devastating impacts, but it was the subsurface that could protect communities during the harsh periods, and hence made life sustainable on Earth. It is not only on the early Earth that subsurface environments could have acted as protected havens for life. The evolution of life has been punctuated by mass extinction events throughout Earth's history due to impacts or global glaciations like the Snowball Earth [26,27,28,29,30,31]. During such catastrophic events, deep subsurface settings could have acted as refuges for microscopic life on Earth. Even if the impacts wiped out substantial portions of the surficial fauna and flora, fossil signatures of life could still be preserved in impact ejecta derived from the subsurface. These ejecta will be discussed later on in this article.

Mars is substantially different from Earth, and the conditions for a possible Martian subsurface biosphere would be markedly different. Even though oceans may have existed on Mars early in its history, the Martian surface has been dry for the majority of the planet's history. The current conditions on the Martian surface are extremely hostile to life. The low atmospheric pressure (~6 hPa), low surface temperature (average of ~240 K), lack of liquid water, UV radiation fluxes three orders of magnitude more damaging to DNA than those on Earth, and the oxidized soil chemistry, make the Martian surface uninhabitable for life as we know it. The subsurface is, on the other hand, much more hospitable for biota, and, in fact, the Martian subsurface might be more hospitable and protective than the Earth's subsurface. First of all, Mars' lower background heat flow and lower gravity allow deeper colonies [14]. Secondly, the thermal heat pulse from a major impact is briefer. On Earth, much of the thermal radiation from minor impacts is absorbed by boiling the oceans: Global thermal excursions are buffered by the heat capacity of the oceans, but at impacts large enough to vaporize the oceans, the thermal buffering serves only to prolong the disaster for thousands of years, while the oceans rain out. Without oceans, thermal buffering does not occur on Mars, but on the other hand, the low Martian escape velocity results in the fact that most of the energetic ejecta escape more easily into space, while massive quantities of less energetic ejecta are globally distributed.

2. EARTH

2.1. The Deep Subsurface Biosphere on Earth

The concept of the deep subsurface biosphere of the lithosphere has been known for several decades, but in the last twenty years, there has been an explosion in the exploration of this biosphere. Much of this is due to the drilling of deep holes, both in terrestrial and marine settings. As early as the 1920s, sulfate reducing bacteria were found in oil well brines at depths greater than 300 m [32,33], and in coal deposits at a depth of 1,089 m [34]. Sediments, as well as sedimentary rock, have also been known, for a long time, to contain microorganisms (e.g., [35,36,37]). Microorganisms have also been found in crystalline hard rock [38,39], and during the 1990s, the knowledge about subsurface microbes in hard rock greatly increased [40]. The depth at which life is being found is constantly increasing, and today, microorganisms have been found at depths that exceeds 5 km [10,41]. Putative fossilized filamentous microorganisms have been found in a variety of subsurface environments, such as hydrous alteration products in volcanic rocks, in oxidation zones of ore deposits, cavities in sediments, in Mississippi-Valley Type ore deposits, and in vein-type hydrothermal mineral deposits in crystalline rocks. In general, fossilized microorganisms are found in rocks that contain low-temperature aqueous alteration assemblages and the microfossils occur in veins and vesicles filled with secondary quartz, carbonates, zeolites, or clays. The fossilized microorganisms usually occur as Fe/Mn-hydroxide filaments upon which later oxides are precipitated [42,43,44].

The sub-seafloor biosphere is much less well-known than its continental counterpart due to obstacles involved in the sampling. It is only during the last two decades that this biosphere has been explored extensively. Sub-seafloor sediments [45,46], as well as hydrothermal vent areas [11,47,48], are known to harbor a vast biosphere of bacteria and archaea, and microorganisms have been found in sediments from depths of 10,897 meters below sea level in the Mariana Trench [49]. Our knowledge of the sub-seafloor biosphere of the basaltic crust, on the other hand, is still today more or less based on a fossil record. Drilled samples of basalt have been shown to contain two types of microfossils: ichnofossils (tracefossils) in volcanic glass [12,50] (Figure 1) and body fossils (fossilized microorganisms) in carbonate filled veins and vesicles [51,52,53] (Figure 2). Fossilized microorganisms have been found at depths of as much as ~950 meters below the seafloor (mbsf). Both ichnofossils in volcanic glass and fossilized microorganisms in the carbonate-filled vugs of basalts are interesting within an astrobiological context. This is because, in the future, they present potential features to search for in rock samples returned from Mars [51,54].

The concept of the deep subsurface biosphere can be problematic when it comes to distinguishing between the ground-surface biosphere, and the subsurface or subterranean biosphere, and there is no general consensus. In our opinion, there is no obvious boundary between the ground-surface biosphere and the subsurface biosphere. Instead, there is merely a gradual transition from the surface biosphere to the subsurface biosphere, such that the influence of the surface decreases with depth. Due to the main focus of this paper on the potential microbial life in deep Martian rocks, we will mainly refer to the subsurface biosphere as microbial life that occurs in the crystal basement or oceanic crust, and not in the overlying sediments. On Earth, subsurface microbial life occurs in most rock types from sedimentary rock to volcanic rock and crystalline granitic rock. On Mars, however, the basement consists of basalt, and we will focus mostly on terrestrial subsurface environments that can be used, more or less, as analogous environments to Martian conditions.

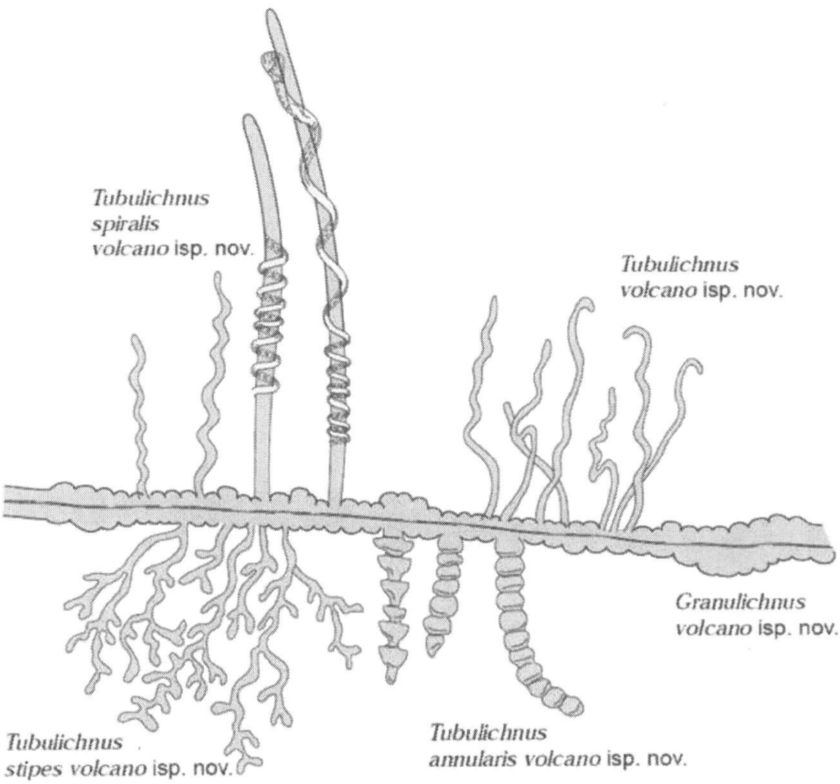

Figure 1. Line drawing showing the various ichnotaxa of tubular and granular ichnofossils found in oceanic volcanic glass worldwide. From left to right: helical shaped or coiled *Tubulohyalichnus spiralis* isp. nov.; simple unornamented and unbranched *Tubulohyalichnus simplus* isp. nov.; on the lower side of the fracture, branched *Tubulohyalichnus stipes* isp. nov.; annulated *Tubulohyalichnus annularis* isp. nov.; on both sides of the fracture, *Granulohyalichnus vulgaris* isp. nov. (Note: this is a schematic juxtaposition of the ichnotaxa; all five have not been observed alongside one another in natural samples; approximate relative scales are illustrated). Modified from McLoughlin *et al.* (2009) [55].

The main prerequisite of the subsurface and sub-seafloor biosphere is the presence of migrating fluids. Microorganisms exist in cracks and fissures in the rock, where fluids migrate and the microorganisms are dependent on the fluids for their transport through the system (Figure 3). Usually, they progress through the system in a passive and non-motile way. Thin films of water are probably enough for the passive transportation of microorganisms by capillary forces. The fluids also contain dissolved elements and compounds that are used by the microorganisms for their metabolism. Since the microorganisms are out of reach from the sun, the base of these ecosystems consists of chemolithoautotrophic microorganisms. Chemolithoautotrophs obtain their energy from the oxidation of inorganic compounds, rather than from sunlight, and use inorganic carbon as a carbon source. Usually, chemolithoautotrophs occupy habitats at the interface between two environments with divergent redox chemistry [56]. In sub-seafloor settings, for example, microbes have been found mostly associated with highly reduced volcanic glass (e.g., [57,58,59]). In contact with oxidized seawater, volcanic glass is out of equilibrium, and redox reactions favorable for microorganisms may occur. The reactions are relatively slow at low temperatures, and the microorganisms take this opportunity to catalyze the reactions and gain metabolic energy in the process. The most accessible compounds for chemolithoautotrophs in subsurface settings would be Fe^{2+}, H_2, H_2S, S, CO, CH_4, and Mn^{2+}, and electron acceptors could be O_2, NO_3^-, Mn^{4+}, Fe^{3+}, SO_4^{2-}, S, and CO_2. Heterotrophs do occur as well, and use organic C in various constellations, like CH_4, as an energy source.

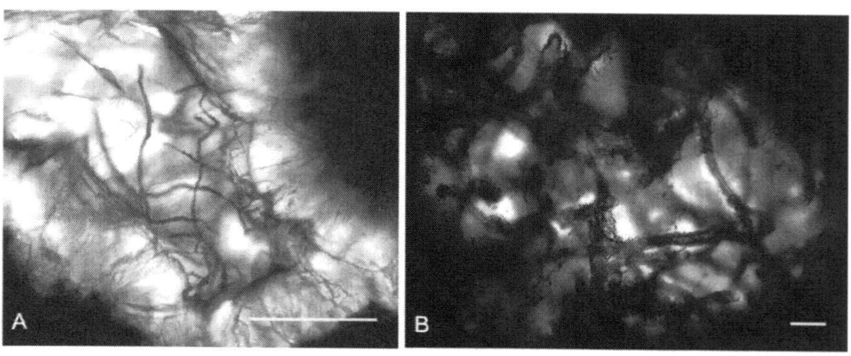

Figure 2. Microphotographs of microfossils in carbonate-filled veins and vesicles in basalts. (A) Microfossils at the interface between zeolites and carbonates in the Ocean Drilling Program (ODP) sample, 1206A-37R-3,72. Scale bar: 100 µm. From [52]. (B) microfossils in carbonate filled veins in basalts from the ODP sample, 1206A-18R-1,104. Scale bar: 10 µm. From [51]. See references [51,52] for more information and discussion regarding the biogenicity of the microfossils.

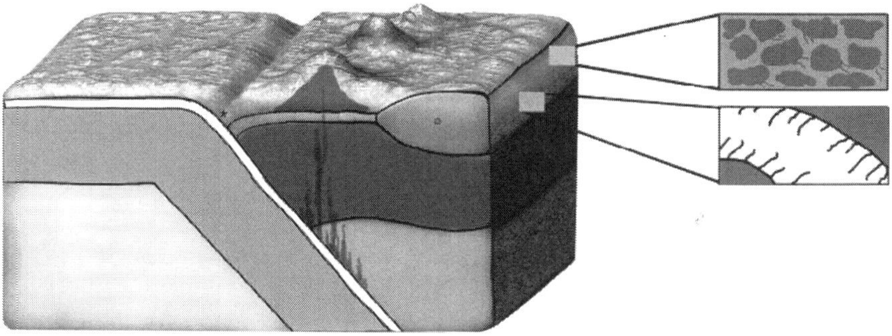

Figure 3. Schematic drawing of the occurrence of microorganisms in the pore spaces of sediments or sedimentary (upper box) rocks, as well as fractures and veins in crystalline rock (lower box). The drawing also indicates the greatest depths at which microorganisms have been observed in the oceanic and continental lithospheres, respectively. * indicates the maximum depth where microorganisms have been found in marine settings: the Mariana Trench at 10,897 meters below sea level. ◦ indicates the maximum depth where microorganisms have been found in continental settings at ~5 km depths.

The Äspö Hard Rock Laboratory (HRL), Sweden, is an underground laboratory focused on the geological and biological issues associated with nuclear waste storage. This underground laboratory is situated 450 m underground, and is a unique location to study the subsurface biosphere, and is probably one of the places on Earth where the subsurface biosphere has been explored most extensively. At Äspö HRL, iron-, sulfide-, manganese-, and methane-oxidizing bacteria have been found in large numbers, as well as nitrate-, iron-, manganese- and sulfate-reducing bacteria, acetogens, and methanogens [41,60].

Usually, the subsurface biosphere is associated with prokaryotes exclusively, but eukaryotes have been found as well. Fungi have been isolated from a wide variety of subsurface environments like soils, deep subseafloor sediments, sedimentary rocks, ground waters, and igneous rock [61]. In soils, fungi are abundant in symbiosis with roots as mycorrhizae, but further down in subsurface settings, fungi are considered to occur more sparsely. Ground water aquifers in sedimentary rocks have been shown to contain a diversity of yeasts and fungi [62,63], as well as ground water from crystalline granite bedrock at Äspö, Sweden [64]. Ekendahl *et al.* [65] isolated five yeasts, three yeast-like fungi, and 17 filamentous fungal strains at depths between 201 and 448 m below sea level at Äspö HRL (Hard Rock Laboratory). Phylogenetic analysis of the 18S rDNA of the five yeasts' isolates revealed their relationships to *Rhodotorula minuta* and*Cryptococcus* spp.

2.2. Sustainability of the Deep Biosphere

The metabolic pathways of chemolithoautotrophs show how well adapted the microorganisms of the subsurface biosphere are to their environment. It is the geochemical conditions of the habitats that define the environmental limitations that the microorganisms have to adapt to. The conditions, at which the subsurface biosphere exists, are normally thought of as extreme, and the microorganisms are often referred to as extremophiles. This is a term that includes a wide array of different microorganisms adapted to certain extreme conditions. Thermophiles are microorganisms with an optimum growth temperature of ~50 °C or above, but with a maximum of 80 °C. Hyperthermophiles are microorganisms with optimum growth temperatures above 80 °C, with a maximum of ~120 °C. Psychrophiles are the opposite, and represent microorganisms that thrive at low temperatures, preferably below 15 °C. Barophilic microorganisms thrive at high pressures, acidophiles at low pHs (preferably below pH 3), and alkaliphiles thrive at high pHs (above pH 9). Combinations of different extremophilic abilities occur. For instance, thermoacidophiles prefer temperatures of up to 80 °C and pH levels below 3.

Adaptation is crucial for the survival and sustainability of the subsurface biosphere, since these organisms are restricted to the subsurface conditions, and are more or less isolated from the surface biosphere. Environmental stress forces the microorganisms to adapt as geochemical conditions change over time in the subsurface settings due to, for example, rock weathering, alterations in fluid flow, or increases in temperature. Microorganisms cope with environmental stress in different ways. Some microorganisms are mixotrophic, which means that they can use a mix of different energy and carbon sources. *Shewanella* spp, for example, a marine bacterium commonly occurring in seafloor basalts, is capable of iron-, manganese- and sulfate reduction [48]. There are also mixotrophs that can alternate between photo- and chemotrophy, litho- and organotrophy, or auto- and heterotrophy. In subsurface settings, microbial communities are forced to adapt to microenvironments with varying redox potentials within a limited spatial scale. The establishment of local niches with varying metabolic pathways increases the enrichment of taxa from diverse metabolic groups, but also the capability of mixotrophy [11].

Environmental changes can sometimes occur instantaneously and many microorganisms are equipped with specific mechanisms to cope with such changes. Various heat shock response mechanisms, such as the thermally stable design of proteins and efficient DNA repair are examples of protective microbial responses to shifts in temperature. Similar response mechanisms have been developed among microorganisms to cope with changes in pressure, acidity, radiation, *etc.* Most of the processes are far from being understood. The organisms have been studied as single isolates, or as part of a limited suite of organisms removed from their natural environments. However, by coupling gene expression studies with analysis of geochemical variations, one may elucidate some of the driving forces of evolution in subsurface ecosystems.

If the microorganisms are subject to unfavorable conditions for a long period of time, they can go into dormancy, which is an inactive period when growth and metabolic activity slows down or completely ceases. This helps an organism to conserve energy. When the environmental conditions are favorable again, the organism can resume from its dormant state and increase its growth rate and metabolic activity. The dormant state can last for very long periods of time. Microorganisms as old as eight million years have been retrieved and grown in a laboratory from ice core samples from Antarctica [66]. However, the sustainability of a biosphere is not maintained by frequent dormant periods. Even if the conditions are considered extreme and the nutrient supply is low in subsurface environments, a stable ecosystem is needed to achieve a steady growth rate. Microorganisms in subsurface environments are known to adapt their growth rate to the supply of nutrients, and thus their metabolic activity is relatively low [67].

Microbial communities can also adapt to dynamic physical and chemical conditions in other ways, such as biofilm formation or motility [68]. Biofilm formation ensures robust stable communities to resist environmental stress. In fact, the production of exopolysaccharides is increased under conditions of stress. Motility, on the other hand, provides a mechanism for microorganisms under environmental stress to find new habitable environments.

2.3. Endoliths

Microorganisms living in the interior of rocks are referred to as endoliths. The microorganisms attached to the external surfaces of rocks are termed epiliths. Endoliths are divided into euendoliths, cryptoendoliths, and chasmoendoliths, depending on their occurrence and interaction with the host rock [69]. Euendoliths are organisms that actively bore in the rock and create etched pits or microtubular cavities. Cryptoendoliths are organisms that inhabit pre-existent cracks and fractures in the rock, and interact with the host rock in the area that they inhabit. Chasmoendoliths are organisms that inhabit pre-existent cracks and fractures without interacting with the host rock.

On Mars, the dominating rock type is basalt, just like the ocean crust on Earth. Future missions that will search for fossilized endolithic life on

Mars will thus be facing basalt as the main substrate. It is therefore interesting that remnants of all types of endolithic life have been found in basalts from Earth. Euendolithic boring is a common feature in volcanic glass from the ocean crust. McLoughlin *et al.* [54] pointed out that this type of ichnofossils will be an important textural biomarker in future astrobiological exploration, and Fisk *et al.* [70] reported microtubular structures in a Martian meteorite similar to biologically formed microtubular ichnofossils from Earth. Putative fossilized cryptoendoliths in carbonate filled vesicles in pillow basalts have been reported by Peckmann *et al.* [53] and Eickmann *et al.* [71] and chasmoendoliths in carbonate and zeolite filled veins and vesicles in deep drilled basalts from the ocean floor have been reported by Ivarsson *et al.* [51,52]. Ivarsson and Holm [72] performed a study on fossilized microorganisms in basalts from the oceanic crust that have implications for the sustainability and the search for a putative Martian subsurface biosphere. They pointed out that fossilized microorganisms and ichnofossils in sub-seafloor basalt are associated with minerals and materials (volcanic glass) that represent different stages of the alteration of the ocean crust. This indicates that the microorganisms have been able to colonize various micro-niches in the sub-seafloor environments as the alteration of the crust proceeded and as the geochemical conditions changed. Palagonitization is the process in which volcanic glass is altered to palagonite (a gel-like, amorphous, yellowish material that replaces glass) and secondary phases like zeolites and smectites. Palagonitization has been described differently by various authors, but according to Honnorez [73], palagonitization takes place in three stages that are distinguished by the presence or absence of glass or secondary minerals. Ichnofossils or fossilized microorganisms are observed in each of the three stages of the palagonitization process [72]. Microbially etched ichnofossils are found in the fresh, unaltered volcanic glass, which represents the first stage. As the alteration proceeds, and the glass, as well as the rest of the basalt, is subject to fluid circulation and related hydrothermal alteration, fossilized filaments are found attached to both altered basalt and altered volcanic glass in the second stage. In samples that represent the third and final stage of palagonitization, and in which the veins are characterized by palagonite, smectite and phillipsite materials, fossilized filamentous microorganisms, as well as globular fossilized microbial cells and fossilized biofilms are found. Considering the microbial interaction with the host rock, the first stage is thus dominated by euendoliths, compared to the second and third stages. These latter stages are dominated by crypto- and chasmoendolithic microorganisms. It appears as if the alteration of the oceanic crust creates micro-habitats for the microorganisms to dwell in, and it is apparent that the microorganisms, by adapting to these new niches, can survive in the systems while their living conditions change with time [72].

3. MARS

3.1. The Martian Subsurface

Theoretically, subsurface life is as plausible on Mars as it is on Earth. The Martian lithosphere consists of basalts, with a high amount of reduced minerals that, when in contact with oxidized fluids, offer conditions with high redox potentials that microorganisms can take advantage of to gain energy. As we have seen in previous sections of this article, basalt systems on Earth harbor extensive microbial ecosystems. Carbon accessible for microorganisms is also present on Mars; the most common form of carbon is probably CO_2 or CH_4 [74,75]. The most plausible source of carbon are volcanic gases from the planet's interior.

The habitability of Martian rocks for endolithic communities is probably comparable to the habitability of rocks on Earth. In fact, Martian rocks may be better suited to host endolithic communities than terrestrial rocks. On Earth, fracturing and the production of micropores where endolithic communities may thrive are a result of cooling and, above all, processes related to plate tectonics. The Martian lithosphere has not been subject to plate tectonics as the terrestrial lithosphere has. However, the present Martian lithosphere has been subject to more impacts than the present terrestrial basement. The terrestrial basement has been recycled many times through plate-tectonics. Fracturing caused by impacts can create new habitats for crypto- and chasmoendoliths and can substantially increase the possibilities for fluid flow and microbial growth in subsurface environments [76,77,78].

Thus, energy, carbon, and habitats do most likely exist on Mars. The major difference between Earth and Mars is the distribution of water. Liquid water is a prerequisite for the existence of sustainable life. Several observations suggest a wetter past on Mars that might even include oceans, but the current situation on Mars is the opposite. At the moment, Mars is a dry planet and water mostly occurs as ice. However, liquid water can occur at great depths, and recent observations indicate that occasional outflows occur on the Martian surface.

3.2. Hydrothermal Activity on Mars

Hydrothermal systems develop wherever a fluid phase coexists with a heat source to drive convective energy. On Earth today, the heat that drives hydrothermal processes originates from internal radioactive decay and core crystallization. Such heat dissipates by mantle convection from the interior of Earth and is finally released by surficial volcanism. Most volcanism on Earth is concentrated in the plate margins, and this is also where the most abundant hydrothermal activity is found. Mars differs fundamentally in this respect, since it never developed plate tectonics [79]. However, crustal volcanism has played an important role in the thermal evolution of the planet. Abundances of radiogenic isotopes (K, U, Th) in Martian meteorites are similar to terrestrial igneous rocks, and it is most likely that radioactive decay provided a long-term heat source to sustain magmatism [80]. Furthermore, morphological evidence of shallow magmatism and volcanism is widespread over the red planet.

Water on Mars in the form of ice is known to exist at the polar caps, in the ground, locally on the surface, and in the atmosphere [74,81,82,83,84]. We also know that liquid water has been abundant both on the surface and as groundwater earlier in Mars' past [85,86]. Thus the conditions, heat and water, that are required for hydrothermal activity exist on Mars.

A number of observations on Mars and of Martian meteorites have been suggested to be the result of hydrothermal processes. Martian meteorites indicate that there has been an active and ongoing process of crustal atmosphere exchange on Mars [87], and hydrothermal convection provides the most compelling mechanism for such exchanges [88]. Martian meteorites also contain trace quantities of minerals like clays, sulfates, carbonates, halides, and ferric oxides that have formed through interactions with aqueous solutions [89]. Another feature that has been assigned to hydrothermal activity is the loss of the early atmosphere. Thermodynamic models indicate that hydrothermal mineralization could have sequestered atmospheric CO_2 in the Martian crust [90]. Furthermore, several authors have suggested that the occurrence of iron and sulfur rich minerals, like hematite and jarosite, indicates aqueous and hydrothermal processes (e.g., [20,91,92,93,94,95]). The origin of some of these minerals is still debated, and alternative interpretations have been presented that involve more surficial processes, such as deposition in stagnant water [96].

Numerous geomorphic observations of channel systems have also served as evidence for hydrothermal activity [20]. The most convincing geomorphic evidence is that of channel systems along the margins of impact craters, and on the slopes of Martian volcanoes. Accumulation of water ice in non-polar regions has also been interpreted as representing hydrothermal activity. The fluid was probably released through melting of the cryosphere or snow caps by volcanically induced heating from underground [97]. In general, hydrothermal activity on Mars is characterized by cryospheric or magmatic systems, compared to hydrothermal processes on Earth that are characterized by meteoric or magmatic systems [20]. Mars contains an extensive subsurface cryosphere that can provide, when heated, a source of water for hydrothermal systems [81]. Hydrothermal systems generated by impacts are widespread in terrestrial impact craters [98,99,100], and an alternative heat source that could have given rise to hydrothermal activity on ancient Mars is the heat generated by impact cratering [15,101,102].

Recent observations by Mars missions are even more compelling and support the idea of past hydrothermal activity on Mars. The rover, Spirit, discovered silica-rich deposits in the Gusev crater. These have been interpreted to be of hydrothermal origin [103]. The silica-rich deposits were found in close association with volcanic materials and ferric sulfates, which are probably also of hydrothermal origin. The deposits resemble the silica deposits sometimes found in the vicinity of fumaroles or hot springs on Earth. No evidence of standing or flowing surface water were observed at the site. This further supports the idea of hydrothermal origin.

Observations of phyllosilicates by the Mars Express spacecraft and the Mars Reconnaissance Orbiter at several sites on Mars have also been used to argue for past hydrothermal activity on the planet [104,105,106]. The presence of minerals, such as nontronite, chamosite, montmorillonite, saponite, kaolinite, chlorite, and illite are indicative of the long-term presence of liquid water and of the alteration of mafic rocks like basalts. These minerals could have been formed on a surface with a warm and humid climate similar to the climate on Earth. However, the most probable explanation is that the minerals originate from hydrothermal processes in the shallow crust.

While hydrothermal systems appear to have existed on the ancient Mars, it is difficult to determine whether hydrothermal activity still exists on the planet or if it has ceased. Liquid water is presently unstable at the surface of Mars today, but extensive groundwater environments may exist at a depth of several km [107]. As a result of overburden pressure and the geothermal gradient, subsurface ice can melt at a given depth, and water can occupy rock pore spaces, forming aquifers. Subsurface zones of liquid water may also exist at more shallow depths in the vicinity of young volcanic centers that have recently erupted through the cryosphere on the high northern plains of Mars. Recent observations of repeated volcanic activation with phases of activity as young as two million years suggest that volcanoes are potentially still active today [97]. The contemporary accumulation and resurfacing of ice caps show that liquid water, possibly a hydrothermal fluid, has been involved. However, no major aquifers have been found at shallow depths by orbiters.

Active hydrothermal systems can be detected by localized anomalies in surface temperature or atmospheric composition [20]. Recently, seasonal concentrations as high as tens of ppb of methane have been detected by Earth-based observations of some localized regions on Mars [75]. The origin of the methane is unclear since the residence time of methane in the

Martian atmosphere is expected to be hundreds of years, and appears to be seasonal. This suggests mechanisms of actively releasing, capturing, or breaking-down methane. This can be explained by biological processes. However, further studies of the distribution of methane and the environment of release on Mars are necessary in order to assess the scenarios for methane's formation.

3.3. Sustainability of a Potential Subsurface Biosphere on Mars

With the knowledge of sustainable microbial ecosystems in subsurface basalt systems on Earth, as well as the potential for deep hydrothermal settings on Mars, it is possible to speculate on a subsurface biosphere on Mars. Subsurface hydrothermal systems on Mars can provide protection against extreme daily temperature variations, UV-radiation, oxidized soil chemistry, low atmospheric pressure (~6 hPa) and impacts. Additionally, the systems can offer heat, liquid water, and nutrients. The redox conditions of basalts with high amounts of reduced elements, like iron and manganese, in combination with the presence of oxidized liquid water, can provide energy to the base of the ecosystems and carbon can be scavenged from volcanic gases. Basalt is known from Earth to support microbial ecosystems for long periods of time and it has been shown that microorganisms can adapt to changing geochemical conditions as the basalt is subject to alteration [72,108,109]. The lower background heat flow and lower gravity on Mars allows for deeper colonies. Thus, the spatial distribution of a subsurface biosphere may be more extended on Mars than on Earth [14].

Another factor that is of interest, concerning a potential Martian subsurface biosphere, is the lack of active plate tectonics. On Earth, volcanic and hydrothermal active areas constantly move and the plate tectonics recycle the crust. It is a dynamic system that constantly causes non-equilibrium conditions in the crust and on the surface. Mars, on the other hand, is characterized by the opposite. No plate tectonics are present to recycle the crust and the volcanic centers are fixed, much like intra-plate volcanism on Earth (except where the plate moves above the active center). This means that a volcanic area on Mars associated with a potential hydrothermal system, as well as a potential subsurface biosphere, would be fixed and spatially isolated. Hydrothermal circulation would be concentrated around the volcanic active area, and probably would not be in contact with another active hydrothermal system. It could have been different on the young Mars, with its warmer climate and liquid water, both on the surface and in the ground. However, today, there is a risk that sporadic hydrothermal systems are isolated. This would have an effect on a potential biosphere: despite the subsurface's capacity to sustain life for long periods of time, eventually, the biosphere would suffer from lack of external contact. Survival of a closed system over billions of years is hard to imagine. However, the present state of Mars may not be a permanent state. The occurrence of water is strongly obliquity-driven and the stability of water has changed periodically [110]. Near-equatorial regions are especially sensitive to obliquity-driven periodicities of water stability, and it has been shown that much of the equatorial regions of Mars had periglacial landscapes indicating surface ice with melt water within the past few million years [111]. Such periods of water stability probably influence the subsurface as well and isolated regions of hydrothermal activity might experience contact with other hydrothermal environments by subsurface throughflow.

Even though the spatial fixation of volcanic active areas and associated hydrothermal systems might result in limited contact with other active hydrothermal systems and limit the input of necessary nutrients and microorganisms, these areas and systems may be able to support a deep biosphere for long periods of time. The volcanic areas, including potential hydrothermal systems, will be extended in both time and space compared to volcanism on Earth. The volcanic areas are larger than on Earth, and thus can support a more extensive hydrothermal system. The largest volcano in the solar system, Olympus Mons, is present on Mars. The volcano is 27 km higher than the mean surface level of Mars, and is 550 km in width. The Martian volcanoes are also active for longer periods of time. Olympus Mons, for example, was active between 3,800 and 100 million years ago. This is equal to 80% of the lifetime of the planet [97] and coincides with a wetter past for the planet. A potential subsurface biosphere that thrives in a hydrothermal system associated with a fixed heat source would have an extensive period of time to evolve in an extensive area of habitat.

4. IMPACT DEPOSITS: A WINDOW TO THE SUBSURFACE

As previously discussed, on Mars, the subsurface is a suitable target to search for life in. However, the sampling and analysis of the Martian subsurface is a tricky business for future planetary missions, due to the heavy drilling equipment needed in order to reach down into the subsurface. Therefore, it has been suggested that sampling and analysis of impact ejecta is one alternative way of studying the subsurface [9]. Impact ejecta are available for sampling at the Martian surface, but they represent geological units from the subsurface (Figure 4). Obviously, the extent and nature of the ejecta depend on the size of impact, but impact cratering is able to excavate large amounts of material from the deep subsurface. For example, in the Haughton impact crater of Canada, material from the crystalline basement, which is located more than 1,880 m below the surface sedimentary rocks, is exposed as crystalline clasts in impact deposits are available today at the surface [112]. It is only through the Haughton impact event that material from the crystalline basement is exposed at the surface. Outside the crater, the basement is buried by almost 2 km of sedimentary rocks.

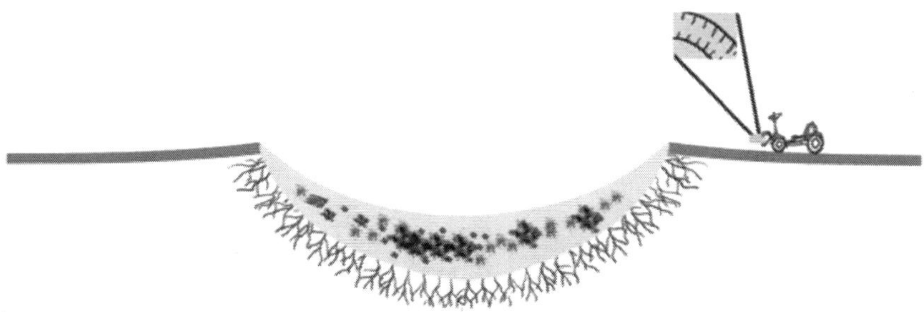

Figure 4. Schematic drawing illustrating a rover at an impact site sampling ejecta that have been excavated from great depths.

Estimates of terrestrial impact frequencies are compiled by French [113]. Based on these impact frequency calculations (references within French [113]), about 1–5 craters of a size >20 km should be produced every million years. These impact frequencies are based on data from the terrestrial record (in stable and well-preserved regions), but also from the cratering record on the moon and on asteroids. It is very hard to estimate impact frequencies, but it is evident that cratering is widespread in our solar system, through, for example, looking at the scarred surface of our moon. The Earth has been reworked by plate tectonics, and many craters are now wiped out. The impact frequency was much higher in the early history of our solar system, but large impacts still happen today, as is evident from, for example, the comet Shoemaker-Levy colliding with Jupiter in 1994. On planetary bodies with little or no crustal-reworking, such as Mars, impact-processed rocks can cover a significant proportion of the surface. This means that impact deposits would be readily available for sampling on the Martian surface.

The Martian surface still is subject to impact cratering and it would be possible to sample relatively fresh excavation material. Mars orbiters observed that between May 1999 and March 2006, 20 impacts created craters 2–150 meters in diameter on Mars [114]. These craters are accompanied by vast ejecta blankets and recent observations further show that the ejecta not only include excavation of rock debris, bui also of ice [115]. The sampling of excavated ice would be highly interesting in an astrobiological context.

Impact cratering is a high-energy event, and involves extremely high pressures and temperatures which could wipe out any pre-impact signatures of life, but the physical and chemical reactions during an impact are far from equilibrium, and the extreme conditions will not apply to all the materials involved in the impact. This gives rise to disordered material next to highly crystalline and shock metamorphosed material [113]. There are several examples of fossil biological signatures which are preserved in impact deposits after an impact event. Organic matter does not necessarily oxidize or change in to highly crystalline forms during an impact event, but substantial amounts can be preserved in a disordered form. For example, Blank *et al.* [116] showed that significant concentrations of organic compounds survived in the shock experiments of amino acids. Fossil biological markers are preserved within lithic clasts of the Haughton impact melt breccia [22], as well as in the carbonate bedrock of the Haughton impact crater [117]. Pre-impact biological markers are also preserved in shale fragments of impact breccias from the Ries impact crater in Germany [118]. Also, experimental impacts show the preservation of fossil biological markers [21].

The examples of the preservation of life from terrestrial impacts all concern a fossil record of life. But since the conditions on Mars today are rather harsh, with an unstable climate, a thin atmosphere, strong irradiation at the surface, and possibly no liquid water, it is more likely that any evidence of life on Mars would occur in the subsurface, and is more likely to be a fossil, which nevertheless would be of a great significance. Therefore, a strong emphasis in Mars exploration is on the search for a fossil record of life [15].

ACKNOWLEDGEMENT

We acknowledge Javier Herbozo at the Swedish Museum of Natural History for the illustrations in Figure 3 and Figure 4. Nicola McLoughlin at the Bergen University is acknowledged for permission to use Figure 1. Anna Neubeck at the Stockholm University is acknowledged for extensive discussions of the manuscript and three anonymous reviewers are acknowledged for constructive comments that greatly improved this paper. Part of this work was funded by the Swedish National Space Board.

REFERENCES

1. Allwood, A.C.; Walter, M.R.; Kamber, B.S.; Marshall, C.P.; Burch, I.W. Stromatolite reef from the Early Archaean era of Australia. Nature 2006, 441, 714–718.

2. Furnes, H.; Banerjee, N.R.; Muehlenbachs, K.; Staudigel, H.; de Wit, M. Early life recorded in Archean pillow lavas.Science 2004, 304, 578–581.

3. Ueno, Y.; Yamada, K.; Yoshida, N.; Maruyama, S.; Isozaki, Y. Evidence from fluid inclusions for microbial methanogenesis in the early Archaean era. Nature 2006, 440, 516–519.

4. Rosing, M.T. C-depleted carbon microparticles in >3700-Ma sea-floor sedimentary rocks from west Greenland. Science1999, 283, 674–676.

5. Han, T.M.; Runnegar, B. Megascopic eukaryoticalgae from the 2.1-billion-year-old Negaunee iron-formation, Michigan. Science 1992, 257, 232–235.

6. Schneider, D.A.; Bickford, M.E.; Cannon, W.F.; Schulz, K.J.; Hamilton, M.A. Age of volcanic rocks and syndepositional iron formations, Marquette Range Supergroup: Implications for the tectonic setting of paleoproterozoic iron formations of the Lake Superior region. Can. J. Earth Sci. 2002, 39, 999–1012.

7. Holland, H.D. The oxygenation of the atmosphere and oceans. Phil. T. Roy. Soc. B. 2006, 361, 903–915.

8. Cockell, C.S.; Kaltenegger, L.; Raven, J.A. Cryptic photosynthesis-extrasolar planetary oxygen without a surface biological signature. Astrobiology 2009, 9, 623–636.

9. Cockell, C.H.; Barlow, N.G. Impact excavation and the search for subsurface life on Mars. Icarus 2002, 155, 340–349.

10. Trimarco, E.; Balkwill, D.; Davidson, M.; Onstott, T.C. In situ enrichment of a diverse community of bacteria from 4–5 km deep fault zone in South Africa. Geomicrobiol. J. 2006, 23, 463–473.

11. Santelli, C.M.; Orcutt, B.N.; Banning, E.; Bach, W.; Moyer, C.L.; Sogin, M.L.; Staudigel, H.; Edwards, K.J. Abundance and diversity of microbial life in ocean crust. Nature 2008, 453, 653–657.

12. Furnes, H.; McLoughlin, N.; Muehlenbachs, K.; Banerjee, N.; Staudigel, H.; Dilek, Y.; de Wit, M.; Van Kranendonk, M.; Schiffman, P. Oceanic pillow lavas and hyaloclastites as habitats for microbial life

through time—A review. In Links between Geological Processes, Microbial Activities and Evolution of Life; Dilek, Y., Furnes, H., Muehlenbachs, K., Eds.; Springer: Berlin, Germany, 2008; pp. 1–68.

13. Rasmussen, B. Filamentous microfossils in a 3,235-million-year-old volcanogenic massive sulphide deposit. Science2000, 405, 676–679.

14. Sleep, N.H.; Zahnle, K. Refugia from asteroid impacts on early Mars and the early Earth. J. Geophys. Res. 1998, 103, 28529–28544.

15. Farmer, J.D.; Des Marais, D.J. Exploring for a record of ancient Martian life. J. Geophys. Res. 1999, 104, 26977–26995.

16. Kminek, G.; Bada, J.L. The effect of ionizing radiation on the preservation of amino acids on Mars. Earth Planet Sci. Lett. 2006, 245, 1–5.

17. Dartnell, L.R.; Desorgher, L.; Ward, J.M.; Coates, A.J. Martian sub-surface ionising radiation: Biosignatures and geology. Biogeosciences 2007, 4, 545–558.

18. Lammer, H.; Lichtenegger, H.I.M.; Kolb, C.; Ribas, I.; Guinan, E.F.; Abart, R.; Bauer, S.J. Loss of water from Mars: Implications for the oxidation of the soil. Icarus 2003, 165, 9–25.

19. Zacny, K.; Bar-Cohen, Y.; Brennan, M.; Briggs, G.; Cooper, G.; Davis, K.; Dolgin, B.; Glaser, D.; Glass, B.; Gorevan, S.; Guerrero, J.; McKay, C.; Paulsen, G.; Stanley, S.; Stoker, C. Drilling systems for extraterrestrial subsurface exploration.Astrobiology 2008, 8, 665–706.

20. Farmer, J.D. Hydrothermal systems on Mars: An assessment of present evidence. In Evolution of Hydrothermal Ecosystems on Earth (and Mars?); Bock, G.R., Goode, J.A., Eds.; Wiley and Sons: Chichester, UK, 1996.

21. Bowden, S.A.; Court, R.W.; Milner, D.; Baldwin, E.C.; Lindgren, P.; Crawford, I.A.; Parnell, J.; Burchell, M.J. The thermal alteration by pyrolysis of the organic component of small projectiles of mudrock during capture at hypervelocity. J. Anal. Appl. Pyrol. 2008, 82, 312–314.

22. Lindgren, P.; Parnell, J.; Bowden, S.A.; Taylor, C.; Osinski, G.; Lee, P. Preservation of biological markers in clasts within impact melt breccias from the Haughton impact structure, Devon Island. Astrobiology 2009, 9, 391–400.

23. The MEPAG next decade science analysis group. Science priorities for Mars sample return. Astrobiology 2008, 8, 489–535.

24. Holm, N.G. Why are hydrothermal systems proposed as plausible environments for the origin of life? Origins. Life. Evol. B. 1992, 22, 5–14.

25. Russel, M.J.; Hall, A.J.; Boyce, A.J.; Fallick, A.E. 100th anniversary special paper: On hydrothermal convection systems and the emergence of life. Econ. Geol. 2005, 100, 419–438.

26. Sleep, N.H.; Zahnle, K.J.; Kasting, J.F.; Morowitz, H. Annihilation of ecosystems by large asteroid impacts on the early Earth. Nature 1989, 342, 139–142.

27. Rampino, M.R. Role of the galaxy in periodic impacts and mass extinctions on the Earth. In Catastrophic Events and Extinctions: Impacts and Beyond; Koeberl, C., MacLeod, K.G., Eds.; GSA Special Paper 356; Geological Society of America: Boulder, CO, USA, 2002; pp. 667–678.

28. Schultz, P.H.; D'Hondt, S. Cretaceous-Tertiary (Chicxulub) impact angle and its consequences. Geology 1996, 11, 963–967.

29. Erwin, D.H.; Bowring, S.A.; Yugan, J. End-Permian mass extinctions: A review. In Catastrophic Events and Extinctions: Impacts and Beyond; Koeberl, C., MacLeod, K.G., Eds.; GSA Special Paper 356; Geological Sociaety of America: Boulder, CO, USA, 2002; pp. 363–384.

30. Berry, W.B.N.; Ripperdan, R.L.; Finney, S.C. Late Ordovician extinction: A Laurentian view. In Catastrophic Events and Extinctions: Impacts and Beyond; Koeberl, C., MacLeod, K.G., Eds.; GSA Special Paper 356; Geological Sociaety of America: Boulder, CO, USA, 2002; pp. 463–472.

31. Hoffman, P.F.; Kaufman, A.J.; Halverson, G.P.; Schrag, D.P. A neoproterozoic snowball Earth. Science 1998, 281, 1342–1346.

32. Bastin, E.S. Microorganisms in oil fields. Science 1926, 63, 21–24.

33. Bastin, E.S.; Greer, F.E. Additional data on sulfate-reducing bacteria in soils and waters of Illinois oil fields. Bull. Amer. Assoc. Petrol. Geol. 1930, 14, 153–159.

34. Lieske, R.; Hoffmann, E. Untersuchungen über den Bakteriengehalt der Erde in grossen Tiefen. Centralbl. F. Bakt, II. Abt. 1929, 77, 305–309.

35. Emery, K.O.; Dietz, R.S. Gravity coring instrument and mechanics of sediment coring. Bull. Geol. Soc. Am. 1941, 52, 1685–1714.

36. Chapelle, F.H.; Zelibor, J.L.J.; Grimes, D.J.; Knobel, L.L. Bacteria in deep coastal plain sediments of Maryland: A possible source of CO2 to groundwater. Water Resour. Res. 1987, 23, 1625–1632.

37. Sinclair, J.L.; Ghiorse, W.C. Distribution of aerobic bacteria, protozoa, algae and fungi in deep subsurface sediments.Geomicrobiol. J. 1989, 7, 15–31.

38. Pedersen, K. Preliminary investigations of deep ground water microbiology in Swedish granitic rock. SKB Tech. Rep.1987, 88, 1–22.

39. Pedersen, K.; Ekendahl, S. Distribution and activity of bacteria in deep granitic groundwaters of southeastern Sweden.Microb. Ecol. 1990, 20, 37–52.

40. Pedersen, K. The deep subterranean biosphere. Earth-Sci. Rev. 1993, 34, 243–260.

41. Pedersen, K. Exploration of deep intraterrestrial microbial life: Current perspectives. FEMS Microbiol. Lett. 2000, 185, 9–16.

42. Hofmann, B.A.; Farmer, J.D.; von Blanckenburg, F.; Fallick, A.E. Subsurface filamentous fabrics: An evaluation of modes of origins based on morphological and geochemical criteria, with implications for exopalaeontology.Astrobiology 2008, 8, 87–117.

43. Kretzschmar, M. Fossile pilze in eisen-stromatolithen von warstein (rheinisches schiefergebirge). Facies 1982, 7, 237–260.

44. Trewin, N.H.; Knoll, A.H. Preservation of devonian chemotrophic filamentous bacteria in calcite veins. Palaios 1999,14, 288–294.

45. D'Hondt, S.; Jørgensen, B.B.; Miller, J.; Batzke, A.; Blake, R.; Cragg, B.A.; Cypionka, H.; Dickens, G.R.; Ferdelman, T.; Hinrichs, K.U.; Holm, N.G.; Mitterer, R.; Spivack, A.; Wang, G.; Bekins, B.; Engelen, B.; Ford, K.; Gettemy, G.; Rutherford, S.D.; Sass, H.; Skilbeck, C.G.; Aiello, I.W.; Guerin, G.; House, C.H.; Inagaki, F.; Meister, P.; Naehr, T.; Niitsuma, S.; Parkes, R.J.; Schippers, A.; Smith, D.C.; Teske, A.; Wiegel, J.; Naranjo Padilla, C.; Solis Acosta, J.L. Distributions of microbial activities in deep subseafloor sediments. Science 2004, 306, 2216–2221.

46. Parkes, R.J.; Cragg, B.A.; Wellsbury, P. Recent studies on bacterial populations and processes in subseafloor sediments: A review. Hydrogeol. J. 2000, 8, 11–28.

47. Toner, B.M.; Santelli, C.M.; Marcus, M.A.; Wirth, R.; Chan, C.S.; McCollom, T.; Bach, W.; Edwards, K.J. Biogenic iron oxyhydroxide

formation at mid-ocean ridge hydrothermal vents: Juan de Fuca Ridge. Geochim. Cosmochim. Ac. 2009,73, 388–403.

48. Lysnes, K.; Thorseth, I.H.; Steinsbu, B.O.; Øvreås, L.; Torsvik, T.; Pedersen, R.B. Microbial community diversity in seafloor basalt from the Arctic spreading ridges. FEMS Microbiol. Ecol. 2004, 50, 213–230.

49. Takami, H.; Ioue, A.; Fuji, F.; Horikoshi, K. Microbial flora in the deepest sea mud of the Mariana Trench. FEMS Microbiol. Lett. 1997, 152, 279–285.

50. Staudigel, H.; Tebo, B.; Yayanos, A.; Furnes, H.; Kelley, K.; Plank, T.; Muehlenbachs, K. The oceanic crust as a bioreactor. In The Subseafloor Biosphere at Mid-Ocean Ridges; Geophysical Monograph Series 144; Wilcock, W.S.D., DeLong, E.F., Kelley, D.S., Baross, J.A., Cary, S.C., Eds.; American Geophysical Union: Washington, DC, USA, 2004; pp. 325–341.

51. Ivarsson, M.; Lausmaa, J.; Lindblom, S.; Broman, C.; Holm, N.G. Fossilized microorganisms from the Emperor Seamounts: Implications for the search for a sub-surface fossil record on Earth and Mars. Astrobiology 2008, 8, 1139–1157.

52. Ivarsson, M.; Lindblom, S.; Broman, C.; Holm, N.G. Fossilized microorganisms associated with zeolite-carbonate interfaces in sub-seafloor hydrothermal environments. Geobiology 2008, 6, 155–170.

53. Peckmann, J.; Bach, W.; Behrens, K.; Reitner, J. Putative cryptoendolithic life in Devonian pillow basalt, Rheinisches Schiefergebirge, Germany. Geobiology 2008, 6, 125–135.

54. McLoughlin, N.; Brasier, M.D.; Wacey, D.; Green, O.R.; Perry, R.S. On biogenicity criteria for endolithic microborings on early Earth and beyond. Astrobiology 2007, 7, 10–26.

55. McLoughlin, N.; Furnes, H.; Banerjee, N.H.; Muehlenbachs, K.; Staudigel, H. Ichnotaxonomy of microbial trace fossils in volcanic glass. J. Geol. Soc. 2009, 166, 159–169.

56. Ehrlich, H.L. Geomicrobiology, 3rd ed.; Marcel Dekker: New York, NY, USA, 1996; p. 719.

57. Furnes, H.; Thorseth, I.H.; Tumyr, O.; Torsvik, T.; Fisk, M.R. Microbial activity in the alteration of glass from pillow lavas from hole 896A. Proc. Oc. Drill. Prog. Sci. Res. 1996, 148, 191–206.

58. Furnes, H.; Staudigel, H. Biological mediation in ocean crust alteration: How deep is the deep biosphere? Earth Planet Sci. Lett. 1999, 166, 97–103.

59. Thorseth, I.H.; Pedersen, R.B.; Christie, D.M. Microbial alteration of 0–30-Ma seafloor basaltic glasses from the Australian Antarctic Discordance. Earth Planet. Sci. Lett. 2003, 215, 237–247.

60. Hallbeck, L.; Pedersen, K. Characterization of microbial processes in deep aquifers of the Fennoscandian Shield. Appl. Geochem. 2008, 23, 1796–1819.

61. Gadd, G.M. Geomycology: Biogeochemical transformations of rocks, minerals, metals and radionuclides by fungi, bioweathering and bioremediation. Mycol. Res. 2007, 111, 3–49.

62. Hirsch, P.; Rades-Rohkohl, E.; Kölbel-Boelke, J.; Nehrkorn, A. Morphological and taxonomic diversity of groundwater micro-organisms. In Progr Hydrogeochem; Matthess, G., Frimmel, F.H., Hirsch, P., Schulz, H.D., Usdowski, E., Eds.; Springer-Verlag: Berlin, Germany, 1992.

63. Krauss, G.; Sridhar, K.R.; Jung, K. Aquatic hyphomycetes in polluted groundwater habitats of central Germany.Microbial. Ecol. 2003, 45, 329–339.

64. Pedersen, K.; Arlinger, J.; Ekendahl, S.; Hallbeck, L. 16S rRNA gene diversity of attached and unattached bacteria in boreholes along the access tunnel to the Äspö Hard Rock Laboratory. FEMS Microbiol. Ecol. 1996, 19, 249–262.

65. Ekendahl, S.; O'Neill, A.H.; Thomsson, E.; Pedersen, K. Characterisation of yeasts isolated from deep igneous rock aquifers of the Fennoscandian Shield. Microb. Ecol. 2003, 46, 416–428.

66. Bidle, K.D.; Lee, S.; Marchant, D.R.; Falkowski, P.G. Fossil genes and microbes in the oldest ice on earth. PNAS 2007,104, 13455–13460.

67. Teske, A.P. The deep subsurface biosphere is alive and well. Trends Microbiol. 2005, 13, 402–404.

68. Edwards, K.J.; Bach, W.; McCollom, T.M. Geomicrobiology in oceanography: Microbe-mineral interactions at and below the seafloor. Trends Microbiol. 2005, 13, 449–456.

69. Golubic, S.; Friedmann, I.; Schneider, J. The lithobiontic ecological niche, with special reference to microorganisms. J. Sediment. Petrol. 1981, 51, 475–478.

70. Fisk, M.R.; Popa, R.; Mason, O.U.; Storrie-Lombardi, M.C.; Vicenzi, E.P. Iron-magnesium silicate bioweathering on Earth (and Mars?). Astrobiology 2006, 6, 48–68.

71. Eickmann, B.; Bach, W.; Kiel, S.; Reitner, J.; Peckmann, J. Evidence for cryptoendolithic life in Devonian pillow basalts of variscan orogens, Germany. Palaeogeogr. Palaeocl. Palaeoecol. 2009, 283, 120–125.

72. Ivarsson, M.; Holm, N.G. Microbial colonization of various habitable niches during alteration of oceanic crust. In Links between Geological Processes, Microbial Activities and Evolution of Life; Dilek, Y., Furnes, H., Muehlenbachs, K., Eds.; Springer: Berlin, Germany, 2008; pp. 69–111.

73. Honnorez, J. Generation of phillipsites by palagonitization of basaltic glass in sea water and the origin of K-rich deep-sea sediments. In Natural Zeolites, Occurrence, Properties, Use; Sand, L.B., Mumpton, F.A., Eds.; Pergamon Press: London, UK, 1978; pp. 245–258.

74. Boynton, W.V.; Ming, D.W.; Kounaves, S.P.; Young, S.M.M.; Arvidson, R.E.; Hecht, M.H.; Hoffman, J.; Niles, P.B.; Hamara, D.K.; Quinn, R.C.; Smith, P.H.; Sutter, B.; Catling, D.C.; Morris, R.V. Evidence for calcium carbonate at the Mars Phoenix landing site. Science 2009, 325, 61–64. [PubMed]

75. Mumma, M.J.; Villanueva, G.L.; Novak, R.E.; Hewagama, T.; Bonev, B.P.; DiSanti, M.A.; Mandell, A.M.; Smith, M.D. Strong release of methane on Mars in northern summer 2003. Science 2009, 323, 1041–1045.

76. Cockell, C.S.; Lee, P.; Broady, P.; Lim, D.S.S.; Osinski, G.R.; Parnell, J.; Koeberl, C.; Pesonen, L.; Salminen, J. Effects of asteroid and comet impacts on habitats for lithophytic organisms—A synthesis. Meteorit. Planet. Sci. 2005, 40, 1901–1914.

77. Cockell, C.S. The origin and emergence of life under impact bombardment. Phil. T. Roy. Soc. B. 2006, 361, 1845–1856.

78. Cockell, C.S.; Osinski, G.R. Impact-induced impoverishment and transformation of a sandstone habitat for lithophytic microorganisms. Meteorit. Planet. Sci. 2007, 42, 1985–1993.

79. Sleep, N.H. Martian plate tectonics. J. Geophys. Res. 1994, 99, 5639–5655.

80. McSween, H.Y.J. What we have learned about Mars from SNC meteorites. Meteoritics 1994, 29, 757–779.

81. Carr, M. Water on Mars; Oxford University Press: New York, NY, USA, 1996.

82. Smith, P.H.; Tamppari, L.K.; Arvidson, R.E.; Bass, D.; Blaney, D.; Boynton, W.V.; Carswell, A.; Catling, D.C.; Clark, B.C.; Duck, T.; DeJong, E.; Fisher, D.; Goetz, W.; Gunnlaugsson, H.P.; Hecht, M.H.; Hipkin, V.; Hoffman, J.; Hviid, S.F.; Keller, H.U.; Kounaves, S.P.; Lange, C.F.; Lemmon, M.T.; Madsen, M.B.; Markiewicz, W.J.; Marshall, J.; McKay, C.P.; Mellon, M.T.; Ming, D.W.; Morris, R.V.; Pike, W.T.; Renno, N.; Staufer, U.; Stoker, C.; Taylor, P.; Whiteway, J.A.; Zent, A.P. H2O at the Phoenix landing site. Science 2009, 325, 58–61.

83. Whiteway, J.A.; Komguem, L.; Dickinson, C.; Cook, C.; Illnicki, M.; Seabrook, J.; Popovici, V.; Duck, T.J.; Davy, R.; Taylor, P.A.; Pathak, J.; Fisher, D.; Carswell, A.I.; Daly, M.; Hipkin, V.; Zent, A.P.; Hecht, M.H.; Wood, S.E.; Tamppari, L.K.; Renno, N.; Moores, J.E.; Lemmon, M.T.; Daerden, F.; Smith, P.H. Mars water-ice clouds and precipitation. Science2009, 325, 68–70. [PubMed]

84. Byrne, S.; Dundas, C.M.; Kennedy, M.R.; Mellon, M.T.; McEwen, A.S.; Cull, S.C.; Daubar, I.J.; Shean, D.E.; Seelos, K.D.; Murchie, S.L.; Cantor, B.A.; Arvidson, R.E.; Edgett, K.S.; Reufer, A.; Thomas, N.; Harrison, T.N.; Posiolova, L.V.; Seelos, F.P. Distribution of mid-latitude ground ice on mars from new impact craters. Science 2009, 325, 1674–1676.

85. Kerr, R.A. On Mars, a Second Chance for Life. Science 2004, 306, 2010–2012.

86. Bibring, J.P.; Langevin, Y.; Mustard, J.F.; Poulet, F.; Arvidson, R.; Gendrin, A.; Gondet, B.; Mangold, N.; Pinet, P.; Forget, F.; The OMEGA Team. Global mineralogical and aqueous mars history derived from Omega/mars express data. Science 2006, 312, 400–404.

87. Jakosky, B.M. Mars volatile evolution: Evidence from stable isotopes. Icarus 1991, 94, 14–31.

88. Jakosky, B.M.; Jones, J.H. Evolution of water on mars. Nature 1994, 370, 328–329.

89. Gooding, J.L. Soil mineralogy and chemistry on Mars: Possible clues from salts and clays in SNC meteorites. Icarus1992, 99, 28–41.

90. Griffith, L.L.; Shock, E.L. A geochemical model for the formation of hydrothermal carbonates on mars. Nature 1995,377, 406–408.

91. Catling, D.C.; Moore, J.M. The nature of coarse-grained crystalline hematite and its implications for the early environment of mars. Icarus 2003, 165, 277–300.

92. Wyatt, M.B.; McSween, H.Y.J. Spectral evidence for weathered basalt as an alternative to andesite in the northern lowlands of mars. Nature 2002, 417, 263–266.

93. Christensen, P.R.; Morris, R.V.; Lane, M.D.; Banfield, J.L.; Malin, M.C. Global mapping of martian hematite mineral deposits: Remnants of water-driven processes on early mars. J. Geophys. Res. 2001, 106, 23873–23885.

94. Klingelhöfer, G. Mössbauer In Situ Studies of the Surface of Mars. Hyperfine Interact. 2005, 158, 117–124.

95. Minitti, M.E.; Rutherford, M.J. Genesis of the mars pathfinder "sulfur-free" rock from SNC parental liquids. Geochim. Cosmochim. Ac. 2000, 64, 2535–2547.

96. Christensen, P.R.; Ruff, S.W. Formation of the hematite-bearing unit in Meridiani Planum: Evidence for deposition in standing water. J. Geophys. Res. 2004, 109, E08003.

97. Neukum, G.; Jaumann, R.; Hoffmann, H.; Hauber, E.; Head, J.W.; Basilevsky, A.T.; Ivanov, B.A.; Werner, S.C.; van Gasselt, S.; Murray, J.B.; McCord, T.; The HRSC Co-investigator Team. Recent and episodic volcanic and glacial activity on Mars revealed by the high resolution stereo camera. Nature 2004, 432, 971–979.

98. Ames, D.E.; Watkinson, D.H.; Parrish, R.R. Dating of a regional hydrothermal system induced by the 1850 Ma Sudbury impact event. Geology 1998, 26, 447–450.

99. Naumov, M.V. Impact-generated hydrothermal systems: Data from Popigai, Kara, and Puchezh-Katunki impact structures. In Impacts in Precambrian Shields; Plado, J., Peasonen, L.J., Eds.; Springer: Berlin, Germany, 2002; pp. 117–171.

100. Osinski, G.R.; Spray, J.G.; Lee, P. Impact induced hydrothermal activity within the Haughton impact structure, arctic Canada: generation of a transient, warm, wet oasis. Meteorit. Planet. Sci. 2001, 36, 731–745.

101. Newsom, H.E. Hydrothermal alteration of impact melt sheets with implications for Mars. Icarus 1980, 44, 207–216.

102. Pope, K.O.; Kieffer, S.W.; Ames, D.E. Impact melt sheet formation on Mars and its implication for hydrothermal systems and exobiology. Icarus 2006, 183, 1–9.

103. Squyres, S.W.; Arvidson, R.E.; Ruff, S.; Gellert, R.; Morris, R.V.; Ming, D.W.; Crumpler, L.; Farmer, J.D.; Des Marais, D.J.; Yen, A.; McLennan, S.M.; Calvin, W.; Bell, J.F., III; Clark, B.C.; Wang, A.; McCoy, T.J.; Schmidt, M.J.; McLennan, P.A., Jr. Detection of silica-richdeposits on mars. Science 2008, 320, 1063–1067.

104. Bishop, J.L.; Noe Dobrea, E.Z.; McKeown, N.K.; Parente, M.; Ehlmann, B.L.; Michalski, J.R.; Milliken, R.E.; Poulet, F.; Swayze, G.A.; Mustard, J.F.; Murchie, S.L.; Bibring, J.P. Phyllosilicate diversity and past aqueous activity revealed at Mawrth Vallis, Mars. Science 2008, 321, 830–833.

105. Mustard, J.F.; Murchie, S.L.; Pelkey, S.M.; Ehlmann, B.L.; Milliken, R.E.; Grant, J.A.; Bibring, J.P.; Poulet, F.; Bishop, J.; Noe Dobrea, E.; Roach, L.; Seelos, F.; Arvidson, R.E.; Wiseman, S.; Green, R.; Hash, C.; Humm, D.; Malaret, E.; McGovern, J.A.; Seelos, K.; Clancy, T.; Clark, R.; Des Marais, D.; Izenberg, N.; Knudson, A.; Langevin, Y.; Martin, T.; McGuire, P.; Morris, R.; Robinson, M.; Roush, T.; Smith, M.; Swayze, G.; Taylor, H.; Titus, T.; Wolff, M. Hydrated silicate minerals on Mars observed by the mars reconnaissance orbiter CRISM instrument. Nature 2008, 454, 305–309.

106. Poulet, F.; Bibring, J.P.; Mustard, J.F.; Gendrin, A.; Mangold, N.; Langevin, Y.; Arvidson, R.E.; Gondet, B.; Gomez, C.; The Omega Team. Phyllosilicates on mars and implications for early martian climate. Nature 2005, 438, 623–627.

107. Clifford, S.M. A model for the hydrologic and climatic behavior of water on mars. J. Geophys. Res. 1993, 88, 2456–2474.

108. McKinley, J.P.; Stevens, T.O.; Westall, F. Microfossils and paleoenvironments in deep subsurface basalt samples.Geomicrobiol. J. 2000, 17, 43–54.

109. Herrera, A.; Cockell, C.S.; Self, S.; Blaxter, M.; Reitner, J.; Thorsteinsson, T.; Arp, G.; Dröse, W.; Tindle, A.G. A cryptoendolithic community in volcanic glass. Astrobiology 2009, 9, 369–381.

110. Laskar, J.; Correia, A.C.M.; Gastineau, M.; Joutel, F.; Levrard, B.; Robutel, P. Long term evolution and chaotic diffusion of the insolation quantities of Mars. Icarus 2004, 170, 343–364.

111. Balme, M.R.; Gallagher, C. An equatorial periglacial landscape on Mars. Earth Planet. Sci. Lett. 2009, 285, 1–15.

112. Osinski, G.R.; Spray, J.G.; Lee, P. Impactites of the Haughton impact structure, Devon island, Canadian high Arctic.Meteorit. Planet. Sci. 2005, 40, 1789–1812.

113. French, B.M. Traces of catastrophe: A handbook of Shock-Metamorphic Effects in Terrestrial Meteorite Impact Structures. In LPI Contribution No. 954; Lunar and Planetary Institute: Houston, TX, USA, 1998; p. 120.

114. Malin, M.C.; Edgett, K.S.; Posiolova, L.V.; McColley, S.M.; Noe Dobrea, E.Z. Present-day cratering rate and contemporary gully activity on Mars. Science 2006, 314, 1573–1577.

115. Byrne, S.; Dundas, C.M.; Kennedy, M.R.; Mellon, M.T.; McEwan, A.S.; Cull, S.C.; Daubar, I.J.; Shean, D.E.; Seelos, K.D.; Murchie, S.L.; Cantor, B.A.; Arvidson, R.E.; Edgett, K.S.; Reufer, A.; Thomas, N.; Harrison, T.N.; Posiolova, L.V.; Seelos, F.P. Distribution of mid-latitude ground ice on Mars fromnew impact craters. Science 2009, 325, 1674–1676.

116. Blank, J.; Miller, G.H.; Ahrens, M.J.; Winans, R.E. Experimental shock chemistry of aqueous amino acid solutions and the cometary delivery of prebiotic compounds. Origins. Life. Evol. B. 2001, 31, 15–51.

117. Parnell, J.; Lee, P.; Osinski, G.R.; Cockell, C.S. Application of organic geochemistry to detect signatures of organic matter in the Haughton impact structure. Meteorit. Planet. Sci. 2005, 40, 1879–1885.

118. Hofmann, P.; Leythaeuser, D.; Schwark, L. Organic matter from the Bunte Breccia of the Ries Crater, southern Germany: Investigating possible thermal effects of the impact. Planet. Space Sci. 2001, 49, 845–851.

CHAPTER 6

Realistic Subsurface Anomaly Discrimination Using Electromagnetic Induction and an SVM Classifier

Juan Pablo Fernández,[1] Fridon Shubitidze,[1, 2] Irma Shamatava,[1, 2] Benjamin E. Barrowes,[1, 3] and Kevin O'Neill[1, 3]

[1]Thayer School of Engineering, Dartmouth College, Hanover, NH 03755, USA

[2]Sky Research, Inc., Hanover, NH 03755, USA

[3]USA ERDC Cold Regions Research and Engineering Laboratory, Hanover, NH 03755, USA

ABSTRACT

The environmental research program of the United States military has set up blind tests for detection and discrimination of unexploded ordnance. One such test consists of measurements taken with the EM-63 sensor at Camp Sibert, AL. We review the performance on the test of a procedure that combines a field-potential (HAP) method to locate targets, the normalized surface magnetic source (NSMS) model to characterize them, and a support vector machine (SVM) to classify them. The HAP method infers location from the scattered magnetic field and its associated scalar potential, the latter reconstructed using equivalent sources. NSMS replaces the target with an enclosing spheroid of equivalent radial magnetization whose integral it uses as a discriminator. SVM generalizes from empirical evidence and can be adapted for multiclass discrimination using a voting system. Our method identifies all potentially dangerous targets correctly and has a false-alarm rate of about 5%.

1. INTRODUCTION

The millions of unexploded ordnance (UXO) strewn about in former battlefields and military practice ranges, of which a significant fraction involve marine or underwater environments, constitute a pressing humanitarian and environmental hazard worldwide [1]. The high false-alarm rates of current sensors and the need to treat every detected anomaly as potentially dangerous result in decontamination costs running into the millions of dollars per acre and extend remediation timescales by decades if not centuries. This state of affairs can only be resolved by developing methodologies that will quickly and reliably identify hazardous items and discriminate them from the morass of innocuous clutter typically found in the field.

The Strategic Environmental Research and Development Program (SERDP) of the United States military supports continuing research that aims to make UXO remediation more efficient and economic. One of SERDP's benchmarks to assess progress is a battery of UXO discrimination blind tests set up in Camp Sibert, a former U.S. Army facility near Gadsden, Alabama. The targets buried in 216 cells—some of which are empty—include unexploded 4.2" mortar shells, mortar explosion byproducts like base plates and partial mortars (i.e., stretched-out half-shells), smaller shrapnel, and unrelated metallic clutter; some examples appear in Figure 1. The different items are distributed in number as shown in Figure 1(d). In 2006, researchers affiliated with Sky Research, Inc. collected data at Camp Sibert using the EM-63, a cart-based step-off time-domain electromagnetic induction (EMI) sensor produced by Geonics Ltd. [2]. In this paper we use those data to demonstrate the performance of a physically complete, fast and clutter-tolerant discrimination approach developed at Dartmouth College and the Cold Regions Research and Engineering Laboratory.

The discrimination process comprises three tasks: localization, characterization, and classification. The secondary field from a visually obscured object depends both on the intrinsic features of the target and on its location and orientation relative to the sensor. Attempts to invert simultaneously for positional and intrinsic parameters often result in slow, ill-posed, computationally expensive optimizations that can easily get stuck in local minima. Our method [3, 4] clears that hurdle by performing the localization step independently at the outset and then using its results to help in the characterization. This permits a fast and accurate determination of the intrinsic parameters of the model. To classify the targets we feed those parameters to an open-source implementation [5] of a support vector machine (SVM) [6], a machine-learning methodology based on statistical learning theory [7, 8] that in the past has been used to perform binary classification [9] and regression [10] and has recently been adapted for multicategory classification [11]. The method has been employed in UXO research, either to classify or regress, in combination with the point-dipole model [1, 12, 13], the Standardized Excitation Approach [14–16], and finite elements [17, 18], and has shown to be competitive in its discrimination ability in relation to neural networks [15, 19] and other statistical methods [20, 21].

Type	Training	Testing	Total
UXO	38	34	72
Partial	12	23	35
Base	5	40	45
Scrap	6	25	31
Clutter	4	22	26
Empty	1	6	7
Total	66	150	216

Figure 1. Representative examples and relative populations of the objects buried at the Camp Sibert cells.(a)4.2" mortar shell, (b) Base plate (c) Half-shell (d) Cell contents

In a previous paper [22], we studied the Camp Sibert data using the same characterization model in combination with nonlinear least squares for the localization step and both template-matching and a Probability Neural Network for classification. We have already noted [4] that the localization procedure described below results in much better discrimination. The SVM-based classification showcased in this paper improves upon the template-matching used before [3, 4] in that it requires less human intervention and is thus faster to run and easier to adapt to other sets of observations. The template-matching procedure made predictions essentially identical to those we report here, perhaps even marginally better, but only after much close monitoring.

SVMs have previously been used for multicategory UXO-related classification [21], though in that reference the authors' choice of forward model and treatment of positional information differ from ours. While they construct parameter libraries at different locations in order to cancel out the geometric effects and enhance classification, we determine those effects separately; that way not only do we recover critically important information but also obtain parameters whose classification is perhaps easier (and thus faster) to perform and still of high quality.

In summary, our procedure aims to be a powerful and efficient discrimination method for UXO. The precise location and orientation estimates supplied by the so-called HAP technique [23] allow an almost instantaneous determination of an unambiguous time-dependent electromagnetic signature, the total NSMS [24]; this in turn can be distilled further using an empirical decay law [25] whose fitting parameters can be mixed into discriminating features that tend to group in well-separated tight clusters, allowing for clear-cut automated classification using the SVM algorithm.

This paper is organized as follows: in Section 2 we introduce the methods we use to locate and characterize scatterers, in Section 3 we briefly present the principles behind SVM classification, in Section 4 we discuss the results we obtain when we apply the combined procedure to the Camp Sibert data, and in Section 5 we conclude.

2. A PROCEDURE TO LOCATE AND CHARACTERIZE OBSCURED TARGETS

The eddy currents and magnetic dipoles induced or realigned by an EMI sensor on and inside a scatterer are distributed nonuniformly and tend to concentrate at some particular points. Under certain conditions, the response of the entire scatterer can be reproduced to arbitrary precision using a set of responding elementary sources—charges, dipoles, or the like—placed at those singularities [26, 27]. This consideration underlies the methods that we use to locate and characterize hidden targets.

2.1. A Dipole-Based Method to Estimate Location

The technique we use to locate an obscured target assumes that the whole scatterer responds as a point dipole. The location and orientation of that dipole are then found by exploiting analytic relations involving a dipole field $\mathbf{H}$ and its associated scalar potential Ψ. (The method originally used the vector potential $\mathbf{A}$ as well and has since been dubbed "HAP" [23].) To construct the potential from the field, one distributes elementary sources on an auxiliary planar layer located between the sensor and the object and finds the sources' amplitudes by fitting measured data.

A point dipole of moment $\mathbf{m}$ located at $\mathbf{r}_d$ generates at the observation point $\mathbf{r}$ a field

$$\mathbf{H}(\mathbf{r}) = \frac{1}{4\pi|\mathbf{r}-\mathbf{r}_d|^3}\left(3\mathbf{m}\cdot(\mathbf{r}-\mathbf{r}_d)\frac{(\mathbf{r}-\mathbf{r}_d)}{|\mathbf{r}-\mathbf{r}_d|^2}-\mathbf{m}\right) = -\nabla\psi(\mathbf{r}), \tag{1}$$

where the scalar potential

$$\psi(\mathbf{r}) = \frac{\mathbf{m}\cdot(\mathbf{r}-\mathbf{r}_d)}{4\pi|\mathbf{r}-\mathbf{r}_d|^3}. \tag{2}$$

Straightforward algebraic manipulation leads to

$$\mathbf{H}\cdot\mathbf{r}_d = -2\psi + \mathbf{H}\cdot\mathbf{r}, \tag{3}$$

which provides a least-squares estimate of $\mathbf{r}_d$ when evaluated at N distinct observation points:

$$\begin{bmatrix} H_x(\mathbf{r}_1) & H_y(\mathbf{r}_1) & H_z(\mathbf{r}_1) \\ H_x(\mathbf{r}_2) & H_y(\mathbf{r}_2) & H_z(\mathbf{r}_2) \\ & \vdots & \\ H_x(\mathbf{r}_N) & H_y(\mathbf{r}_N) & H_z(\mathbf{r}_N) \end{bmatrix} \begin{bmatrix} x_d \\ y_d \\ z_d \end{bmatrix} = \begin{bmatrix} -2\psi(\mathbf{r}_1)+\mathbf{H}(\mathbf{r}_1)\cdot\mathbf{r}_1 \\ -2\psi(\mathbf{r}_2)+\mathbf{H}(\mathbf{r}_2)\cdot\mathbf{r}_2 \\ \vdots \\ -2\psi(\mathbf{r}_N)+\mathbf{H}(\mathbf{r}_N)\cdot\mathbf{r}_N \end{bmatrix}. \tag{4}$$

For this particular implementation of the HAP method, it is important to note that the EM-63 sensor measures only the vertical component of the field. To construct the other two components and the scalar potential we assume that the field is produced by a surface distribution of magnetic charge $q(s')$ spread on a fictitious plane located just below the ground (see Figure 2). The positions $\mathbf{r}_{s'}$ of the sources are fixed and known by construction, and the field can be expressed as the matrix-vector product

$$H_z(\mathbf{r}) = \int \frac{q(s')}{4\pi} \frac{z - z_{s'}}{\left|\mathbf{r} - \mathbf{r}_{s'}\right|^3} ds' \equiv \overleftrightarrow{\mathbf{Z}}_z \cdot \mathbf{q} \tag{5}$$

by employing a quadrature scheme. To determine the array $\mathbf{q}$ of charges, one minimizes the difference between model predictions and collected data $\mathbf{H}_z^{meas}$ at a set of known points:

$$\mathbf{q} = \operatorname{argmin} \frac{1}{2} (\overleftrightarrow{\mathbf{Z}}_z \cdot \mathbf{q} - \mathbf{H}_z^{meas})^2 = \left[\overleftrightarrow{\mathbf{Z}}_z^T \cdot \overleftrightarrow{\mathbf{Z}}_z\right]^{-1} \left[\overleftrightarrow{\mathbf{Z}}_z^T \cdot \mathbf{H}_z^{meas}\right], \tag{6}$$

where each matrix row corresponds to a different measurement point and each column to a subsurface of the underground virtual source layer. The potential is then found from

$$\psi(\mathbf{r}) = \int \frac{q(s')}{4\pi \left|\mathbf{r} - \mathbf{r}_{s'}\right|} ds' \equiv \overleftrightarrow{\mathbf{Z}}_\psi \cdot \mathbf{q} . \tag{7}$$

This method and its adaptation to monostatic sensors like the EM-63 used for the Camp Sibert test are discussed in further detail in [23]. One last point is worth reiterating: the HAP method replaces the scatterer with a point dipole, and is thus based on a rather drastic simplification; yet it provides acceptable location estimates because the sources within the target that produce the scattered field tend to concentrate at a set of "scattered field singularities" [26, 27]. The locations of these singularities change at every measurement point, since the primary field of the sensor also changes; the HAP method takes these variations into account and outputs an average location as a result.

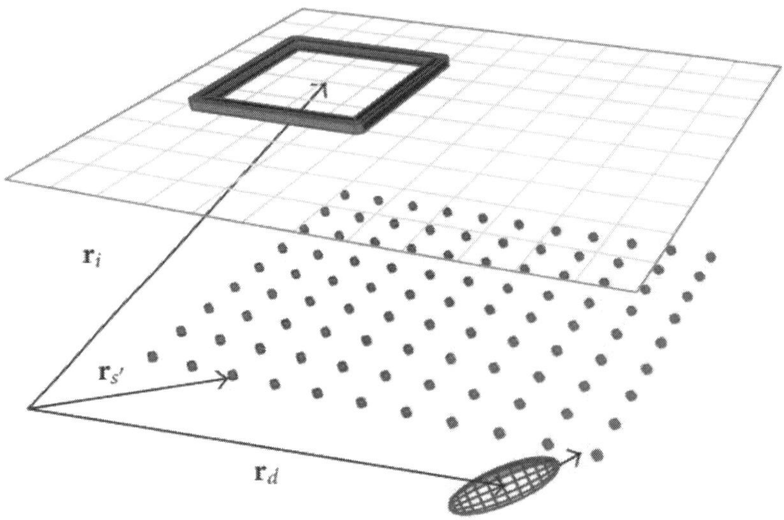

Figure 2. Determining the location and orientation of a buried target. The method assumes the object is a point dipole and exploits an analytic relation between the field measured at $\mathbf{r}_i$ and the scalar potential at the same point to find the location $\mathbf{r}_d$. The potential is constructed using a layer of equivalent magnetic sources placed between the sensor and the object; $\mathbf{r}_{s'}$ is a typical location on the layer.

2.2. The Normalized Surface Magnetic Source Model

To encapsulate the electromagnetic signature of a target, we use the fast and robust normalized surface magnetic source (NSMS) model [24]. The particular version we use here associates a scatterer with a surrounding prolate spheroid on which a continuum of radially oriented dipoles are distributed. The strengths of these dipoles—normalized by the normal component of the primary field to take monostaticity into account—are determined as those that best reproduce actual measurements. The composite dipole moment, referred to as the "total NSMS" and denoted by Q, varies significantly for different targets but is remarkably consistent for different specimens of the same object.

We divide the spheroid S into patches (or belts to exploit the azimuthal symmetry) and assign

$$\mathbf{H}^{sc}(\mathbf{r}) = \oint_S \frac{\Omega(s')}{4\pi R_{s'}^3}\left[\frac{3(\hat{\xi}_{s'}\cdot\mathbf{R}_{s'})\mathbf{R}_{s'}}{R_{s'}^2} - \hat{\xi}_{s'}\right]\left[\hat{\xi}_{s'}\cdot\mathbf{H}^{pr}(s')\right]ds'$$

$$\equiv \overleftrightarrow{\mathbf{Z}}\cdot\Omega, \qquad (8)$$

where $\mathbf{R}_{s'}$ is a vector that points from the location $\mathbf{r}_{s'}$ of the s'-th infinitesimal patch on the spheroid to the observation point $\mathbf{r}$ and $\hat{\xi}_{s'}$ is the unit vector normal to the patch. To factor out the particulars of location and orientation we have introduced the normalized surface polarization distribution $\Omega(s')$. The integral can again be transformed to a matrix-vector product via numerical quadrature; each column of $\overleftrightarrow{\mathbf{Z}}$ corresponds to a different source element and each row to a measurement point. The amplitude array Ω is determined by minimizing in a least-squares sense the difference between measured data with a known object-sensor configuration (as in the case of the Camp Sibert training data) and the predictions of (8).

Once $\Omega(s')$ is found, one can define a total polarizability by integrating over the whole spheroid. The resulting quantity

$$Q = \oint_S \Omega(s')ds', \qquad (9)$$

a global magnetic capacitance of sorts, has been shown to be intrinsic to the object and can be used, on its own or combined with other quantities, in discrimination processing. Figure 3 shows Q for all 216 Camp Sibert anomalies.

Our analysis of the time dependence of Q has been presented elsewhere [3, 4, 22] but is worth summarizing here. At early times, where higher frequencies are involved, the skin depth $\delta \propto f^{-1/2}$ is small and the induced eddy currents are superficial. As time passes and lower frequencies start to dominate, the currents diffuse into the object, making the late-time response involve the whole volume of the scatterer rather than just its surface. Thus a smaller but solid body like the base plate of Figure 1(b) has a relatively weak early response that dies down slowly, while a large but essentially hollow object like the partial mortar of Figure 1(c) has a strong initial response that decays quickly. The unexploded 4.2" mortar is large and compact and has a substantial early response that takes a while to die off. Our aim is to use these characteristics of Q to highlight quantitatively the differences between the various targets.

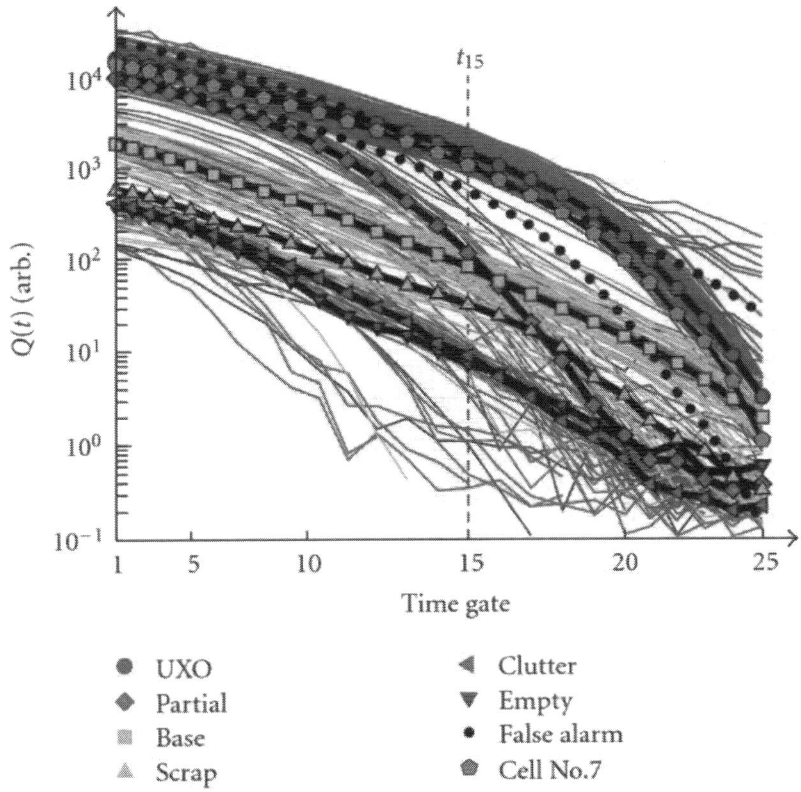

Figure 3. Total NSMS amplitudes Q(t) of the 216 Camp Sibert anomalies as a function of time. The 25 time gates are distributed approximately logarithmically between 191 μs and 25 ms. The Q-values are shown in the same colors and with the same markers as in subsequent figures, with the black line denoting the median Q for each group. The 15th time channel, corresponding to t = ms, is highlighted; the ratio $R \equiv Q(t_{15})/Q(t_1)$ has been found to be a robust classifier. The gray pentagons correspond to Cell no. 7, which we study in detail in the main text. The thinner dotted lines show the two false alarms from Figure 6, also depicted on Figure 5.

The information contained in Q as a function of the time t can be summarized further by fitting to it an empirical decay law first proposed by Pasion and Oldenburg [25]:

$$Q(t) = kt^{-\beta}e^{-\gamma t}. \tag{10}$$

Various combinations of these fitting parameters can be used as inputs to classifier programs, of which the support vector machine (SVM) is an example.

3. SUPPORT VECTOR MACHINES FOR SUBSURFACE OBJECT CLASSIFICATION

A support vector machine learns from data: when fed a series of answered training examples, it attempts to make sense of them by weighing the available empirical evidence, with no need for an underlying model, and applies this knowledge to make predictions about unseen cases. The examples can be any combination of model parameters expected to contain evidence of the essence of an object. In the simplest instance of binary classification, each n-dimensional example $\mathbf{x}_i$ has an associated yes/no attribute $y_i = \pm 1$; the SVM performs the classification by finding a hyperplane that divides the parameter space into two distinct regions, each of which ideally contains points from only one of the categories. During the learning or training process the machine readjusts the hyperplane parameters to accommodate every training vector until it strikes an optimal balance between fitting accuracy and model simplicity. All information about the hyperplane is contained in a subset of the examples—the *support vectors* that give the method its name—which are then combined to specify a predicting function.

The SVM algorithm uses two different strategies to tackle the nonseparability of realistic data. On one hand, it projects the examples into a space of higher dimensionality by means of a *kernel* function [28]. The separating surface thus found is flat by construction in the new space but can be curved and even multiply connected in the original. On the other hand, the technique tries to control overfitting—and thus concentrate on essentials rather than on details, resulting in better generalization—by having an adjustable penalty on misclassifications. This penalty is represented by a single scalar parameter, the *capacity* of the machine [29].

During training, an SVM solves the constrained quadratic optimization problem [30]

$$\max_{\alpha} \quad \sum_i \alpha_i - \frac{1}{2}\sum_{i,j} \alpha_i y_i \mathbf{x}_i^T \mathbf{x}_j y_j \alpha_j$$
$$\text{s.t.} \quad \sum_i \alpha_i y_i = 0, \quad 0 \le \alpha_i \le C, \tag{11}$$

whose solution is a vector of coefficients α_i that measure the information content of the examples and are nonzero only for the support vectors. The coefficients are prescribed not to exceed the capacity C, which limits the influence of potentially problematic points on the final result.

The projection to higher dimensions occurs by substituting the scalar products

$$\mathbf{x}_i^T \mathbf{x}_j \longrightarrow K(\mathbf{x}_i, \mathbf{x}_j) \equiv \phi(\mathbf{x}_i)^T \phi(\mathbf{x}_j) \tag{12}$$

for some mapping $\phi(\mathbf{x})$. The function K is the kernel we mentioned earlier. It is not necessary to know ϕ to find K: any function that combines two vectors into a scalar and fulfills the (not very restrictive) set of conditions spelled out in Mercer's theorem [31] can be used as a kernel. Some kernels stretch out the examples into the added dimensions in such a way that gaps open up between the examples which permit a flat separating surface to pass through. In this paper, we use the radial basis function (RBF) kernel

$$K(\mathbf{x}_i, \mathbf{x}_j) = \exp\left(-\frac{(\mathbf{x}_i - \mathbf{x}_j)^T(\mathbf{x}_i - \mathbf{x}_j)}{2\sigma^2}\right), \tag{13}$$

which surrounds every example with a surface that in a sense "repels" the separating hyperplane. The Gaussian width σ is a second adjustable parameter and usually has a scale on the order of the average separation between points. In [32] it was found that polynomial kernels may outperform the RBF kernel in some electromagnetic inverse problems. We find that the linear kernel makes similar predictions and runs faster than the RBF, though the difference in run time is negligible for the number of training data and example features that we use in this study.

Once α is known, the SVM can predict the attribute of an unknown example using the function [29, 33]

$$f(\mathbf{x}) = \operatorname{sgn}\left(\sum_{i \in SV} \alpha_i y_i K(\mathbf{x}_i, \mathbf{x}) \right). \tag{14}$$

There are several ways to generalize the SVM procedure to perform multiclass categorization. These have been reviewed in [11], whose authors conclude that the methods more suitable for practical use perform several binary classifications instead of attempting to separate all classes at once. In this work we adopt a one-against-one approach [34] in which the system carries out $\binom{k}{2} = k(k-1)/2$ optimizations and obtains the same number of decision functions of the form (14). When given an example to predict, the algorithm proceeds by ballot: it evaluates the decision functions one by one on the example and adds a vote to the one category (out of two) in which it is predicted to be. At the end, the example is assigned to the category with the most votes; should there be a tie between two classes, the program arbitrarily selects that with the smallest label.

4. RESULTS

4.1. Data Acquisition

The Camp Sibert blind-test data were collected over 216 test cells, each of which was a square plot of side 5 m and contained at most one anomaly. The targets of interest were 4.2″ mortar shells like the one in Figure 1(a), which were to be discriminated from explosion byproducts represented by the base plates and partial shells of Figures 1(b) and 1(c). Other sites had smaller shrapnel or non-UXO related scrap instead, and a few were essentially empty. We were given the ground truth for 66 of these cases, which we used to build a catalog of expected total NSMS values that were then tested on the 150 other cells. The EM-63 took data over 26 channels that span in approximately logarithmic fashion a lapse of time between 180 μs and 25 ms. In our analysis, we use 25 of these channels, starting with the second. Measurements were taken at grids of between 400 and 700 points that crisscrossed each cell; each grid row was separated some 50 cm from its neighbors, and within each track the spacing between consecutive measurements was on the order of 5 cm. The EM-63 was always placed 30 cm above the ground. Figure 4 shows a typical experimental situation, corresponding to Cell no. 7 of the study and containing 668 measurement points.

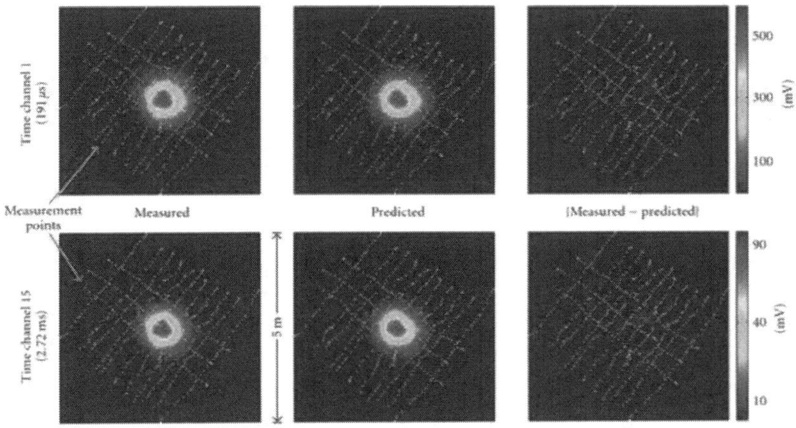

Figure 4. A typical cell from the Camp Sibert EM-63 blind test, no. 7 in this case, consists of a square plot of side 5 m. The white dots show the measurement points distributed on a 668-point grid; the separation between rows is about 50 cm and that between consecutive points is some 5 cm. The contour plots show the measured (left column) and HAP/NSMS-predicted (middle column) near-field distributions for this cell, as well as the mismatch between the two (right column), for two time gates. The first time channel (top row) is taken 191 μs after shutdown; the 15th (bottom) is centered at 2.72 ms.

4.2. Target Location and Characterization

For each data set we run the HAP method of Section 2.1 to locate the target and a fully three-dimensional implementation of the NSMS model of Section 2.2 to characterize it. Consider again the example cell shown on Figure 4. To find the target, we take a fictitious $5\,\mathrm{m} \times 5\,\mathrm{m}$ flat square surface concentric with the plot and located 30 cm below the sensor (i.e., at ground level) and divide it into 11×11 patches, each of which is assumed to contain a magnetic-charge distribution of uniform density. We take the measured field data (as seen for example on the left column of Figure 4) and use (5) to determine $\mathbf{q}$, which in turn allows us to determine $\psi(\mathbf{r})$ using (7) and construct the matrices of (4) to find the location. We do this separately for every time channel and get consistent location estimates from gate to gate, which lends credence to their precision. For the case of Cell no. 7, we obtain a target depth of 55 cm, acceptably close to the ground truth of 60 cm.

To compute the NSMS amplitude $Q(t)$ we surround the target with a prolate spheroid with semiminor axis $a = 5$ cm and elongation $e \equiv b/a = 4$. This spheroid is divided into 7 azimuthal belts, each of which is assumed to contain a radial-magnetic-dipole distribution of constant density. The spheroid is placed at the location estimated by the HAP method and the orientation given by the dipole moment $\mathbf{m}$ obtained from (2) and (7). With all the pieces in place, we can proceed to apply (8) to find Ω and (9) to extract $Q(t)$ for the target, which eventually reveals it as the UXO of Figure 5(a). This $Q(t)$ curve appears as a line of pentagons in Figure 3, which also depicts the resulting total NSMS values for the rest of the targets. We determine the Pasion-Oldenburg parameters k, β, and γ for each anomaly by a direct nonlinear least-squares fit of (10) and by linear (pseudo)inversion of its logarithm; both procedures gave consistent results. In general, we obtain good fits to the measured fields [4]; Figure 4 shows that the discrepancy between the actual data and the model prediction runs only to a few percent.

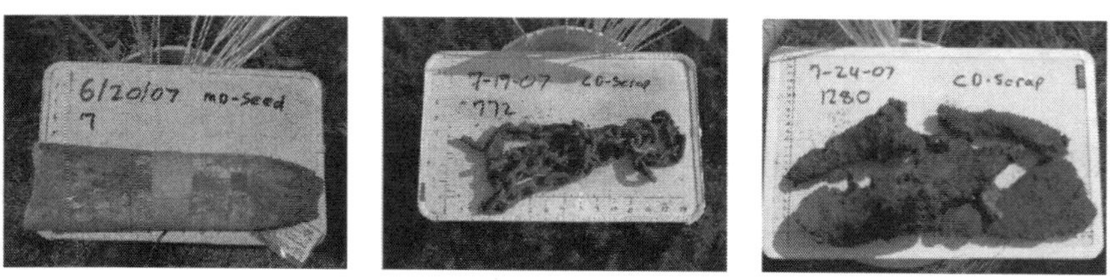

Figure 5. (a) Unexploded shell from Cell no. 7, and (b), (c) the two false alarms obtained by the SVM classifier using *k* and *R* as discriminators.Shell at no. 7Twisted chainCrumpled debris

We have previously found [3, 4] that the ratio of Q at the 15th time channel to Q at the first time channel, which involves a fixed superposition of β and γ, shows discernible clustering for this particular data set when combined with the third parameter k. (The 15th time channel, centered at about 2.7 ms, was chosen because it takes place late enough to show the behavior described above but early enough that all targets still have an acceptable signal-to-noise ratio; nearby time channels produce similar results.) The values of $R \equiv Q(t_{15}) / Q(t_1)$ for the 4.2″ mortars are particularly well grouped and for the most part noticeably distinct from those of the others, suggesting that this two-dimensional feature space may be used to perform dependable classification. This suggestion is confirmed by our SVM analysis.

4.3. SVM Classification

We use a Gaussian RBF kernel for the SVM analysis. The kernel width turns out not to have much influence on the outcome; we usually set it so that a unit in a typical x- or y-axis in a log plot (Figure 6, say) comprises 100Δ Gaussian widths, where Δ is the dimensionality of the feature space. To find the capacity, we train the SVM with a subset of the training data and a given C, scramble the training set, and use a new subset of the data for testing. We then vary C, setting it to a high value initially and then lowering it, and keep the lowest capacity with which the machine identifies all dangerous items in the test. The procedure is rather ad hoc but effective for the data at hand, given the small sample sizes, the low dimensionality of the feature spaces, and the speed of the SVM implementation. A more systematic search for C and γ using five-fold cross-validation [5] recommends slightly higher capacities that result in identical predictions.

For R and k as features, we find the best SVM performance using $C = 10$. The results are displayed (for testing data only) in Table 1 and shown pictorially (for both training and testing) in Figure 6. The matrix element c_{ij} in the table denotes an item of category i that was identified by the SVM as belonging to category j; in other words, the rows of this contingency table correspond to the ground truth and the columns to predictions. The small markers in the plot show the ground truth (hollow for training data and filled for the tests), while the large markers point out the items for which the SVM makes wrong predictions. For example, a small yellow upright triangle surrounded by a large cyan square is a piece of scrap (clutter unrelated to UXO) incorrectly identified as a base plate. The UXO, with their high initial amplitudes and slow decay, are clustered at the top right corner. We see that there are only two false alarms (i.e., objects identified with UXO that were in fact something else) and that all potentially dangerous items have been identified correctly.

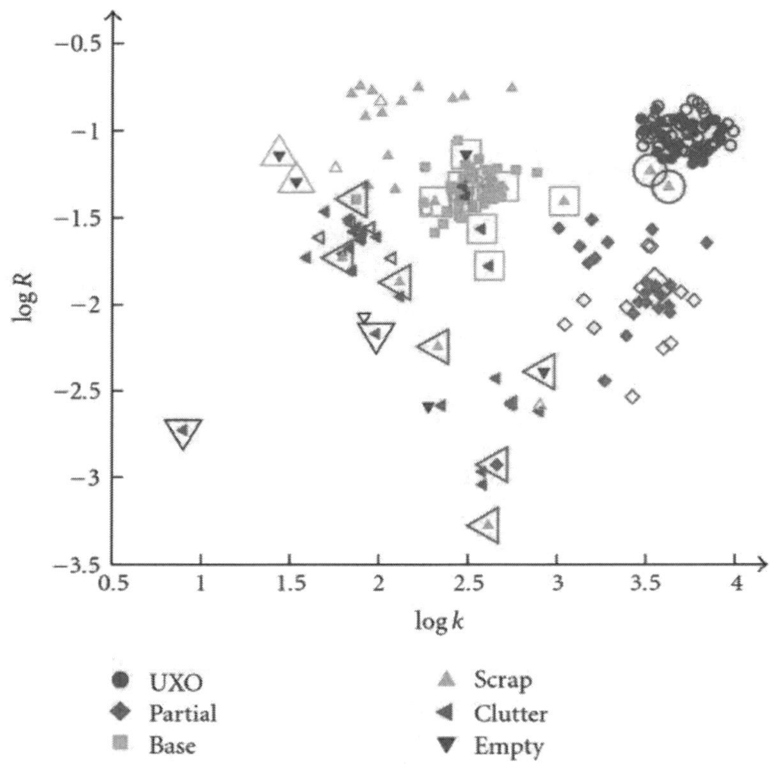

Figure 6. Result of the SVM classification for the Camp Sibert anomalies using the logarithms of k and $R \equiv Q(t_{15})/Q(t_1)$. The SVM has been trained with capacity $C = 10$ and kernel width $\sigma = 1/200$. The small markers denote the ground truth for both training (hollow) and testing (solid) cells. The larger markers highlight the cases where there is a disagreement between the ground truth and the SVM prediction.

Table 1. SVM classification of Camp Sibert anomalies using k and R with $C = 10$.

$k, R; C = 10$		SVM prediction					
		UXO	Partial	Base	Clutter	Scrap	Empty
Ground truth	UXO	34	0	0	0	0	0
	Partial	0	22	0	1	0	0
	Base	0	0	39	1	0	0
	Clutter	0	0	4	19	0	2
	Scrap	2	0	3	4	13	0
	Empty	0	1	1	1	2	1

The false alarms, two pieces of non-UXO clutter, appear on Figures 5(b) and 5(c). They are seen to be similar to the 4.2″ mortars in size and metal content (cf. Figure 5(a)), which makes their k and R values lie closer to the tight UXO cluster than to any other anomaly in Figure 6. Here we note that, as can be seen in Figure 1(d), the training data provided by the examiners was somewhat biased toward UXO, while clutter and scrap samples were underrepresented (this was not the case with the testing data and should not be expected in future tests). If we switch training and testing data in the SVM analysis, we can achieve perfect discrimination without varying the capacity—though in this case we have more training data than tests. This highlights the importance of having a diverse collection of representative samples to use during the training stage.

We can repeat the analysis using other two-dimensional combinations of the Pasion-Oldenburg parameters. Combining k and γ yields results similar to those of k and R, as Figure 7 and Table 2 show. Figure 8 and Table 3 show the classification resulting from the use of β and γ as discriminators. The table shows that we can obtain reasonable discrimination, with all the UXO once again correctly identified, but the increased number of false alarms and the very high capacity needed (four orders of magnitude larger than the previous ones) indicate that this combination of parameters may not be optimal and that this machine is prone to overfitting. A glance at the figure shows that the clustering is much less clear-cut than in the previous cases, partly because the range of β is rather small. In fact, combining k and β greatly reduces the performance, since the small β -range and the close similarity in k of the UXO and the partial mortars causes an overlap between the two categories that cannot be disentangled.

Table 2. SVM classification of Camp Sibert anomalies using γ and k with $C = 9$.

$\gamma, k; C = 9$		SVM prediction					
		UXO	Partial	Base	Clutter	Scrap	Empty
Ground truth	UXO	34	0	0	0	0	0
	Partial	5	17	0	1	0	0
	Base	0	0	39	0	1	0
	Clutter	0	0	4	15	5	1
	Scrap	2	1	3	5	11	0
	Empty	1	1	2	2	0	0

It is helpful and straightforward to increase the dimensionality of the feature space. Figure 9 shows the discrimination obtained by running the SVM using all three Pasion-Oldenburg features. The capacity $C = 9$ here, and increasing it changes the results only slightly. The number of false alarms increases: we get the same two pieces of scrap from before, and now a few of the partial mortars are identified as UXO by the algorithm, due in part to the small range of β and in part to the large gap between the UXO and the other anomalies, clearly visible in the figure, which again calls out for more and more-diverse training information.

Table 3. SVM classification of Camp Sibert anomalies using γ and β with $C = 10^5$.

$\gamma, \beta; C = 10^5$		SVM prediction					
		UXO	Partial	Base	Clutter	Scrap	Empty
Ground truth	UXO	34	0	0	0	0	0
	Partial	0	14	5	2	1	1
	Base	3	0	37	0	0	0
	Clutter	0	1	5	14	1	4
	Scrap	3	1	1	3	13	1
	Empty	2	0	0	3	1	0

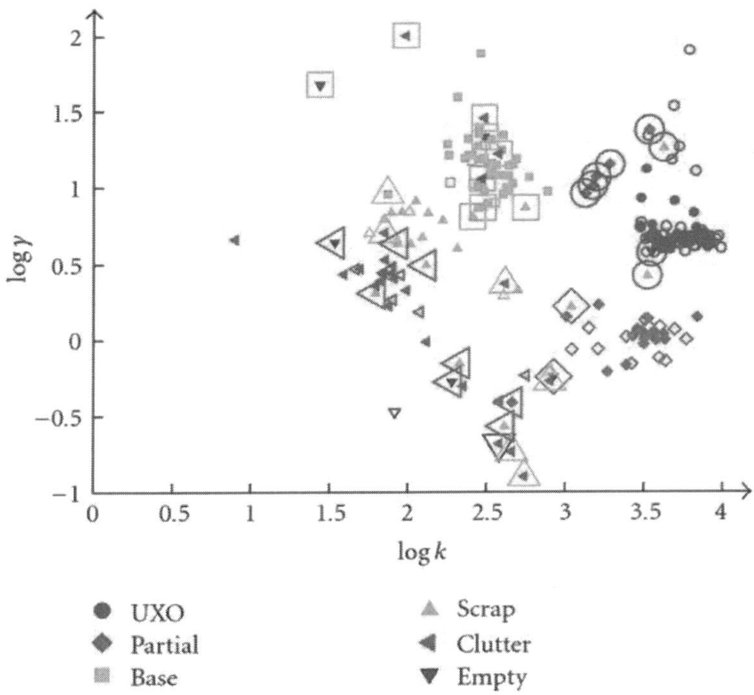

Figure 7. Result of the SVM classification for the Camp Sibert anomalies using the logarithms of the Pasion-Oldenburg parameters $k$and γ. The SVM here has a capacity $C = 9$. The small markers denote the ground truth for both training (hollow) and testing (solid) cells. The larger markers show the wrong SVM predictions.

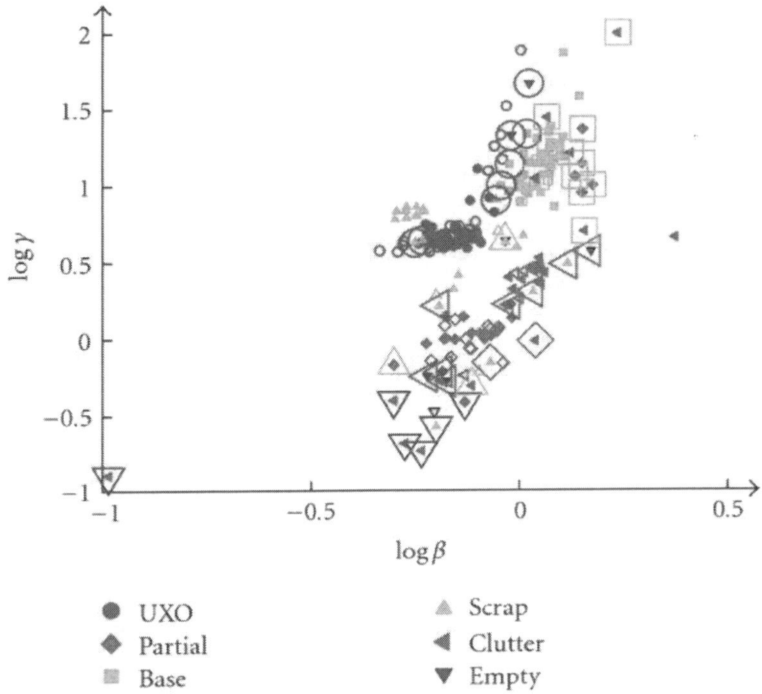

Figure 8. Result of the SVM classification for the Camp Sibert anomalies using the logarithms of the Pasion-Oldenburg parameters β and γ. The SVM capacity $C = 10^5$. The small markers denote the ground truth for both training (hollow) and testing (solid) cells. The larger markers highlight the wrong predictions made by the SVM.

Finally, it is possible to dispense with the Pasion-Oldenburg model altogether and run an SVM using the "raw" $Q(t)$ as input. The feature space has dimensionality $\Delta = 25$. We scale the values by $Q(t_1)$ and take the logarithm. We find $C = 20$ to be the optimal value. Table 4 shows the results. The performance is slightly inferior to that of R versus k; the usual two false alarms are there, along with a few new ones. All the UXO are identified correctly. We can also use the logarithm of Q without any scaling (though the SVM internally rescales the feature space to $[0,1]^{\Delta}$). A capacity $C = 1$ suffices here. The results appear on Table 5. All dangerous items are once more identified as such.

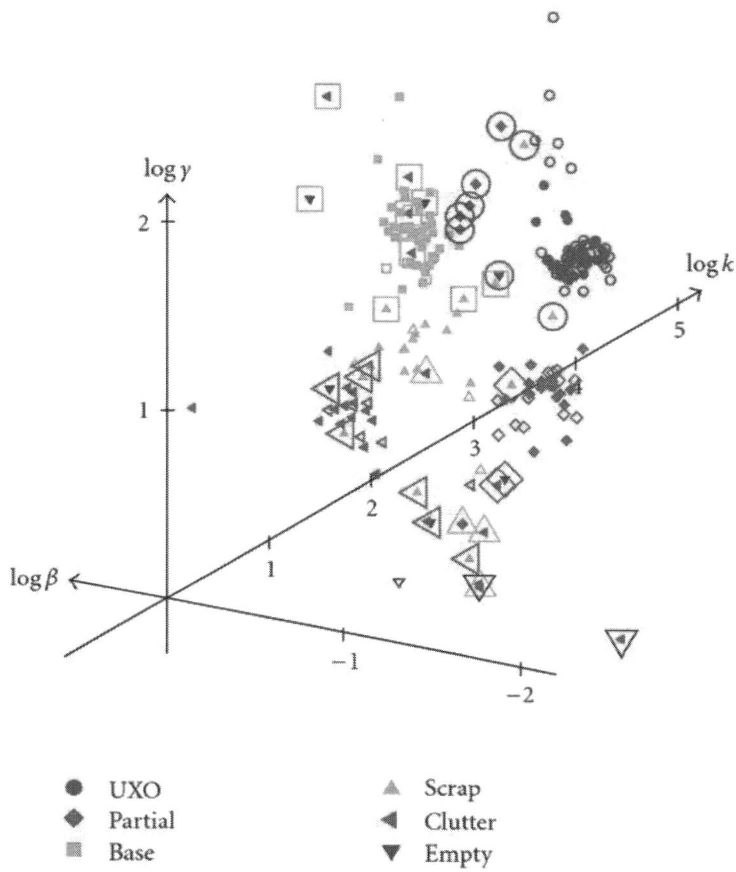

Figure 9. SVM classification of the Camp Sibert anomalies using the logarithms of k, β, and γ. The SVM has $C = 9$. The small markers denote the ground truth for both training and testing cells. The larger markers highlight the cases where there is a disagreement between the ground truth and the SVM prediction.

Table 4. SVM classification of Camp Sibert anomalies using all Q values (scaled by $Q(t_1)$) with $C = 20$.

$Q/Q(t_1)$; $C = 20$		SVM prediction					
		UXO	Partial	Base	Clutter	Scrap	Empty
Ground truth	UXO	34	0	0	0	0	0
	Partial	0	15	0	7	1	0
	Base	3	0	34	3	0	0
	Clutter	0	2	3	14	4	2
	Scrap	3	1	3	3	12	0
	Empty	2	2	1	0	1	0

Table 5. SVM classification of Camp Sibert anomalies using all Q values (unscaled) with $C = 1$.

All Q; C = 1		SVM prediction					
		UXO	Partial	Base	Clutter	Scrap	Empty
Ground truth	UXO	34	0	0	0	0	0
	Partial	5	17	0	1	0	0
	Base	0	0	39	0	1	0
	Clutter	0	2	4	18	1	0
	Scrap	2	2	3	2	13	0
	Empty	0	1	1	2	2	0

5. CONCLUSION

In this paper, we have applied the NSMS model to EM-63 Camp Sibert discrimination data sets. First the locations of the objects were inverted for by the fast and accurate dipole-inspired HAP method. Subsequently, each anomaly was characterized at each time channel through its total NSMS strength. Discrete intrinsic features were selected and extracted for each object using the Pasion-Oldenburg decay law and then used as input for a support vector machine that classified the items.

Our study reveals that the ratio of an object's late response to its early response can be used as a robust discriminator when combined with the Pasion-Oldenburg amplitude k. Other mixtures of these parameters also result in good classifiers. Moreover, we can use Q directly, completely obviating the need for the Pasion-Oldenburg fit. In each case, the classifier runs by itself and does not require any human intervention. The SVM can be trained very quickly, even when the feature space has more than 20 dimensions, and it is a simple matter to add more training data on-the-fly. It is also possible to use already processed data to classify examples as yet unseen.

We should stress that none of our classifications yielded false negatives: all UXO were identified correctly in every instance. (This is due in part to the clean, UXO-intensive training data provided by the examiners and may change under different conditions.) The number of false alarms (false positives) varies with the classification features, but is in general low and can be as low as 2 out of 36 reported positives. Figures 6 and 5 show, among others, how these false alarms come to be: some of the clutter items have a response that closely resembles that of UXO. While this will inevitably arise, it may still be possible to make the SVM more effective—and perhaps get close to reaching 100% accuracy—by including some of these refractory cases during the training. That said, there will certainly be cases in the field where the nonuniqueness inherent to noisy inverse scattering problems will cause the whole procedure to fail and yield dubious estimates. In those cases it will be necessary to assume the target is dangerous and dig it out.

In a completely realistic situation, where in principle no training data are given and the ground truth can be learned only as the anomalies are excavated, one can never be sure that the data already labeled constitute a representative sample containing enough of both dangerous and innocuous items. This difficulty is mitigated by two facts: (1) usually at the outset we have some idea of the kind of UXO present in the field and (2) the (usually great) majority of detected anomalies will not be UXO and thus random digging will produce a varied sampling of the clutter present. Methods involving *semi* supervised learning exploit this gradual revealing of the truth and have been found to perform better at UXO discrimination than supervised learning methods like SVM when starting from the point dipole model [35, 36]. (*Active* learning methods, which try to infer which anomalies would contain the most useful information and could thus serve to guide the anomaly unveiling, show further, though fairly minor, improvement.) Combining this more powerful learning procedure with the excellent performance of the HAP/NSMS method may enhance the discrimination protocol and should be the subject of further research.

In summary, the results presented here show that our search and characterization procedure, whose effectiveness is apparent from several recent studies [3, 4, 37, 38], can be combined with an SVM classifier to produce a UXO discrimination system capable of correctly singling out dangerous items from among munitions-related debris and other natural and artificial clutter. In future investigations, we will continue to hone these algorithms and use them on other blind tests, including some already carried out in saltwater instead of soil.

ACKNOWLEDGMENT

This work was supported by the Strategic Environmental Research and Development Program through Grants no. MM-1572 and no. MM-1573.

REFERENCES

1. Byrnes J (Ed): Unexploded Ordnance Detection and Mitigation, NATO Science for Peace and Security Series B: Physics and Biophysics. Springer, Dordrecht, The Netherlands; 2009.

2. McNeill JD, Bosnar M: Application of TDEM techniques to metal detection and discrimination: a case history with the new Geonics EM-63 fully time-domain metal detector. Geonics LTD, Mississauga, Canada;http://www.geonics.com/ 2000

3. Shubitidze F, Fernández JP, Barrowes BE, Shamatava I, O'Neill K: Normalized surface magnetic source model applied to camp sibert data: discrimination studies. Proceedings of the Applied Computational Electromagnetics Symposium (ACES '09), March 2009, Monterey, Calif, USA

4. Shubitidze F, Fernández JP, Shamatava I, Pasion LR, Barrowes BE, O'Neill K: Application of the normalized surface magnetic source model to a blind unexploded ordnance discrimination test. Applied Computational Electromagnetics Society Journal 2010, 25(1):89-98.

5. Chang C-C, Lin C-J: LIBSVM: a library for support vector machines. 2009, http://www.csie.ntu.edu.tw/~cjlin/libsvm/

6. Boser BE, Guyon IM, Vapnik VN: A training algorithm for optimal margin classifiers. In Proceedings of Annual 5th ACM Workshop on Computational Learning Theory. Edited by: Haussler D. ACM Press, London, UK; 1992:144-152.

7. Vapnik VN: An overview of statistical learning theory. IEEE Transactions on Neural Networks 1999, 10(5):988-999.

8. Cherkassky V, Mulier F: Learning from Data: Concepts, Theory, and Methods. John Wiley & Sons, New York, NY, USA; 1998.

9. Cortes C, Vapnik V: Support-vector networks. Machine Learning 1995, 20(3):273-297.

10. Smola AJ, Schölkopf B: A tutorial on support vector regression. Statistics and Computing 2004, 14(3):199-222.

11. Hsu C-W, Lin C-J: A comparison of methods for multiclass support vector machines. IEEE Transactions on Neural Networks 2002, 13(2):415-425.

12. Zhang Y, Collins L, Yu H, Baum CE, Carin L: Sensing of unexploded ordnance with magnetometer and induction data: theory and signal processing. IEEE Transactions on Geoscience and Remote Sensing 2003, 41(5, part 1):1005-1015.

13. Fernández JP, Barrowes B, O'Neill K, et al.: Evaluation of SVM classification of metallic objects based on a magnetic-dipole representation. In Detection and Remediation Technologies for Mines and Minelike Targets XI, April 2006, Bellingham, Wash, USA, Proceedings of SPIEEdited by: Broach JT, Harmon RS, Holloway JH Jr.. 6217:

14. Chen X: Inverse problems in electromagnetics, Ph.D. dissertation. Massachusetts Institute of Technology, Cambridge, Mass, USA; June 2005.

15. Zhang B, O'Neill K, Kong JA, Grzegorczyk TM: Support vector machine and neural network classification of metallic objects using coefficients of the spheroidal MQS response modes. IEEE Transactions on Geoscience and Remote Sensing 2008, 46(1):159-171.

16. Fernández JP, Sun K, Barrowes B, et al.: Inferring the location of buried UXO using a support vector machine. In Detection and Remediation Technologies for Mines and Minelike Targets XII, April 2007, Bellingham, Wash, USA, Proceedings of SPIE Edited by: Harmon RS, Broach JT, Holloway JH Jr.. 6553: 65530B.1-65530B.9.

17. Massa A, Boni A, Donelli M: A classification approach based on SVM for electromagnetic subsurface sensing. IEEE Transactions on Geoscience and Remote Sensing 2005, 43(9):2084-2093.

18. Bermani E, Boni A, Caorsi S, Massa A: An innovative real-time technique for buried object detection. IEEE Transactions on Geoscience and Remote Sensing 2003, 41(4):927-931.

19. Caorsi S, Anguita D, Bermani E, Boni A, Donelli M, Massa A: A comparative study of NN and SVM-based electromagnetic inverse scattering approaches to on-line detection of buried objects. Applied Computational Electromagnetics Society Journal 2003, 18(2):65-75.

20. Tarokh AB, Miller EL, Won IJ, Huang H: Statistical classification of buried objects from spatially sampled time or frequency domain electromagnetic induction data. Radio Science 2004., 39(4):Article ID RS4S05, 2004.

21. Aliamiri A, Stalnaker J, Miller EL: Statistical classification of buried unexploded ordnance using nonparametric prior models. IEEE Transactions on Geoscience and Remote Sensing 2007, 45(9):2794-2806.

22. Shamatava I, Shubitidze F, Barrowes B, Fernández JP, Pasion LR, O'Neill K: Applying the physically complete EMI models to the ESTCP camp sibert pilot study EM-63 data. In Detection and Sensing of Mines, Explosive Objects, and Obscured Targets XIV, May 2009, Bellingham, Wash, USA, Proceedings of SPIE Edited by: Harmon RS, Broach JT, Holloway JH Jr.. 7303: 73030O1-73030O-11.

23. Shubitidze F, Karkashadze D, Barrowes B, Shamatava I, O'Neill K: A new physics-based approach for estimating a buried object's location,

orientation and magnetic polarization from EMI data. Journal of Environmental and Engineering Geophysics 2008, 13(3):115-130.

24. Shubitidze F, O'Neill K, Barrowes BE, et al.: Application of the normalized surface magnetic charge model to UXO discrimination in cases with overlapping signals. Journal of Applied Geophysics 2007, 61(3-4):292-303.

25. Pasion LR, Oldenburg DW: A discrimination algorithm for UXO using time domain electromagnetics. Journal of Environmental & Engineering Geophysics 2001, 6: 91-102.

26. Kyurkchan AG, Sternin BY, Shatalov VE: Singularities of continuation of wave fields. Physics-Uspekhi 1996, 39(12):1221-1242.

27. Zaridze R, Bit-Babik G, Tavzarashvili K, Economou DP, Uzunoglu NK: Wave field singularity aspects in large-size scatterers and inverse problems. IEEE Transactions on Antennas and Propagation 2002, 50(1):50-58.

28. Aizerman MA, Braverman EM, Rozonoer LI: Theoretical foundations of the potential function method in pattern recognition learning.Automation and Remote Control 1964, 25: 821-837.

29. Burges CJC: A tutorial on support vector machines for pattern recognition. Data Mining and Knowledge Discovery 1998, 2(2):121-167.

30. Mangasarian OL, Musicant DR: Lagrangian support vector machines. Journal of Machine Learning Research 2001, 1(3):161-177.

31. Mercer J: Functions of positive and negative type and their connection with the theory of integral equations. Philosophical transactions of the Royal Society of London. Series A 1909, 209: 415-446.

32. Bermani E, Boni A, Kerhet A, Massa A: Kernels evaluation of SVM-based estimators for inverse scattering problems. The phenomenon of resonance radiation of electromagnetic Research 2005, 53: 167-188.

33. Cristianini N, Shawe-Taylor J: An Introduction to Support Vector Machines and other Kernel-Based Learning Methods. Cambridge University Press, Cambridge, UK; 2000.

34. Kreßel U: Pairwise classification and support vector machines. In Advances in Kernel Methods: Support Vector Learning. Edited by: Schölkopf B, Burges JC, Smola AJ. MIT Press, Cambridge, Mass, USA; 1999:268-255.

35. Zhang Y, Liao X, Carin L: Detection of buried targets via active selection of labeled data: application to sensing subsurface UXO. IEEE Transactions on Geoscience and Remote Sensing 2004, 42(11):2535-2543.

36. Liu Q, Liao X, Carin L: Detection of unexploded ordnance via efficient semisupervised and active learning. IEEE Transactions on Geoscience and Remote Sensing 2008, 46(9):2558-2567.

37. Shubitidze F, Barrowes B, Shamatava I, Fernández JP, O'Neill K: APG UXO discrimination studies using advanced EMI models and TEMTADS data. In Detection and Sensing of Mines, Explosive Objects, and Obscured Targets XIV, May 2009, Bellingham, Wash, USA, Proceedings of SPIE Edited by: Harmon RS, Broach JT, Holloway JH Jr.. 7303:

38. Shamatava I, Shubitidze F, Barrowes B, Fernández JP, O'Neill K: Physically complete model applied to BUD time-domain EMI data. InDetection and Sensing of Mines, Explosive Objects, and Obscured Targets XIV, May 2009, Bellingham, Wash, USA, Proceedings of SPIEEdited by: Harmon RS, Broach JT, Holloway JH Jr.. 7303: 73030N1-73030N-9.

CHAPTER 7

Audio-frequency magnetotelluric imaging of the Hijima fault, Yamasaki fault system, southwest Japan

Satoru Yamaguchi[1]*, Yasuo Ogawa[2], Kiyoshi Fuji-ta[1], Naoto Ujihara[3], Hiroo Inokuchi[4], and Naoto Oshiman[5]

[1]Department of Earth and Planetary Sciences, Faculty of Science, Kobe University, Nada, Kobe 657-8501, Japan

[2]Volcanic Fluid Research Center, Tokyo Institute of Technology, Tokyo 152-8551, Japan

[3]Departments of Earth and Planetary Sciences, Graduate School of Science and Engineering,

Tokyo Institute of Technology, Tokyo 152-8551, Japan

[4]School of Human Science and Environment, University of Hyogo, Himeji 670-002, Japan

[5]Disaster Prevention Research Institute, Kyoto University, Uji 611-0011, Japan

ABSTRACT

An audio-frequency magnetotelluric (AMT) survey was undertaken at ten sites along a transect across the Hijima fault, a major segment of the Yamasaki fault system, Japan. The data were subjected to dimensionality analysis, following which two-dimensional inversions for the TE and TM modes were carried out. This model is characterized by (1) a clear resistivity boundary that coincides with the downward projection of the surface trace of the Hijima fault, (2) a resistive zone (>500 Ω m) that corresponds to Mesozoic sediment, and (3) shallow and deep two highly conductive zones (30–40 Ω m) along the fault. The shallow conductive zone is a common feature of the Yamasaki fault system, whereas the deep conductor is a newly discovered feature at depths of 800–1,800 m to the southwest of the fault. The conductor is truncated by the Hijima fault to the northeast, and its upper boundary is the resistive zone. Both conductors are interpreted to represent a combination of clay minerals and a fluid network within a fault-related fracture zone. In terms of the development of the fluid networks, the fault core of the Hijima fault and the highly resistive zone may play important roles as barriers to fluid flow on the northeast and upper sides of the conductive zones, respectively.

KEYWORDS

Hijima fault; Yamasaki fault system; audio-frequency magnetotelluric; conductivity structure

1. INTRODUCTION

A common feature of magnetotelluric surveys is the appearance of changes in the apparent resistivity and/or phase value across the surface trace of an active fault (e.g., Ogawa and Honkura, 1997; Unsworth *et al.*, 1997; Yamaguchi *et al.*, 2001; Ritter *et al.*, 2005). A prominent feature of such changes, and an important factor in characterizing faults, is a distinctive conductive zone termed the fault zone conductor (FZC). For example, Unsworth *et al.* (1999) reported different imaging characteristics for FZCs located near the ground surface within locked and actively creeping segments of the San Andreas fault, USA. Yamaguchi *et al.* (2002) imaged contrasting near-surface FZCs within segments of both the large displacement and the little displacement along the Nojima fault, Japan, which were initiated at the time of the 1995 Hyogo-ken Nanbu earthquake.

A second type of FZC has been found in deep regions near active faults. Ogawa and Honkura (1997) performed an audio-frequency magnetotelluric (AMT) survey across the Atera fault, a 66-km-long active fault in Central Japan, and found conductors located near the fault at a depth of 0.5–2.0 km. In a later publication, Ogawa *et al.* (2002) reported the presence of a deep conductor at mid-crustal depths (15–20 km) near the Itoigawa-Shizuoka Tectonic line, a structure that is much longer than the Atera fault and which crosses Central Japan. The authors of these earlier studies proposed that the conductors may represent localized zones of fluids that occupy areas of enhanced porosity in fracture zones associated with active faults. A similar conductor was found in the lower crust along a major tectonic fault in India (Normada-Son Lineament; Patro *et al.*, 2005). It is interesting that conductors of varying sizes but with the same mechanism of formation have been found near faults.

The Yamasaki fault system of southwest Japan is a typical left-lateral strike-slip fault system that extends for over 80 km along a general strike of N60°W–S60°E (Fig. 1(a)). Many micro-earthquakes have been recorded along this fault system (Shibutani, 2004; Fig. 1(b)), along with large historical earthquakes, such as the magnitude 7.1 Harima earthquake of 868 AD (Okada *et al.*, 1987).

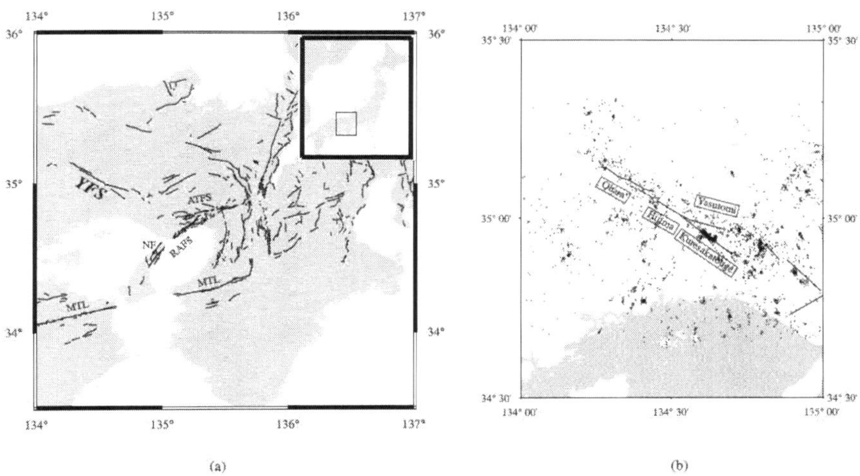

Figure 1. (a) Location of the Yamasaki fault system and nearby faults within the study area (after Yamaguchi *et al.*, 2007). YFS: Yamasaki fault system, MTL: median tectonic line, NF: Nojima fault, RAFS: Rokko-Arima fault system, ATFS: Arima-Takatsuki fault system. (b) Epicenters of micro-earthquakes centered around the Yamasaki fault system. The hypocenters were recorded by the Disaster Prevention Research Institute, Kyoto University, between June 1976 and December 2003 (Shibutani, 2004). Only the epicenters of earthquakes with magnitudes >1.0 are shown. The Ohara, Hijima, Yasutomi, Kuresakatouge, Biwako, Miki, and Kusatani faults make up the Yamasaki fault system. The open star symbol on the Hijima fault indicates the location of the study area.

The Earthquake Research Committee of Japan evaluates the probability of earthquake occurrence at major active faults in Japan, estimating the probability of earthquakes of a given magnitude occurring in the following 30 years. For the southeastern part of the Yamasaki fault system (Biwako and Miki faults), the committee has estimated a maximum probability of 5% for an earthquake with a magnitude of ~7.3; this places the system in the high group of earthquake occurrence probability. For the northwestern part of the system (the Ohara, Hijima, Yasutomi, and Kuresakatouge faults), the maximum probability of a magnitude ~7.7 event is 1%, corresponding to the slightly high group of earthquake occurrence probability (Earthquake Research Committee, 2003, 2007).

An FZC that is 1,000 m wide and with an apparent resistivity of 100–400 Ω m has been detected along the Yasutomi fault of the Yamasaki fault system (Electromagnetic Research Group for the Active Fault (ERGAF), 1982); however, the base of the zone has yet to be determined. Handa and Sumitomo (1985) conducted an ELF-MT (extremely low frequency magnetotelluric) survey around the Yasutomi fault with the aim of determining the resistivity structure beneath the fault. Their model is characterized by a large conductive zone (< 1,000 Ω m) that is 6 km wide and 3 km deep, including the surface fault trace; however, the precise nature of the resistivity structure beneath the Yamasaki fault system has yet to be established.

In this paper, we report a detailed two-dimensional (2-D) geoelectrical model across the Hijima fault of the Yamasaki fault system. The model was developed based on data collected during an AMT survey that was carried out with the aim of placing geophysical constrains on the nature of the Yamasaki fault system. This fault system is located close to large cities and is considered to be a high earthquake risk.

2. OBSERVATIONS

An AMT survey was undertaken in January 2006 at ten sites along a transect across the Hijima fault (Fig. 2(a)). A dense site spacing of 250 m in the center of the profile was supplemented by more widely spaced sites near the margins. In performing the survey, we used the AMT systems developed and manufactured by Phoenix Geophysics Ltd. (Toronto, Canada). Five components (H_x, H_y, H_z, E_x, and E_y) were simultaneously measured for periods of 3–17hat each site. Measurements at either Site 1 or Site 2 were made for the entire survey period to enable remote-reference processing of the time series data (Gamble *et al.*, 1979).

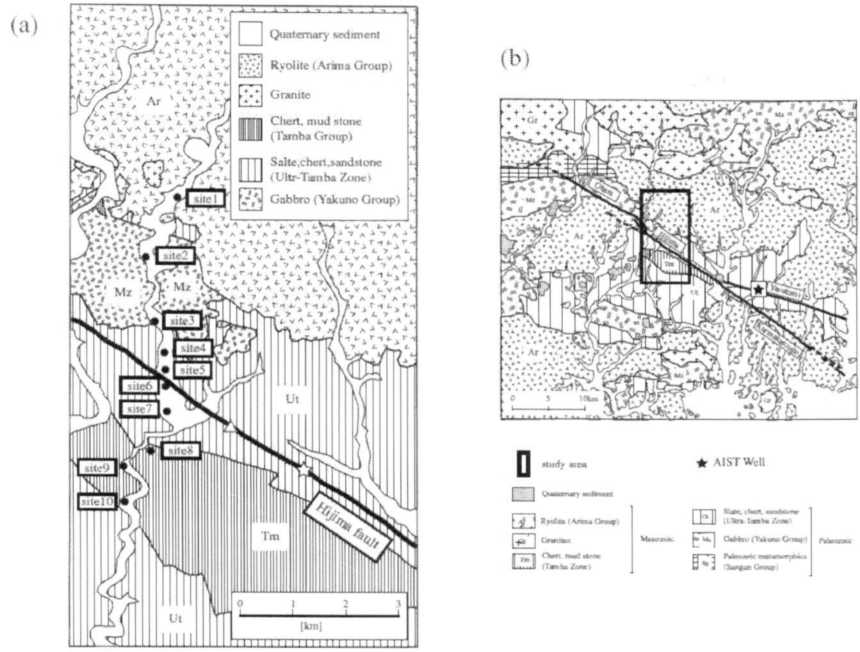

Figure 2. (a) Simplified geological map (after Hyogo-ken, 2000) of the study area showing the distribution of observation sites. Solid circles denote audio-frequency magnetotelluric (AMT) sites. Solid line indicates the surface trace of the Hijima fault. (b) Simplified geological map of the area around the Yamasaki fault system (after Hyogo-ken, 2000). Black star indicates the location of the AIST well.

3. DATA ANALYSIS

We adopted phase tensor analysis (Caldwell *et al.*, 2004) to estimate the dimensionality of the resistivity structure beneath the study area and to determine the regional strike in the case that the resistivity structure was 2-D. Figure 3 shows a β-frequency section along a model profile. All of the data have small values (within ±5°), with just two exceptions: (1) a high-frequency band (>400 Hz) at Site 3 and (2) a mid-frequency range (~200–5000 Hz) at Site 9 and Site 10. Small β values indicate that the resistivity structure beneath the study area is 1-D or 2-D. We then estimated the regional strike of the resistivity structure. Figure 4 shows the phase tensor major axis (a) of three frequency bands ($>10^3$, 10^3–10^2, and $<10^2$ Hz) with 90° ambiguity at all sites. Strikes of N60°W–S60°E and N30°E–S30°W are dominant for the high-frequency band, whereas those of N30°W–S30°E and N60°E–S60°W are dominant for the low-frequency band; thus, the study area is not strictly 2-D. However, it may be acceptable to assume that the resistivity structure of the study area is 2-D and that the regional strike is N30°W-S30°EorN60°E-S60°W. This assumption is based on the fact that these strikes are determined not only from the structure of a shallow part of the structure but also from a part that extends from shallow to deep, and the main aim of this paper is to determine the structure down to deep levels around the Hijima fault. Ultimately, we adopted N30°W–S30°E as the regional strike because this orientation is closer to the geological strike than N60°E-S60°W.

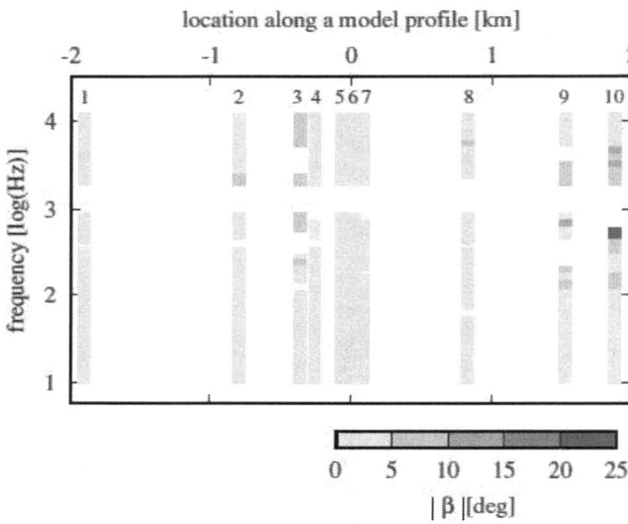

Figure 3. β-frequency section along the model profile. Site number is shown above each column.

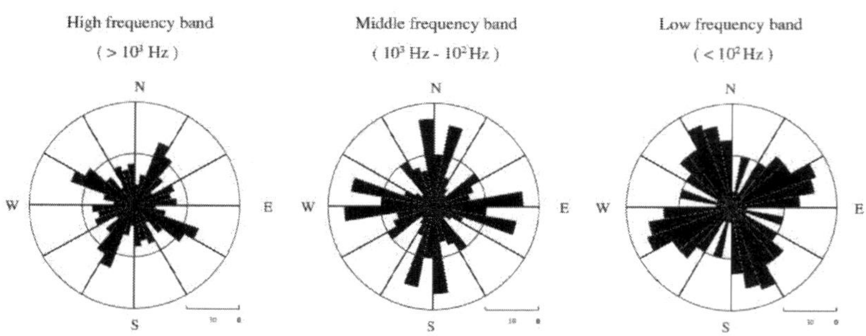

Figure 4. Rose diagrams showing the regional strikes determined from phase tensor analysis. The figure shows the major axes of the phase tensor (a) for all sites in the frequency bands of $>10^3$,10^3-10^2,and $<10^2$ Hz.

In this study, the transverse electric (TE)-mode impedance corresponds to the electric current flowing along the regional strike, while the transverse magnetic (TM) mode data correspond to electric currents flowing normal to the regional strike. Magnetotelluric (MT) responses (apparent resistivity and phase values) of both modes at each site are shown in Fig. 5. The apparent resistivity shows a smooth variation with frequency at all sites. In contrast, the phase value of the high-frequency band in TM mode at Site 3 is slightly scattered and small compared with those at other sites and those of the mid-frequency band (200–5,000 Hz) in TE mode at Site 10 show a degree of scatter.

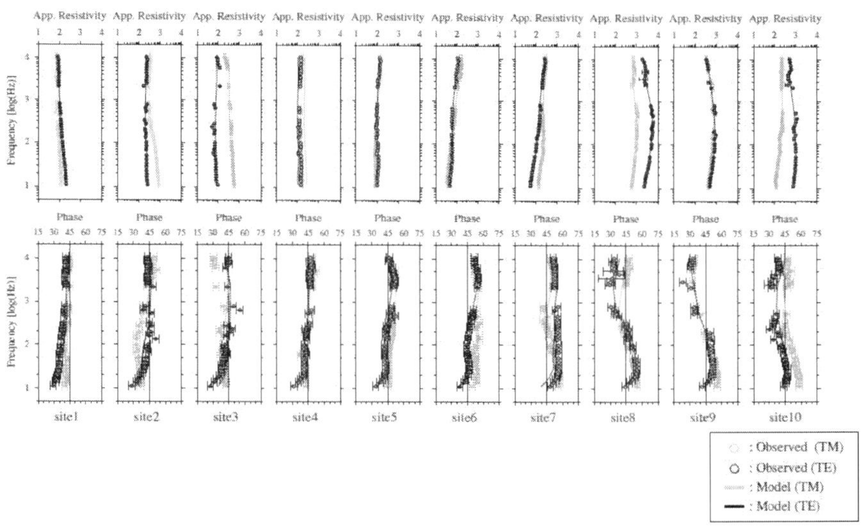

Figure 5. Observed MT responses (apparent resistivity and phase value) at all sites compared with those of the optimum model. The upper panels show observed apparent resistivity in TM (gray circles) and TE (black circles) modes compared with those of the optimum model (solid lines). The lower panels show phase values.

Qualitative insights into variations in the subsurface structure along a model profile can be obtained from pseudo-sections of the apparent resistivity and phase data of the TM and TE modes (Fig. 6). The surface trace of the Hijima fault is marked by low apparent resistivity, whereas high resistivity is recognized in the southwestern part of the fault (0–2 km) in both the TE and TM modes. In both modes, the phase value is > 45° along the surface trace of the fault, and it increases with decreasing frequency. These results suggest the existence of two conductive zones: a shallow zone located near the fault and a deep zone located to the southwest of the fault. The northeastern part of the profile (−3 to −2 km) is characterized by low apparent resistivity and a uniform phase value of ~30° for the entire frequency range, indicating the occurrence of a near-surface conductive zone.

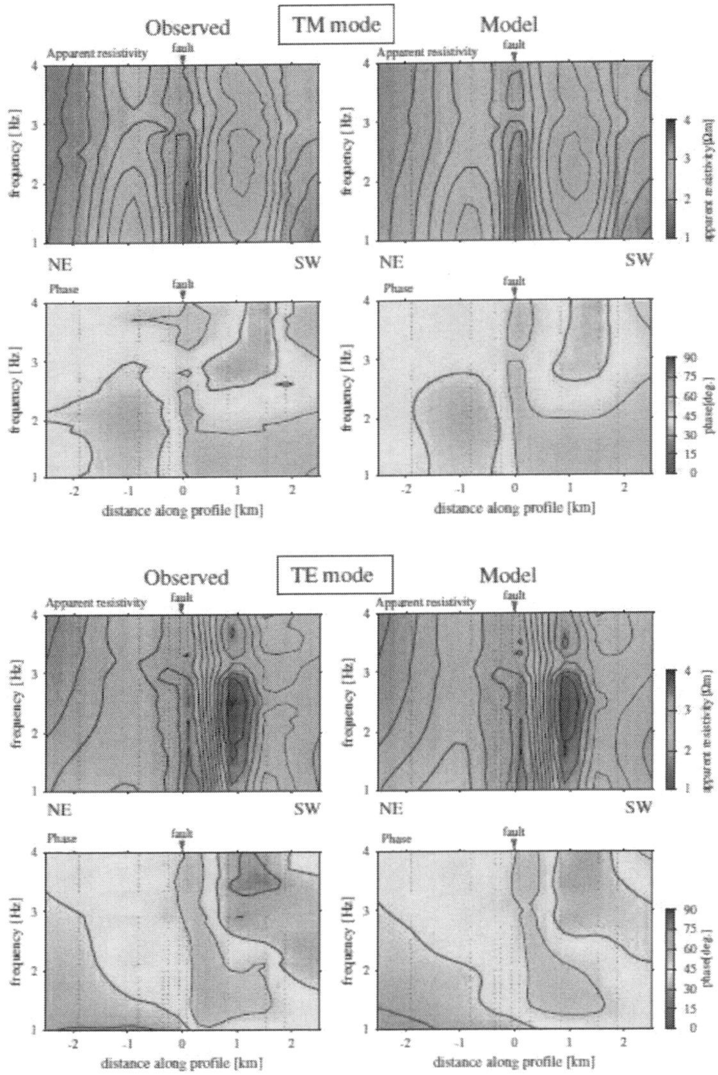

Figure 6. Observed and calculated pseudo-sections of apparent resistivity and phase for TM and TE modes. The four uppermost panels show the observed responses (left column) and calculated responses of the optimum model (right column) for the TM mode. The four lowermost panels show the observed responses (left column) and calculated responses of the optimum model (right column) for the TE mode.

4. MODEL ANALYSIS AND RESULTS

All of the studied sites are projected on a model profile oriented at N60°E–S60°W, and the apparent resistivity and phase data of both the TM and TE modes are inverted using the inversion code of Ogawa and Uchida (1996), in which a static shift is also a model parameter. The initial model was a uniform halfspace of 100 Ω m. An error floor of 10% in apparent resistivity and an equivalent value for phase (2.87°) were applied to the observed MT impedances. All data errors less than the prescribed error were replaced by the error floor; this ensures a uniform fit to the data at all frequencies. After 20 iterations, the ABIC (Akaike's Bayesian Information Criterion) minimum model was adopted as the most plausible model (Fig. 7(b); denoted the model *the optimum model* hereafter), with a root mean square misfit of 0.74. The responses and pseudo-section of the optimum model compared with the observed data are shown in Figs. 5 and 6, respectively. We calculated the variance of the resistivity of each block and found that the area shallower than 2.3 km between Site 1 and Site 10 is well resolved, with the exception of the near-surface area between Site 1 and Site 2. We identified the following features of the optimum model.

1 A clear resistivity boundary between Site 5 and Site 6 below a depth of 600 m that coincides with the downward projection of the surface trace of the Hijima fault.
2 A conductive zone (C1; <100 Ω m and minimum resistivity of 40 Ω m) that is 350 m wide and 50–500 m deep beneath the surface trace of the Hijima fault.
3 A conductive zone (C2; <100 Ω m and minimum resistivity of 30 Ω m) that is 1,700 m wide and 800–1,800 m deep, truncated to the northeast by the Hijima fault.

4 A highly conductive zone (C3) near the surface between Site 1 and Site 2, with a thickness of 200 m and minimum resistivity of 30 Ω m. However, this zone is poorly resolved because of the large distance between Site 1 and Site 2.

5 A resistive zone (R; >500 Ω m and maximum resistivity of 1,000 Ω m) at a depth of 100–600 m beneath the area between Site 7 and Site 9.

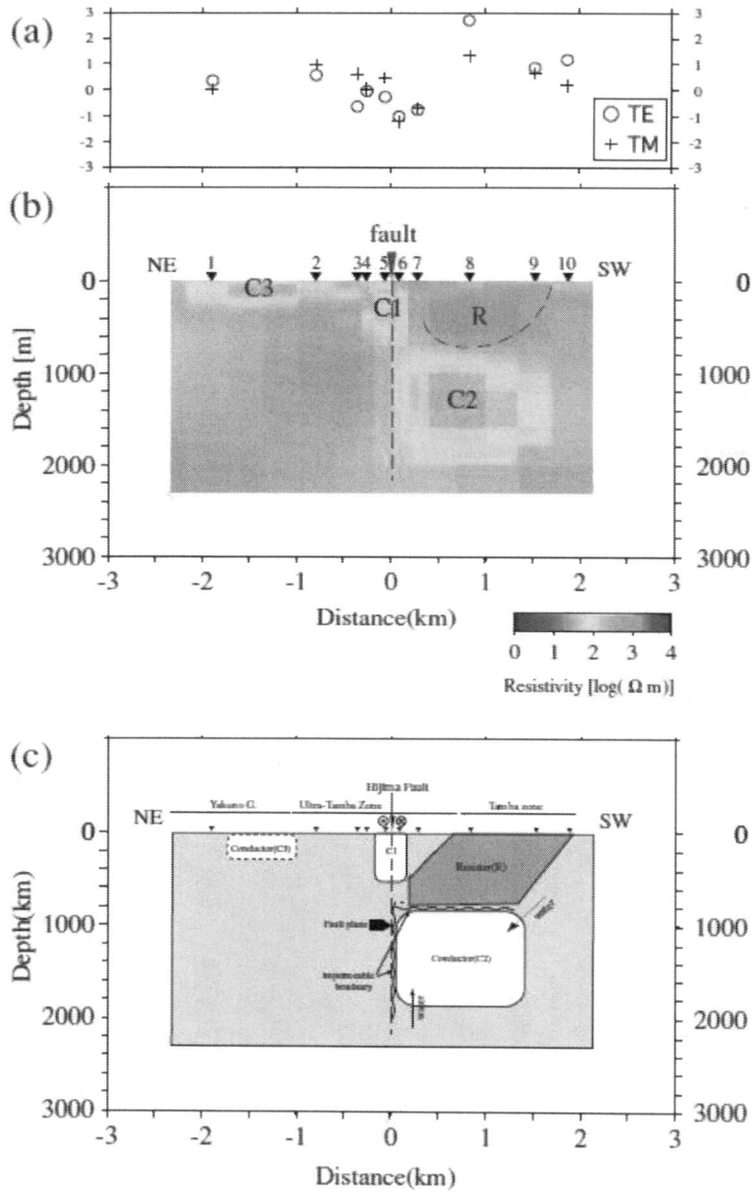

Figure 7. (a) Static shift coefficients estimated as a function of location for the TM and TE modes. (b) Final 2-D resistivity model. Black and red inverted triangles represent audio-frequency magnetotelluric (AMT) sites and the surface trace of the Hijima fault, respectively. C1–C3 and R indicate the characteristic conductive and resistive zones in the optimum model, respectively. (c) Simplified interpretation of the optimum model.

5. DISCUSSION

5.1 Appropriate ranges of resistivity within the characteristic conductive and resistive zones

The optimum model is characterized by two large conductive zones (C1 and C2) and one resistive zone (R). To determine the appropriate range of resistivity in each zone, we constructed representative resistivity models in which the resistivity in each zone was replaced with a number of resistivity values. The forward responses were computed and then compared with the optimum model response and observed data.

(a) Appropriate range of resistivity within zone C1 We constructed representative resistivity models in which we replaced the resistivity in zone C1 with one of nine set values (20, 40, 60, 80, 100, 120, 140, 180, 220 Ω m). We found clear differences in MT responses between the representative models and the optimum model for both apparent resistivity and the phase value at Site 5–Site 7 in the TE mode and Site 5–Site 6 in the TM mode. Figure 8 shows the results obtained at Sites 5 and 6. For the case of 20 Ω m, plots of apparent resistivity in the TE mode at both sites are concave, with a minimum at 100 Hz and values of <100 Ω m for the entire frequency range. The phase trends are convex, with peaks at 3,000 Hz at Site 5 and 6,000 Hz at Site 6. In TM mode, the apparent resistivity shows a minimum value of 40 Ω m at Site 5 and 5 Ω m at Site 6 at the lowest frequency; with increasing frequency, the values increase up to 200 Ω m. For frequencies <4,000 Hz, the phase value shows a monotonous increase with increasing frequency, reaching 60° at Site 5. The phase curve for Site 6 shows two peaks and one trough, with values >60° for the entire frequency range. This representative model is unable to explain the observed responses, as the MT responses of the model differ markedly. The cases of 40 and 60 Ω m show responses more similar to the observed responses than those of the representative models for 20 Ω m; however, the apparent resistivity in TE mode and the phase value in TM mode at Site 6 are inconsistent with the observed responses (not shown). We therefore conclude that these models are also unsuitable.

(b) Appropriate range of resistivity within zone C2 We constructed representative resistivity models in which we replaced the resistivity in zone C2 with one of five resistivity values (20, 50, 100, 200, 300 Ω m). We observed a clear difference in response between the representative models and the optimum model in phase values at Site 3-Site 10 (TE mode) and at Site 5–Site 7 (TM mode). Figure 9 shows the MT responses at Site 6–Site 8 in both modes. For the case of 20 Ω m, at Site 6–Site 8, the phase value in TE mode is much larger than the observed value for frequencies <100 Hz. A large difference is also recognized in the phase value of the TM mode at Site

8 for frequencies <200 Hz. We therefore conclude that the model with a resistivity of 20 Ω m is unable to explain the observed values. The model with a resistivity of 50 Ω m, however, yields a response that is closer to the observed values, with insignificant differences; this model is therefore considered to be appropriate.

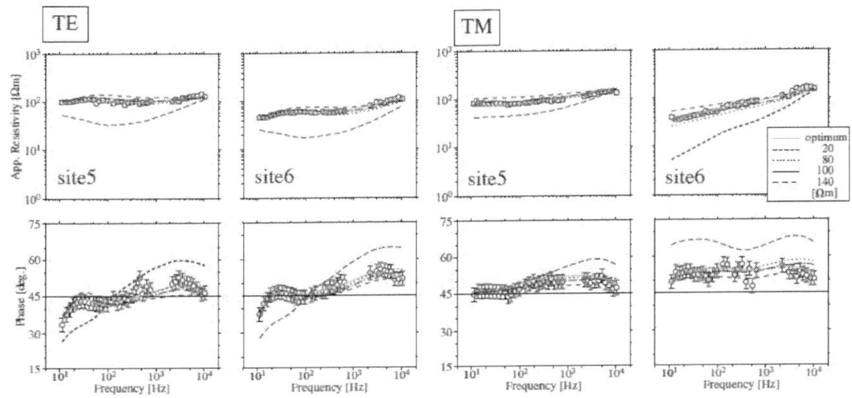

Figure 8. Comparison of MT responses between the optimum model and the models with assumed values of resistivity in zone C1. Solid gray lines indicate the responses of the optimum model. The responses of the representative models are shown by thick broken lines (20 Ω m), dotted lines (80 Ω m), dash-dotted lines (100 Ω m), and thin broken lines (140 Ω m). Circles with error bars indicate observed responses.

In contrast, in the case of 100 Ω m, the phase value of the TE mode at Site 6–Site 8 is unable to explain the large phase value in the frequency band <200 Hz. This same feature is also recognized in the phase value of the TM mode at Site 7. We therefore conclude that the model with a resistivity of 100 Ω m is unable to explain the observed values. The models with values of 200 and 300 Ω m show larger differences than that of the model in the case of 100 Ω m. We therefore conclude that only the 50 Ω m model yields appropriate resistivity in zone C2.

(c) Appropriate range of resistivity within zone R We constructed representative resistivity models in which we replaced the resistivity in zone R with one of seven different values (300, 500, 750, 1000, 1500, 2000, and 2500 Ω m). For the case of 300 Ω m, a clear difference in response between the representative models and the optimum model in phase values is observed at Site 7–Site 9 (Fig. 10). The phase values at Site 8 are significantly larger than the observed values for frequencies of 200–5,000 Hz (TE mode) and 200–2,000 Hz (TM mode). For the cases of 2,000 and

2,500 Ω m, the phase value of TM mode at Site 8 is systematically larger—although not much larger—than the observed value for frequencies of 20–200 Hz. We conclude that the models with resistivities of 500–1,000 Ω m yield the appropriate resistivity values for zone R.

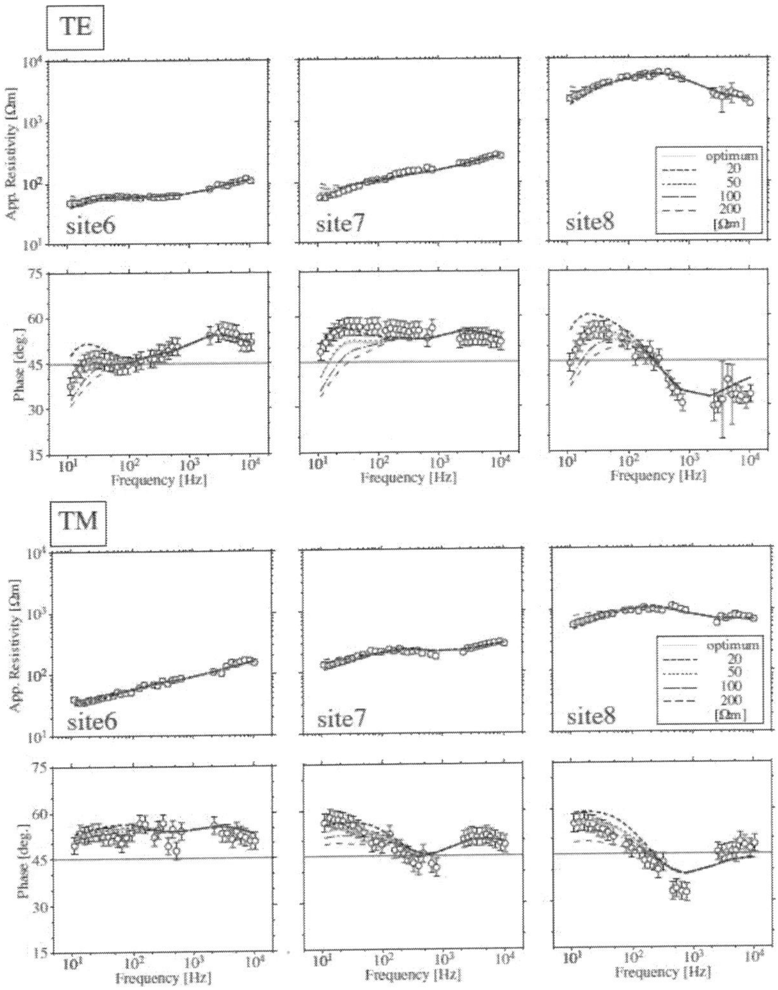

Figure 9. Comparison of MT responses between the optimum model and the models with assumed values of resistivity in zone C2. Solid gray lines indicate the responses of the optimum model. The responses of the representative models are shown by thick broken lines (20 Ω m), dotted lines (50 Ω m), dash-dotted lines (100 Ω m), and thin broken lines (200 Ω m). Circles with error bars indicate observed responses.

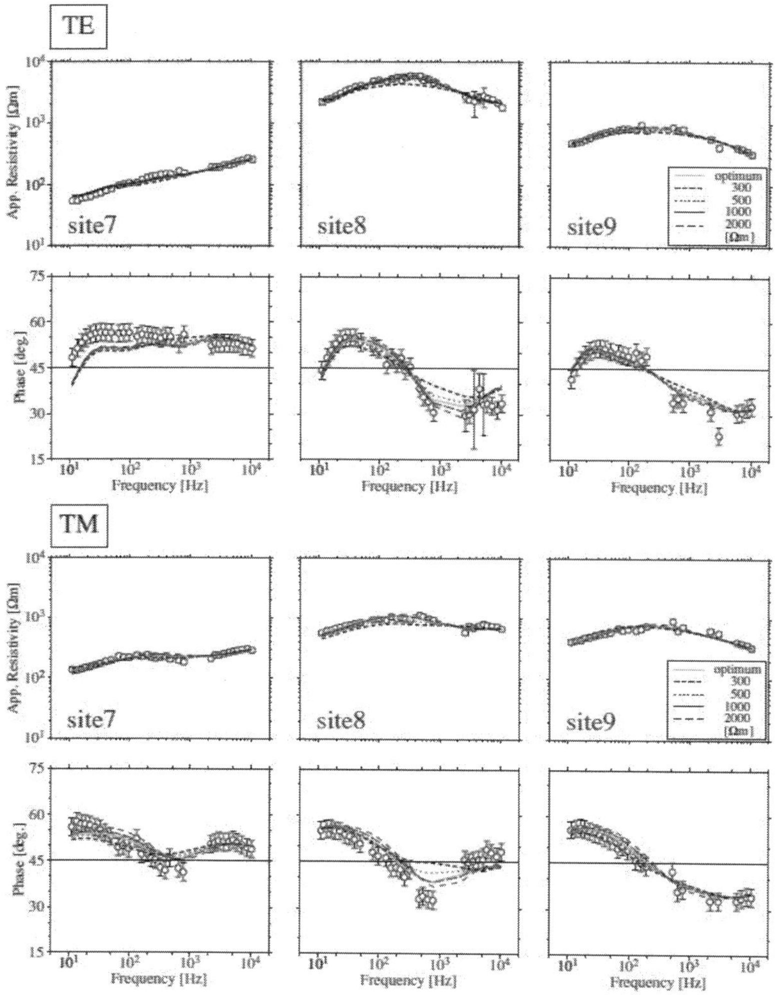

Figure 10. Comparison of MT responses between the optimum model and the models with assumed values of resistivity in zone R. Solid gray lines indicate the responses of the optimum model. The responses of the representative models are shown by thick broken lines (300 Ω m), dotted lines (500 Ω m), dash-dotted lines (1000 Ω m), and thin broken lines (2000 Ω m). Circles with error bars indicate observed responses.

5.2 Appropriate range of the lower boundary of the conductive zone C2

Separate tests were carried out for those cases in which the boundary is shallower and deeper than their counterparts in the optimum model.

(a) Deeper case We constructed representative models in which we placed the base of the zone at depths of 2.0 and 2.3 km, respectively. We chose 2.3 km as the maximum depth because areas deeper than this are poorly resolved, as stated above. We assigned 50 Ω m to an extended area of zone C2 by deepening the base of the zone. The test results indicate that we cannot resolve the appropriate depth of the lower boundary of zone C2 at depths greater than that of the optimum model because a small difference in response between the representative models and the optimum model is only observed near the lowest range of our measurements for both representative models.

(b) Shallower case We constructed representative models in which the base of C2 is placed at 1.6, 1.4, and 1.2 km. We assigned 200 Ω m to an area that is changed from being within zone C2 to outside the zone by raising the base of the zone. Figure 11 shows the MT responses of the representative models at Site 6 and Site 7. No significant difference in response between the representative models and the optimum model is observed in the case of 1.6 km. In the case of 1.4 km, the model shows a negligible decrease in the phase value at Site 7 in TE mode for low frequencies. The model with a base at 1.2 km shows a significant difference. We conclude that, in the shallowest case, the lower boundary of zone C2 is located at a depth of 1.4 km—but it may occur at greater depths.

5.3 Sharp resistivity boundary coincident with the downward projection of the surface trace of the Hijima fault

A clear resistivity boundary occurs below a depth of 600 m, coinciding with the downward projection of the surface trace of the Hijima fault; this boundary is interpreted to represent the subsurface fault plane (Fig. 7(c)). Takemura and Suzuki (1996) presented a structural section (oriented N15°E–S15°W) across the Hijima fault, showing a vertical fault plane extending to a depth of ~500 m. Our result indicates that the fault plane extends to a depth of 1.4 km along the resistivity boundary, possibly deeper.

5.4 Shallow conductive zone

A conductive zone similar to zone C1 described herein has been reported along the Yasutomi fault (ERGAF, 1982). This conductive zone comprises a three-part structure: an inner conductive belt (1,000 m wide and 100–400 Ω m) that contains highly conductive streaks (~20 Ω m) near the surface trace of the fault, and an outer area that is somewhat resistive but still low (~ 1,000 Ω m) compared with the surrounding area (>10,000 Ω m). The inner conductive zone along the Yasutomi fault is comparable with zone C1 along the Hijima fault in terms of its location relative to the fault, width, and resistivity. These observations suggest that highly conductive zones are a common occurrence along the Yamasaki fault system.

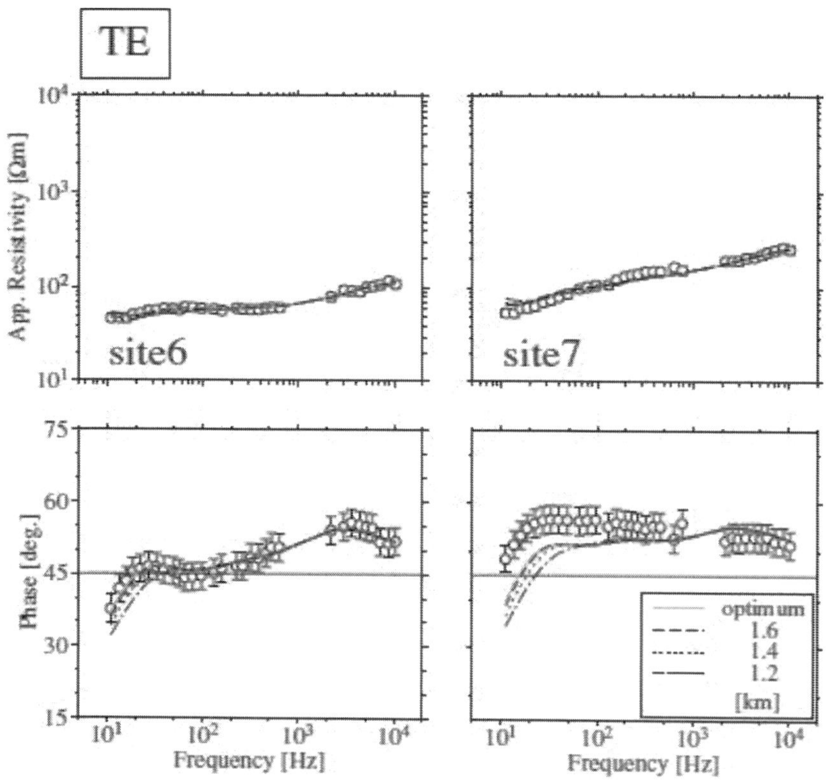

Figure 11. Comparison of MT responses between the optimum model and the models with assumed depths of the lower boundary of zone C2. Solid gray lines indicate the responses of the optimum model. The responses of the representative models are shown by broken lines (1.6 km), thin dotted lines (1.4 km), and dash-dotted lines (1.2 km). Circles with error bars indicate observed responses.

The high conductivity in zone C1 results from the presence of clay minerals and a fluid network within a fault-related fracture zone, as proposed by ERGAF (1982) and as also reported for the San Andreas fault, USA (Unsworth *et al.*, 1997,1999; Bedrosian *et al.*, 2002). Clay minerals are a common component of a fault gouge, and the typical resistivity of clay-rich sedimentary rocks is 5–30 Ω m (Palacky, 1987). An additional contributor is a fluid network within a fault-related fracture zone, as this can enhance the conductivity of the fault zone. In systems in which electrical conductivity is dominated by an interconnected pore fluid, Archie's law (Archie, 1942) is often used to convert the bulk resistivity of rock (ρ_0) to that of porosity (Φ). When we assumed all pores were saturated:

$$\rho_0 = \rho_w \Phi^{-m} \tag{1}$$

where ρ_w is the resistivity of pore fluid and m is an empirical constant.

In order to assess porosity in zone C1, we used the resistivity of groundwater in a well at Yasutomi along the Yasutomi Fault as the reference value; this well is located very close to the study area (Fig. 2(b)). The National Institute of Advanced Industrial Science and Technology (AIST) of Japan have been monitoring the crustal strain field and groundwater pressure in the well since 1991 (Koizumi *et al.*, 2000). The resistivity of groundwater sampled within the well at depths of 254–265 m is reported to be 43.1 Ω m (N. Koizumi, 2007, personal communication). Assuming that these fluids are representative of those along the fault zone, the ρ_w would then be 43.1 Ω m. To obtain ρ_0= 100 Ω m in the shallow conductive zone would require a porosity of 57% for $m = 1.5$, which is the value for slightly cemented sandstone (Schön, 1998). It is worth noting that this estimation is the maximum porosity value.

5.5 Deep conductive zone

As with zone C1, the high conductivity measured in zone C2 may reflect the combined influences of a fluid network and the presence of clay minerals. Two other mechanisms are generally cited in explaining high conductivity in the upper crust: high temperatures and the presence of highly conductive materials (e.g., graphite and metallic minerals). The former mechanism is unlikely to play a significant role in the case described here because heat flow in the study area is <60 mW/m^2 (Furukawa *et al.*, 1998), which is among the lowest values in the area shown in Fig. 1(a). The significance of the latter mechanism cannot be assessed due to a lack of data.

For a fluid network to form in zone C2, it is important that the fault core of the Hijima fault and the resistive zone R act as barriers to fluid flow, which permeates from the surface along north-dipping strata (Takemura and

Suzuki, 1996) or is derived from depth via the highly permeable damage zone that exists along the fault. The northeastern end of zone C2 is truncated along a plane that coincides in location with the downward projection of the surface trace of the Hijima fault; this sharp boundary suggests the presence of an impermeable zone along the Hijima fault. Ritter *et al.* (2003) described a similar conductor along the Araba fault of the Dead Sea Transform, Jordan. These authors proposed a model entailing a strong lateral contrast in conductivity where a highly conductive layer at a depth of 1.5–3 km is truncated at a position coinciding with the downward projection of the surface trace of the fault, which then acts as an impermeable barrier at depth. Caine *et al.* (1996) proposed a conceptual model of fault zone architecture and related permeability structure that consists of three components: the fault core, a damaged zone, and the protolith. In this model, most of the displacement is accommodated in the fault core, which acts as a barrier to fluid movement. The broad and highly permeable damaged zone represents a network of subsidiary structures that bound the fault core, while the protolith consists of undeformed country rocks.

In our study area, the resistive zone R above zone C2 may play an important role as a cap rock that defines the upper boundary of C2. If such a cap rock were not to exist, fluid in C2 (meteorological water and/or groundwater) would not be confined to the zone; rather, it would be dispersed throughout a wide area, with part of it reaching the surface.

5.6 Possibility of along-strike variation in the resistivity structure

In our explanation of the enhanced conductivity observed in zone C2, the resistive zone R plays an important role as a cap rock; as such, the immediate geological setting of the Hijima fault is an important factor in the development of a resistivity structure along the fault. The Mesozoic sediments (chert and mudstone of the Tamba zone) that correspond to the resistive zone are only exposed on the southern side of the Hijima fault; Paleozoic sediment (sandstone and slate of the Ultra-Tamba zone) occur around the Mesozoic sediments. In contrast, Paleozoic sediments occur on both sides of the Yasutomi fault; rhyolite and gabbro are distributed along the Ohara fault; rhyolite and Paleozoic sediment are distributed along the Kuresakatouge fault (Fig. 2(b)). These variations in geological setting along the strike of the Yamasaki fault system are expected to cause variations in resistivity structure. To clarify the factors that determine the resistivity structure of a fault system, it is important to identify both similar and contrasting features of resistivity structure along different fault systems. In a previous seismological study, Shibutani (2004) showed that both seismic activity and the *b*-value vary along the Yamasaki fault system. We have also proposed the existence of along-strike variations in structure and/or condition (e.g., distribution of fluid) beneath the fault system. Despite these differences, however, both the Hijima fault and Yasutomi fault are characterized by a shallow conductive zone along the surface fault trace.

Follow-up MT and resistivity surveys of those that have previously been undertaken along many profiles across the fault system would clarify the nature of along-strike variations in the subsurface structure along the Yamasaki fault system.

6. CONCLUSIONS

An audio-frequency magnetotelluric survey was undertaken along a profile across the Hijima fault, Yamasaki fault system, to image the subsurface structure. The detailed 2D inversion model for both TE and TM modes shows an electrical model of the Hijima fault. First, a clear resistivity boundary is recognized at a depth greater than 600 m at a position coinciding with the downward projection of the surface trace of the Hijima fault. This boundary represents the subsurface fault plane, indicating that the Hijima fault is near-vertical in orientation to at least a depth of 1.4 km, possibly deeper. Second, we imaged a highly conductive zone (<100 Ω m) that is 350 m wide and 50–500 m deep, including the surface trace of the Hijima fault. We interpreted this zone as reflecting the presence of clay minerals and a fluid network within a fracture zone generated by fault movement. A similar conductive zone occurs along the Yasutomi fault, suggesting that shallow conductive zones are common along the Yamasaki fault system. Finally, we newly identified a deeper conductive zone (<100 Ω m) that is 1,700 m wide and 800–1,800 m deep. The fault-core of the Hijima fault and a highly resistive zone form impermeable boundaries on the northeast side and upper surface of the zone, respectively. The fluid in this deep zone, which permeates downward from the surface or is derived from deeper levels via the damaged zone of the fault, occurs as a fluid network, enhancing conductivity along with the presence of clay minerals.

ACKNOWLEDGEMENTS

We thank the many private landowners in the area around the Hijima fault for allowing us to take measurements upon their land. This study was supported by the Joint Research Project (Exploratory Subject: 17H-1) of the Disaster Prevention Research Institute, Kyoto University. Financial support was also provided by the Research Center for Urban Safety and Security, Kobe University. We gratefully acknowledge Dr. T. Shibutani (Kyoto

University) for providing hypocentral data, and Dr. N. Koizumi (AIST) for providing electrical conductivity data for groundwater in the well at Yasutomi station. We thank Mr. S. Kato (The Museum of Nature and Human Activities, Hyogo) for providing much useful geological and geomorphological advice that proved helpful in selecting the observation sites. The constructive comments from Dr. S. Takakura and Dr. P. K. Patro are deeply appreciated. Most of the figures were drawn using Generic Mapping Tools (Wessel and Smith, 1998).

REFERENCES

1. Archie, G. E., The electrical resistivity log as an aid in determining some reservoir characteristics, Trans. Am. Inst. Mineral. Met., **146**, 54–62, 1942.

2. Bedrosian, P. A., M. J. Unsworth, and G. Egbert, Magnetotelluric imaging of the creeping segment of the San Andreas Fault near Hollister,Geophys. Res. Lett, **29**(11), 1506,

3. Caine, J. S., J. P. Evans, and C. B. Forster, Fault zone architecture and permeability structure, Geology, **24**, 1025–1028, 1996.

4. Caldwell, T. G., H. M. Bibby, and C. Brown, The magnetotelluric phase tensor, Geophys. J. Int., **158**, 457–169, 2004.

5. Earthquake Research Committee, Long-term Evaluation of the Yamasaki Fault System, p.38, 2003 (in Japanese).

6. Earthquake Research Committee, National Seismic Hazard Map of Japan 2007, p.52, 2007 (in Japanese).

7. Electromagnetic Research Group for the Active Fault, Low electrical resistivity along an active fault, the Yamasaki fault, J. Geomag. Geoelectr, **34**, 103–127, 1982.

8. Furukawa, Y., H. Shinjoe, and S. Nishimura, Heat flow in the southwest Japan arc and its implication for thermal processes under arcs,Geophys. Res. Lett, **25**, 1087–1090, 1998.

9. Gamble, T. D., W. M. Goubau, and J. Clarke, Magnetotellurics with a remote magnetic reference, Geophysics, **44**, 53–68, 1979.

10. Handa, S. and N. Sumitomo, The geoelectric structure of the Yamasaki and the Hanaori faults, southwest Japan, J. Geomag. Geoelectr,**37**, 93–106, 1985.

11. Hyogo-ken, The Report of an Activity of the Yamasaki Fault System, 2000 (in Japanese).

12. Koizumi, N., A. Cho, M. Takahashi, and E. Tsukuda, Groundwater level observation of Geological Survey of Japan in and around the Kinki district, Japan for earthquake prediction research, Proc. Hokudan Internat. Symp. and School on Active Faulting, pp. 183–186, 2000.

13. Ogawa, Y. and Y. Honkura, An audiomagnetotelluric view of the Atera fault, J. Geomag. Geoelectr, **49**, 1065–1071, 1997.

14. Ogawa, Y. and T. Uchida, A two-dimensional magnetotelluric inversion assuming Gaussian static shift, Geophys. J. Int., **126**, 69–76, 1996.

15. Ogawa, Y., S. Takakura, and Y. Honkura, Resistivity structure across Itoigawa-Shizuoka tectonic line and its implications for concentrated deformation, Earth Planets Space, **54**, 1115–1120, 2002.

16. Okada, A., M. Ando, and T. Tsukuda, Trenching study for Yasutomi Fault of the Yamasaki Fault System at Anji, Yasutomi Town, Hyogo Pref., Japan, J. Geogr, **96**(2), 1–17, 1987 (in Japanese with English abstract).

17. Palacky, G., Resistivity characteristics of geologic targets, in Electromagnetic Methods in Applied Geophysics, edited by Naibighian, M. N., Tulsa, Oklahoma, Society of Exploration Geophysicists, v. 1, P. 53, 1987.

18. Patro, B. P. K., T. Harinarayana, R. S. Sastry, M. Rao, C. Manoj, K. Naganjaneyulu, and S. V. S. Sarma, Electrical imaging of Narmada-Son Lineament Zone, Central India from magnetotellurics, Phys. Earth Planet. Inter, **148**, 215–232, 2005.

19. Ritter, O., T. Ryberg, U. Weckmann, A. Hoffmann-Rothe, A. Abueladas, Z. Garfunkel, and DESERT Research Group, Geophysical images of the Dead Sea transform in Jordan reveal an impermeable barrier for fluid flow, Geophys. J. Int., **30**(14), 1741,

20. Ritter, O., A. Hoffmann-Rothe, P. A. Bedrosian, U. Weckmann, and V. Haak, Electrical conductivity images of active and fossil fault zones, 165–186, in High-Strain Zones: Structure and Physical Properties, GSA Special Publication 245, 2005.

21. Schön, J. H., Physical Properties of Rocks; Fundamentals and Principles of Petrophysics (Handbook of Geophysical Exploration. Section I, Seismic Exploration, vol. 18), p.408, Pergamon, Netherland, 1998.

22. Shibutani, T., Seismic activity around the Yamasaki fault for the last 29 years and the significance of an earthquake near Yumesaki Town (12 July 2004, Mj=3.9), Abstracts of the Seismological Society of Japan, 2004, Fall Meeting, A036, 2004.

23. Takemura, S. and S. Suzuki, The geology and tectonics of the Ultra-Tamba zone, western Hyogo Prefecture, Southwest Japan, J. Geol. Soc. Jpn, **102**, 1–12, 1996 (in Japanese with English abstract).

24. Unsworth, M. J., P. E. Malin, G. D. Egbert, and J. R. Booker, Internal structure of the San Andreas fault at Parkfield, California, Geology,**25**, 359–362, 1997.

25. Unsworth, M., G. Egbert, and J. Booker, High-resolution electromagnetic imaging of the San Andreas fault in Central California, J. Geophys. Res., **104**, 1131–1150, 1999.

26. Yamaguchi, S., T. Murakami, and H. Inokuchi, Resistivity mapping using the VLF-MT method around surface fault ruptures of the 1995 Hyogoken Nanbu earthquake, Japan, The Island Arc, **10**, 296–305, 2001.

27. Yamaguchi, S., S. Sutoh, H. Hashimoto, H. Murakami, and N. Takagi, 2-D resistivity structure of the southern part of the Nojima fault and relationship to the activity ofthe faults, Zisin (J. Seismol. Soc. Jpn.), **55**, 143–151, 2002 (in Japanese with English abstract).

28. Yamaguchi, S., H. Murakami, H. Iwamoto, K. Takemoto, K. Kitada, I. Shiozaki, N. Oshiman, and S. Katoh, Two-dimensional resistivity structure of the fault associated with the 2000 Western Tottori Earthquake, Earth Planets Space, **59**, 1211–1217, 2007.

29. Wessel, P. and W. H. F. Smith, New, improved version of generic mapping tools released, EOS Trans. AGU, **79**, 579, 1998.

CHAPTER 8

An experimental investigation of geochromatography during secondary migration of petroleum performed under subsurface conditions with a real rock

Steve Larter,a Berni Bowler,*a Ed Clarke,b Colin Wilson,b Brian Moffatt,b Barry Bennett,a , Gareth Yardleyc and Dan Carruthersc

a Fossil Fuels and Environmental Geochemistry (Postgraduate Institute) NRG, University of Newcastle, UK.

b BG International, Reading, Berkshire, UK

c Dept of Petroleum Engineering, Heriot-Watt University, Edinburgh, UK

ABSTRACT

An understanding of the size of petroleum secondary migration systems is vital for successful exploration for petroleum reserves. Geochemists have suggested that compositional fractionation of petroleum accompanying the migration process (geochromatography) can potentially be used to infer distances petroleum may have travelled and the ratio of oil in the reservoir to that lost in the carrier. To date, this has been attempted by measuring concentrations and distributions of specific steranes, and aromatic oxygen and nitrogen compounds in reservoired oils which have been proposed to respond to migration rather than to source maturity or other effects. We report here an experiment involving oil migration through an initially water wet siltstone under realistic subsurface carrier bed or reservoir conditions (48 MPa, 70°C) where source facies and maturity effects are eliminated. We show that geochromatography does indeed occur even for initially water-saturated rocks and that the migration fractionations observed for alkylcarbazoles, benzocarbazoles and alkylphenols are very similar to those seen in field data sets. In contrast, sterane based migration parameters show no compositional fractionation under these conditions.

1. INTRODUCTION

Petroleum may migrate up to several hundred kilometers from source to reservoir.[1] The physics of fluid flow dictates that this migration takes place in a focussed manner, with only a small fraction of the carrier bed being used.[2] This makes direct measurement of secondary petroleum migration conduits, by drilling or geophysics, virtually impossible. Consequently, petroleum geochemists have sought to quantify petroleum migration processes by chemical analysis of reservoired oil using the concentrations and distributions of several compound types as migration tracers.

Geochemical migration tracers are components of petroleum systems which partition from petroleum to water/rock, or *vice versa*, during migration. Of all the potential components, it is those that partition from petroleum to the carrier bed medium that constitute the largest group studied to date. Numerous studies have assessed the usefulness of sterane isomers,[3] petroleum alkylcarbazoles and benzocarbazoles, [4–9] alkylbenzoquinolines[10] and petroleum alkylphenols.[11, 12] Atmosphere derived noble gases which partition from pore water into oil have also been studied[13] and migration applications of both tracer types reported (Western Canada,[14] Oman,[15] North Sea[13]).

The migration of petroleum from source to reservoir usually results in compositional changes brought about by oil/water/rock partition and distribution processes, collectively termed 'geochromatography'.[16, 17] Large compositional fractionations accompany primary migration as oil leaves the source rock but fractionations occurring during secondary migration are much more subtle [5–7] and may be overprinted in reservoired oils with influences attributable to facies and maturity variation in source kitchen areas.[7–9, 14, 18, 19] By choosing petroleum systems in the North Sea and from the petroleum systems of the Western Canada sedimentary basin where such effects were minimised, Larter *et al.*[7] suggested that for benzocarbazoles, migration in clastic carrier beds over distances of more than 100 km resulted in the reduction of absolute concentrations of benzocarbazoles in oils by over an order of magnitude, with accompanying decreases in the ratio of the benzo[*a*]carbazole to benzo[*c*]carbazole isomers, though they did note that in some petroleum systems no fractionations were observed. Li *et al.* similarly suggested,[5, 14] from studies of oils from China and Western Canada, that decreases in absolute alkylcarbazole concentrations in oils and the accompanying increase in the relative proportions of more alkylated carbazoles compared to methylcarbazoles were again related to secondary migration. Li *et al.*[5] further suggested that changes in the isomer distributions of methyl and dimethyl carbazoles were also migration related. It is difficult to be certain of such effects from field studies, where migration, source facies and maturity effects on oil composition may be combined; consequently, numerous laboratory geochromatography studies have been made.

Demonstrable compositional fractionations during simulated oil migration are numerous,[17, 20, 21] but most experiments use unrealistic artificial "carrier beds" or migrating phases other than oil or unrealistic flow sequences. Although useful, these approaches do not simulate continued oil injection into a consolidated, initially water saturated rock at high temperature and pressure. Here, we discuss a more realistic experiment in which oils of constant initial composition were migrated through a rock under subsurface conditions. The lithology chosen was a siltstone in the microdarcy permeability range typical of portions of carrier systems involved in vertical petroleum migration. There is very little data available for oil flow through such low permeability lithologies even though rocks such as siltstones often form critical parts of carrier systems and seals and may significantly affect basin scale fluid flow. The experiment eliminates source facies and maturity influences on oil composition, allowing detailed investigation of petroleum migration fractionation effects in isolation.

2. METHODS

The main components of the experiment were as follows:

i. rock selection and preliminary testing of coreplugs for permeability/sorption properties;

ii. computerised tomography (CT) scanning of selected core (before and after core-flooding);

iii. core-flooding with synthetic brine to give water saturation;

iv. core-flooding with oil;

v. chemical analysis of produced fluid samples (brine, oil).

2.1 Core selection

The siltstone chosen was from the Cadley Hill A coal mine section of the Westphalian coal measures, East Midlands, UK. An unfractured block was selected and a full length (1000 × 5 cm) core cut and trimmed. Basic data for the core and the initial oil composition are shown in Table 1.

Table 1. Properties of the core and geochemical characteristics of the oil used in the experiment [a]

Siltstone, Cadley A mine, Leics.	
Porosity	*ca.* 12%
Permeability (brine)	0.01–0.02 mD
Total carbon content	0.9%
Total organic carbon content	0.4%
Clay content (estimated XRD)	37%
Oil	
Saturated hydrocarbons	52.4%
Aromatic hydrocarbons	29.4%
Resins (NSO cmpds)	14.5%
Asphaltenes	3.7%
Carbazole	0.9 $\mu g\ g^{-1}$
1-Methylcarbazole	1.6 $\mu g\ g^{-1}$
Benzo[*a*]carbazole	0.5 $\mu g\ g^{-1}$
Benzo[*c*]carbazole	0.2 $\mu g\ g^{-1}$
Total C_0–C_2 alkylcarbazoles	20.2 $\mu g\ g^{-1}$
Sum C_2 alkylcarbazoles	15.4 $\mu g\ g^{-1}$
Phenol	28.3 $\mu g\ g^{-1}$
o-Cresol	13.7 $\mu g\ g^{-1}$
m-Cresol	3.3 $\mu g\ g^{-1}$
p-Cresol	5.9 $\mu g\ g^{-1}$
2-Ethylphenol	3.5 $\mu g\ g^{-1}$
4-Ethylphenol	2.4 $\mu g\ g^{-1}$
2-Isopropylphenol	0.8 $\mu g\ g^{-1}$
3-Isopropylphenol	1.0 $\mu g\ g^{-1}$

[a]Prefixes C_0, C_1, C_2, C_3 refer to the number of alkyl carbon substituents on the basic phenol or carbazole nucleus.

2.2 CT scanning

X-ray computerised tomography (CT) scanning of the core with a medical CT scanner before water-flooding and after oil-flooding showed pervasive petroleum flow to have occurred, with no obvious focussing evident (Fig. 1). An iodide enhanced brine was used to improve fluid contrasts. This pervasive flow throughout the core was confirmed subsequently by geochemical analysis of residual core extracts, the results of which will be discussed elsewhere.

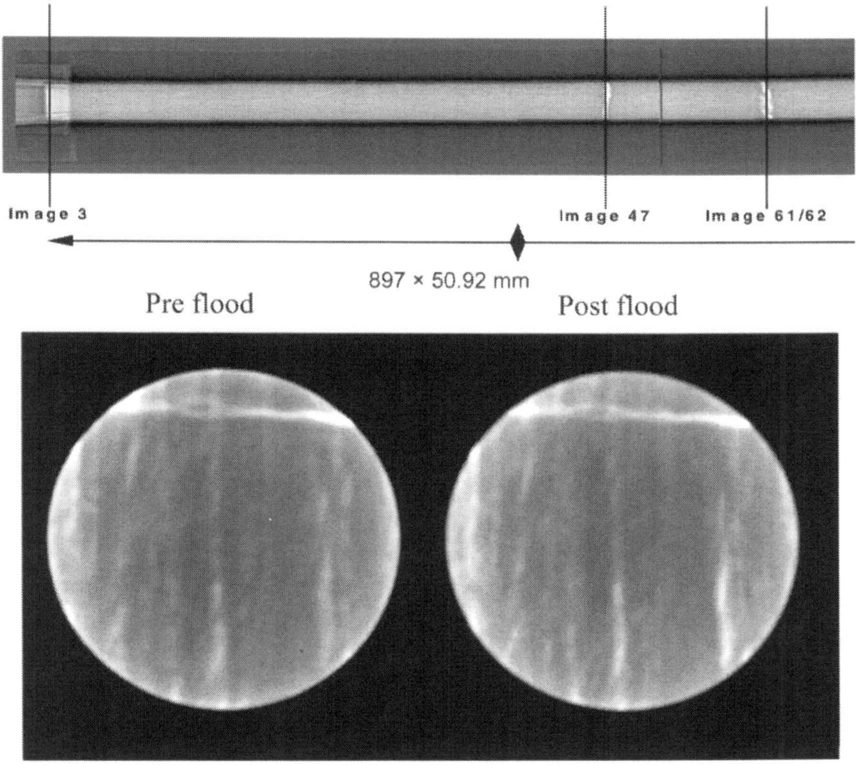

Figure 1. X-ray tomograms of the whole core prior to flooding with fluid (top), and of the same cross section of the core (bottom), before (left) and after (right) flooding with brine then oil. The bright regions in the CT scans are calcite-filled fractures in the core.

2.3 Core-flood

The experimental set-up is shown diagrammatically in Fig. 2. The Viton jacketted core was installed vertically with flow from base to top, parallel to bedding. The oil-flood commenced after initial saturation of the siltstone core with a synthetic N. Sea formation water was achieved.

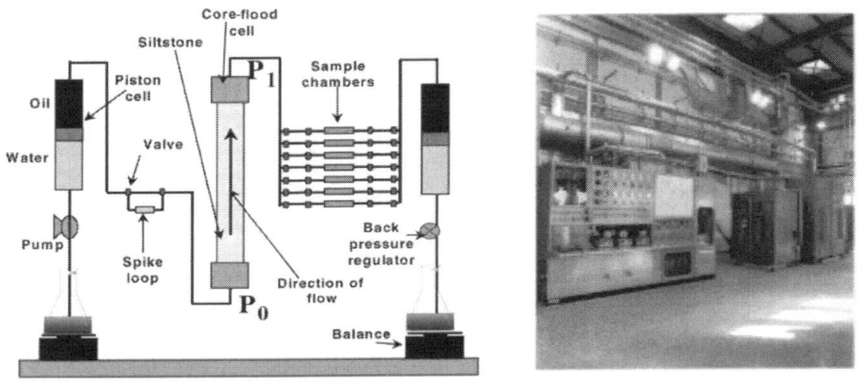

Figure 2. Diagram of the core-flood experimental set-up, with the actual lab set-up shown right. The core-flood cell was maintained at 7000 psi confining pressure and 70°C. Pressure differential across P_0/P_1 = 700 psi.

A typical 41° API gravity, non-biodegraded North Sea black oil was filtered (0.45 μm mesh sieve) and homogenised prior to sub-sampling. A 400 ml volume of oil was taken for the main charge. Five additional sub-samples were taken and mixed with different aliphatic hydrocarbon standards (at 10 mg g^{-1} oil) to form five distinct spiked oils of essentially the same composition as the main oil but which could be differentiated by gas chromatography to track oil movement through the core. For practical reasons no hydrocarbon solution gas was present, although the oils were saturated with nitrogen prior to injection; this is thought unlikely to affect the results observed.

After stabilisation of the core under water-saturated conditions, oil was flowed at typical subsurface conditions *i.e.*, 70°C, 48.3 MPa (7000 psi) confining pressure controlled by a pressure differential of *ca.* 4.8 MPa (700 psi) along the core. The target flow rate for the oil, determined by time constraints and core-plug screening work, was set and maintained at 2.4 ml per day (0.01 ml h^{-1}).

The spiked oils were injected peridodically and "produced fluid" flowing through the outlet valve was collected in 1.6 ml pressurised sample chambers, minimising cross-contamination during sampling. Injected oil volumes are used for reference in this paper as they could be determined more reliably than eluted oil volumes. The spiked oil data will be discussed elsewhere.

2.4 Chemical analysis of eluted oils

The produced fluid samples were comprised largely of oil and a small amount (<20%) of water, both at "oil breakthrough" and during the remaining time of the experiment. Periodically the sample loops were depressurised and the contents decanted into glass vials where the two phases readily separated. Samples of the oil were analysed for bulk composition using Iatroscan,[22] and for alkylcarbazoles, benzocarbazoles and alkylphenols using solid-phase extraction (SPE) and GC-MS methods.[7, 23, 24] Briefly, a hydrocarbon fraction and non-hydrocarbon fraction were obtained using C18 modified non-endcapped silica SPE cartridges and elution with hexane to give the hydrocarbon fraction and with dichloromethane (DCM) to give the phenol and carbazole fractions. The hydrocarbon fraction was further fractionated into saturated hydrocarbons and aromatic hydrocarbons by silver ion SPE, as described by Bennett and Larter.[25] A subsample of the SPE DCM fraction was derivatised using BSTFA after addition of internal standards, and phenol and the C_1–C_3 alkylphenols were analysed by GCMS in selected ion monitoring (SIM) mode.[23] Carbazole and benzocarbazole analyses were performed by direct GCMS-SIM analysis of the SPE DCM fraction.[7]

3. RESULTS AND DISCUSSION

Oil was flowed for 89 days; oil breakthrough occurred after 58 d, when 121 ml of oil (57% of core pore volume) had been injected into the core. Prior to oil breakthrough, the samples of produced fluid collected in the sample loops consisted of the brine used to achieve initial water-saturation. These were analysed for dissolved BTEX compounds (benzene, toluene, ethylbenzenes and xylenes) by purge and trap-GC but none were found at the > 10 ppb level. At oil breakthrough, water production dropped to approximately 20% of produced fluid volume before steadily reducing to less than 10% by the end of the experiment. Oil saturation was 57% of the pore volume at breakthrough increasing to approximately 65% at the end of the experiment, after 205 ml of oil had been injected (97% pore volume).

The first oil samples produced from the core were a clear orange colour, contrasting markedly with the black colour of the initial oil (Fig. 3). Subsequently produced oil samples became progressively darker as the experiment continued. The removal of colour in the first-eluted oil samples is a dramatic demonstration of (geo)chromatography in its original sense. Analysis of the core extracts showed that most of the pigmented compounds are removed near the core inlet (Fig. 4); these coloured polar species will be discussed elsewhere. Changes in the bulk composition of the produced oil were small, the resin content, for example, reducing from *ca.* 15% in the original oil to *ca.* 10% in the first-eluted oil. This was accompanied by a similarly small increase in saturated hydrocarbons content. By the end of the experiment, the bulk composition and colour of the produced oil samples were similar to those of the original oil.

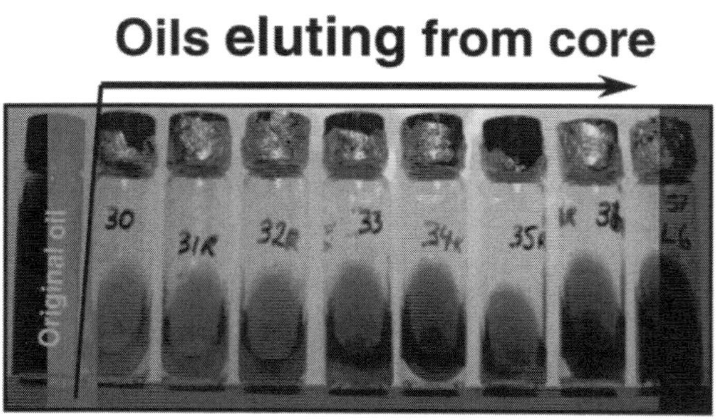

Figure 3. Geochromatography in action! The colour of the eluted oil changes during the experiment; the earliest eluting oils (vials 30–36) are light orange, subsequent samples rapidly becoming darker through the experiment, the final eluted oil (vial 57) being of similar colour to the original oil.

3.1 Fractionation of migration tracers

Comparison of the sterane based "Biomarker migration index" of Siefert and Moldowan[3] (ratio of C_{29} sterane 5α, 14β, 17β (S + R)/[5α, 14β, 17β (S + R) + 5α, 14α, 17α (S + R)] isomers) and a typical biomarker maturity related ratio (the ratio of the 20S/(20S + 20R) isomers for the regular C_{29} steranes) in the original and migrated oils showed no statistically detectable variation even with the first produced orange oils. A similar absence of variation accompanied most hydrocarbon biomarker ratios used for facies and maturity assessment.

Figure 4. Dichloromethane–methanol (2:1) extracts of 1 cm thick slices along the core show that most coloured material removed from the injected oil has been retained near the inlet.

In contrast, non-hydrocarbons containing nitrogen and oxygen showed substantial changes in concentration and composition as the experiment proceeded, Fig. 5 shows the normalised concentrations of representative prospective migration tracers in the sequentially eluted oils. Nitrogen compounds (carbazoles and benzocarbazoles) show reduced concentrations in the early eluted oils, with no benzocarbazoles being detectable in the first samples of oil produced from the core. Thereafter, concentrations rise to values near 80% of the initial oil by the end of the experiment. The larger scatter in the benzocarbazole data reflects analytical error associated with the low concentrations in the oils (peak height data were used in this case for quantitation to minimise interference from closely eluting compounds in the GCMS analysis). This elution profile is typical of other aromatic heterocyclic compounds, such as 9-fluorenone,[26] which show similar elution characteristics to carbazole. In contrast, alkylphenols, represented here by *p*-cresol, are severely depleted in all the oils collected, and the total C_0–C_3alkylphenols concentrations in the eluted oils reach only 10% of initial oil concentrations by the end of the experiment.

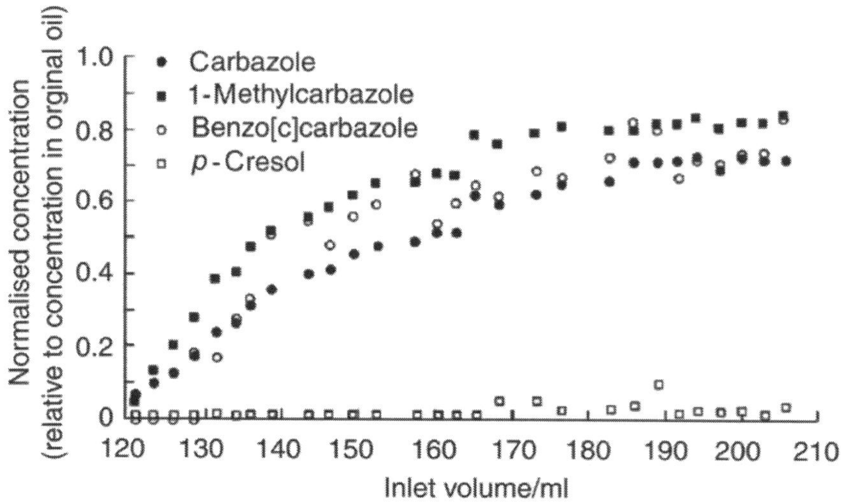

Figure 5. Variation in concentrations of selected representative non-hydrocarbons in oil samples collected at the core outlet during the core flood experiment. Concentration data has been normalised to values obtained for the original (input) oil (see Table 1). Note that phenols (as exemplified here by *p*-cresol) show much stronger depletion than carbazoles. The first oil was eluted from the core after *ca.* 120 ml of oil had been injected.

Distributions of different isomeric and homologous carbazoles and phenols also show systematic trends in the migrated (eluted) oils. Fig. 6shows relative abundances of higher to lower alkylcarbazole homologues in the migrated oils. The first oil samples from the core are characterised by enrichment in higher alkyl homologues (*cf.* C_2 carbazoles; sum of dimethyl- and ethylcarbazoles) relative to lower homologues. As more oil is flowed, the homologue ratios decrease towards the initial value for the oil. Except in the very first eluted oils, no significant fractionations of isomeric methyl or dimethylcarbazoles were observed.

Benzocarbazoles are very strongly sorbed by the siltstone and isomeric fractionations are also seen, with the first oils to contain both benzocarbazole [*a*] and benzocarbazole [*c*] having a reduced [*a*]/[*c*] ratio compared to that of the initial oil, as shown in Fig. 7.

In contrast to the alkylcarbazoles, where concentrations attained 80% of that in the initial oil by the end of the experiment, alkylphenol concentrations reached only 10% of that in the original oil, indicating much more extensive removal of phenols by the core. Fractionations between alkylphenol homologues and isomers were marked, the early eluting oils being enriched in phenol relative to higher homologues, possibly due to the higher water content of the earlier-eluted fluids. Substantial isomeric ratio variations were also seen in the alkylphenols between *ortho*-isomers (where the OH group on the phenol molecule is shielded by an adjacent alkyl substituent) and *meta*- and *para*-isomers (where there is no alkyl substituent in the *ortho* position, the OH group thus being exposed). The ratios of ortho-substituted to *meta*- and *para*-substituted isomers for the C_3 and C_2 alkylphenols in eluted oils seem to increase to a maximum and then gradually decrease with further oil flow to values close to those seen in natural oils. The C_1 phenols, however, show only a rise in this ratio to high values unlike those seen in natural crude oils. These variations suggest involvement of non-equilibrium competitive sorption–desorption processes in controlling the relative abundance of phenol isomers in the migrated oils and these will be discussed elsewhere.

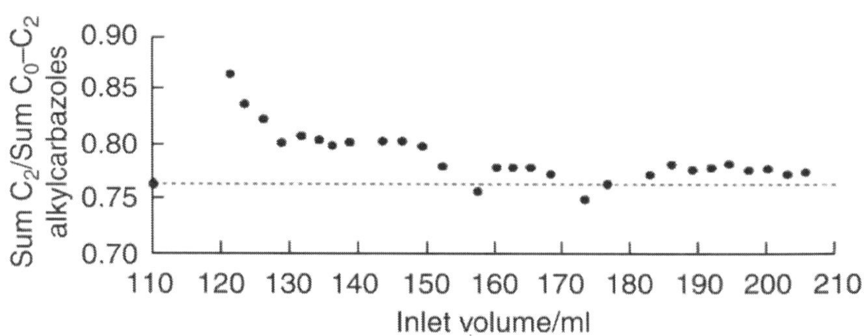

Figure 6. Variation in the abundance ratio of higher alkylated carbazole homologues (C_2 alkylcarbazoles) relative to total (C_0–C_2) alkylcarbazoles in outlet oil during the coreflood [$C_2/(C_0 + C_1 + C_2)$ ratio]. The value for the original oil is shown at inlet volume = 110 ml and by the dotted horizontal line. The first eluted oil at 120 ml shows significant relative enrichment in C_2 alkylcarbazoles.

3.2 Geological implications

The experiment clearly shows that even with short rapid transits through initially water saturated rocks, hydrophobic components such as carbazoles and benzocarbazoles can be effectively removed from migrating petroleum by partition–sorption processes. It is clear that at least in the case of the alkylphenols the core and oil are far from equilibrium at the end of the experiment, with isomer ratios in migrated/eluted oils very different from those in normal crude oils. Even for the (more hydrophobic) nitrogen compounds, equilibrium was only just being approached, with carbazole and benzocarbazole concentrations and isomer distributions reaching only *ca.* 20% of values in the initial oil by the end of the experiment. While these differences in equilibration time are significant on a lab timescale they are unlikely to be evident in field data sets.

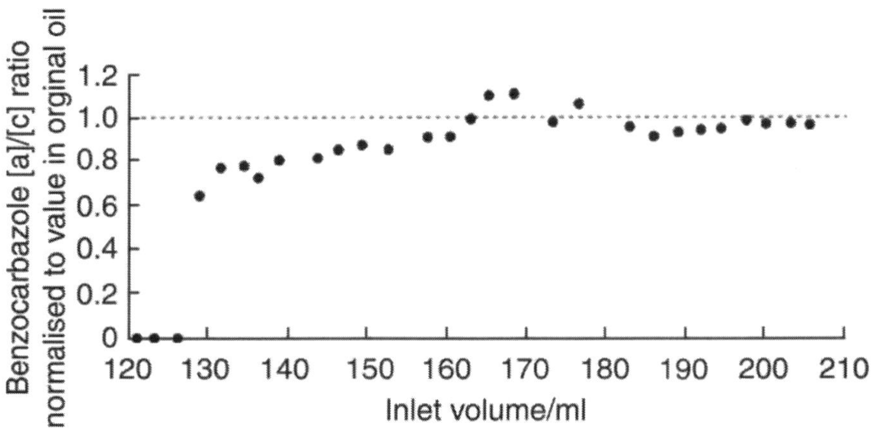

Figure 7. Variation in benzocarbazole [*a*]/[*c*] isomer ratios in outlet oil samples, determined from peak heights in *m/z* 217 mass chromatograms of the SPE DCM fraction. The values are normalised to the value in the original oil. The benzocarbazoles were completely depleted in the first three oil samples collected, and subsequent samples to *ca.* 160 ml showed statistically significant depletion of benzo[*a*]carbazole relative to benzo[*c*]carbazole compared to later eluted oils. The horizontal dotted line is the reference (injected) oil value at 1.0.

The dramatic removal of alkylphenols from migrated oils in the experiment confirms field observations[11, 12, 27] that phenol concentrations in migrated oils are rapidly attenuated over migration distances of only a few tens of kilometers through sandstones. Petroleum phenols can sorb to both minerals and solid organic matter in sediments,[11] and rapid wettability changes have been reported in initially water wet sandstones, brought about by synthetic oils containing small surlactant molecules such as phenols.[28] The rapid sorption of hydrophilic compounds such as the phenols might be expected, as they would readily partition from oil to water and then to solid phases.[11] It seems however that even with free water phases initially present in the rock, large hydrophobic molecules such as benzocarbazoles, with estimated oil–water partition coefficients (*P*) in the range of 10^4-10^5,[29] are also very quickly removed. While oil wet kerogen in the core may be responsible for some of this sorption, prior sorption of surfactant moieties such as alkylphenols, with very low subsurface oil–water partition coefficients ($P << 50$)[30] may also aid in the rapid sorption of carbazoles to organic and mineral surfaces. Molecular dynamics studies suggest that, indeed, prior sorption of alkyphenols to silica and calcite mineral surfaces effectively aids in subsequent sorption of hydrophobic moieties (van Duin and Larter – in preparation). The rapid oil wetting of pore surfaces in such fine grained rocks has major implications for cap rock sealing mechanisms and this will be discussed elsewhere. Similar synergistic processes may well aid in the removal of hydrophobic pollutants from water to mineral surfaces within aquifers.

The reduction in oil alkylcarbazole and benzocarbazole concentrations during the experiment, the enhancement in the $C_2/(C_0 + C_1 + C_2)$ alkylcarbazole ratio and the reduction in the benzocarbazole [*a*]/[*c*] ratio is similar to the compositional changes seen in migrated oil field data sets[5, 7, 14, 15] and suggests that these compounds do respond to secondary petroleum migration. The fractionations observed are similar to those seen in normal phase chromatographic separations in the laboratory,[31] and it is likely that both mineral and organic matter sorption processes contribute to the effects seen.

Source facies and maturity effects sometimes exert a major role in controlling the nitrogen compound composition of source rock petroleums.[6–9, 19] We suggested previously[7] that for some clastic oil source rocks, primary migration fractionations effectively removed any source related maturity influence on benzocarbazole composition from the secondary migrated oil, but in a study of the carbonate sourced Sonde de Campeche petroleum system of Mexico, where prolific vertical petroleum migration dominates, there is further evidence that maturity effects do sometimes dominate the carbazole and benzocarbazole compositions in the reservoired oils.[8, 9] In this case, the benzocarbazole [*a*]/[*c*] ratios and concentrations in the oils increase in the vitrinite reflectance defined maturity range up to 0.9% *R*o (*R*o, vitrinite reflectance; source temperatures *ca.* 140°C). Terken and Frewin[15] confirm that maturity effects on benzocarbazole distributions are sometimes present in oil datasets from around the world but in many cases they are not.

Our experience is that for shale-sourced petroleum systems with oils generated in the maturity range up to 0.9% *R*o, the concentration of benzocarbazoles in the migrated petroleum is typically an order of magnitude lower, or less, than that in the mature source petroleum.[27, 32,33] Thus, while mature Kimmeridge Clay Formation source rocks typically have benzocarbazole concentrations of 15–60 ppm in their source petroleum, North Sea oils derived from this source rock typically have only a few ppm of benzocarbazoles present.[33] In contrast, Horsfield and coworkers[8, 9] report that for the Sonde de Campeche carbonate-sourced petroleum system, both reservoired oil and source rock petroleum have very similar concentrations of these compounds in this same maturity range. This suggests greater primary migration fractionations in clastic source rock systems compared to carbonate source rocks which may variably affect inherited source-related nitrogen compound signals in the expelled oils.

For a series of petroleum reservoirs along a migration pathway where maturity gradients are present from field to field, it is often the case that the fields nearest the source kitchen contain the most mature petroleum and those more distal contain less mature petroleum, In this common scenario, the effects of any transmitted source maturity effect and secondary migration fractionation of benzocarbazole concentrations and distributions, as demonstrated here, would appear to have the potential to operate in concert, with further migrated oils tending to be less mature and more migration fractionated, and *vice versa.* Both effects would result in greater reductions in concentrations and benzocarbazole [*a*]/[*c*] ratios in the further migrated and least mature oils. This may explain the very low values of isomer ratios and concentrations seen in some migrated oils. While our core flood experiment confirms that these compounds do have the potential to work as migration indicators, a thorough maturity/source facies assessment of fluids in a possible migration suite is, as always, necessary.[5, 8, 14, 15]

Simple chromatographic theory[17] indicates that for equilibrium sorption of a tracer to a rock, the tracer concentration in a reservoir is not a simple function of distance migrated but depends critically on sorption properties of the carrier system and on the relationship between the volume of carrier rock accessed by the oil and the volume of oil migrated through it to reservoirs. Our theoretical studies[34] suggest that the concentration of tracer in a migrated oil in local equilibrium with its carrier rock correlates inversely and linearly with the ratio of carrier bed pore volume transgressed to oil volume transmitted through it (*i.e.*, volume carrier/volume oil transmitted).

This simple result can be thought of in terms of high and low throughput carrier systems. The petroleum in a low throughput carrier system comprised of small reservoirs charged *via* a comparatively long migration pathway (*i.e.*, volume carrier/volume reservoired oil is large) will be characterised by large reductions in tracer concentration during migration. Such examples would be the highly fractionated petroleums on the 2nd White Speckled Shale (source) – Cardium/Viking Formation (carrier/reservoir) petroleum systems of the Western Canada foreland sedimentary basin,[7] the Bakken oils in the Canadian portion of the Williston basin with source kitchens in the USA,[35] and the Dhahaban sourced oils of Oman which have also travelled well over 200 km from their source basin.[15] In contrast, in prolific basins (*i.e.*, with high throughput carriers), large volumes of oil will be charged through the same carrier volume, ultimately saturating the carrier to equilibrium saturation with the feed oil, with little net tracer removal seen in the final reservoired oil. An example would be the Kimmeridge Clay Formation (source) – Hugin Formation/Paleocene/Eocene sand (carrier/reservoir) petroleum systems of the Tertiary reservoirs in the Viking Graben, North Sea,[36] where many reservoir volumes of oil have been charged through a single focussed pathway. The Sonde de Campeche system[9] is probably a somewhat similar setting.

Even when maturity, facies and in-reservoir alteration complications can be resolved, in many prolific petroleum systems, where large volumes of petroleum pass through the same carrier system, it is likely that little or no net tracer fractionation may be observed. Large fractionations are most likely to be observed in basins with relatively small fields at the end of long migration routes, as seen in foreland basins[1] or where carriers involve highly sorptive siltstones or mudstones. Thus, large reductions in benzocarbazole ratios and concentrations are seen in reservoired petroleums in the foreland basin settings of Western Canada and Oman,[7, 14, 15] and for vertically migrated oils in the Gulf of Mexico and in the Kekeya Field of the Tarim Basin[37] which must have migrated through kilometers of mudstones.[36]

4. CONCLUSION

A core-flood experiment in which oil is flowed through water saturated siltstone under subsurface conditions of temperature and pressure confirms that geochromatographic fractionation of petroleum does occur during migration through water wetted rocks. While hydrocarbon fractions showed little fractionation, the more polar non-hydrocarbon fraction showed substantial compositional change. The systematic removal of carbazoles, benzocarbazoles and phenols during migration under subsurface conditions is similar to that inferred previously from field data. As expected from physicochemical properties, phenols were removed much more actively than carbazoles.

While little fractionation of sterane based migration parameters in migrated oils was observed, there was a systematic increase in the proportion of C_2 alkylcarbazoles compared to lower homologues, and a reduction in the benzocarbazole [*a*]/[*c*] ratio. The variations in homologue and isomer ratios were accompanied by reductions in absolute concentrations of these nitrogen compounds. These observations support field data reported earlier and suggest that while maturity effects may have a local influence on nitrogen compound distributions in reservoired oils, these species do track secondary migration processes *via* both concentration and isomer ratio variation.

Alkylphenols are very strongly absorbed by the siltstone and the phenol distributions in the migrated oils do not resemble those we see in nature. Non-equilibrium sorption–desorption processes within the timescale of the experiment are indicated, with rapid removal of phenols possibly resulting in the subsequent enhanced removal of more hydrophobic components such as alkylcarbazoles.

ACKNOWLEDGEMENTS

The research was supported by the EU FP4 SMACCERS project. Andy Aplin, Yunlai Yang, Maowen Li (GSC-Calgary), Kim Noke, Trevor Whitfield, Rob Hunter and Paul Donohoe are thanked for comments and technical assistance.

REFERENCES

1. Creaney S, Allan J: Classic Petroleum Provinces. Edited by: Brooks J. 1990, Geological Society of London Special Publication 50, London, 189

2. England WA, Mackenzie AS, Mann DM, Quigley TM: The movement and entrapment of petroleum fluids in the subsurface. J Geol Soc London. 1987, 144: 327-

3. Seifert WK, Moldowan JM: Applications of steranes, terpanes and monoaromatics to the maturation, migration and source of crude oils. Geochim Cosmochim Acta. 1978, 42: 77

4. Li M, Larter SR, Stoddart D, Bjorøy M: Liquid chromatographic separation schemes for pyrrolic and pyridinic nitrogen aromatic heterocycle fractions from crude oils suitable for rapid characterisation of geochemical samples. Anal Chem. 1992, 64: 1337

5. Li M, Larter SR, Stoddart D, Bjorøy M: The Geochemistry of Reservoirs. Edited by: England WA, Cubitt J. 1995, Geological Society Special Publication, London, 103-

6. Li M, Yao H, Stasiuk LD, Fowler MG, Larter SR: Effect of maturity and petroleum expulsion on pyrrolic nitrogen compound yields and distributions in Duvernay Formation petroleum source rocks in central Alberta. Org Geochem. 1997, 26: 731

7. Larter SR, Bowler BFJ, Li M, Chen M, Brincat D, Bennett B, Noke K, Donohoe P, Simmons D, Kohnen M, Allan J, Telnaes N, Horstad I: Molecular indicators of secondary oil migration distances. Nature. 1996, 383: 593

8. Clegg H, Wilkes H, Oldenburg T, Santamaria Orozco D, Horsfield B: Influence of maturity on carbazole and benzocarba zole distributions in crude oils and source rocks from the Sonda de Campeche, Gulf of Mexico. Org Geochem. 1998, 29: 183

9. Horsfield B, Clegg H, Wilkes H, Santamaria Orozco D: Effect of maturity on carbazole distributions in petroleum systems: new insights from the Sonda de Campeche, Mexico, and Hils syncline. Naturwissenschaften. 1998, 85: 288

10. Yamamoto M: Fractionation of azaarenes during oil migration. Org Geochem. 1992, 19: 389

11. Taylor P, Larter S, Jones M, Dale J, Horstad I: The effect of oil–water–rock partitioning on the occurrence of alkylphenols in petroleum systems. Geochim Cosmochim Acta. 1997, 61: 1899

12. Galimberti R, Ghiselli C, Chiaramonte MA: Acidic polar compounds in petroleum: a new analytical methodology and applications as molecular migration indexes. Extended abstracts of the 19th Intl. Meeting on Organic Geochemistry, Istanbul, Turkey. 1999, TUBITAK Marmara Research Center, Gebze-Kocaeli, 147-6–10 Sept 1999

13. Ballentine CJ, O'Nions RK, Coleman ML: A Magnus Opus: Helium, neon and argon isotopes in a North Sea oilfield. Geochim Cosmochim Acta. 1996, 60: 831

14. Li M, Yao HX, Fowler MG, Stasiuk LD: Geochemical constraints on models for secondary petroleum migration along the upper Devonian Rimbey-Meadowbrook reef trend in central Alberta, Canada. Org Geochem. 1998, 29 (1–3): 163

15. Terken JMJ, Frewin NL: The Dhahaban petroleum system of Oman. Bulletin AAPG. 2000, 84 (4): 523-

16. Nagy B: Review of the chromatographic plate theory with reference to fluid flow in rocks and sediments. Geochim Cosmochim Acta. 1960, 19: 289

17. Krooss BM, Brothers L, Engel MH: Petroleum Migration. Edited by: England WA, Fleet AJ. 1991, Geological Society Special Publication, 149-

18. Clegg H, Horsfield B, Stasiuk L, Fowler M, Vliex M: Geochemical characterisation of organic matter in Keg River Formation (Elk point group, Middle Devonian), La Crete Basin, Western Canada. Org Geochem. 1997, 26: 627

19. Clegg H, Wilkes H, Horsfield B: Carbazole distributions in carbonate and clastic source rocks. Geochem Cosmochim Acta. 1997, 61: 5335

20. Greibrokk T, Lundanes E, Norli HR, Dyrstad K, Olsen SD: Experimental simulation of oil migration – distribution effects on organic compound groups and on metal/metal ratios. Chem Geol. 1994, 116: 28

21. Bonilla JV, Engel MH: Chemical and isotopic redistribu tion of hydrocarbons during migration. Laboratory simulation experiments. Org Geochem. 1986, 10: 181

22. Karlsen DA, Larter SR: Analysis of petroleum fractions by TLC-FID: applications to petroleum reservoir description. Org Geochem. 1991, 17: 603

23. Bennett B, Bowler BFJ, Larter SR: Determination of C_0–C_3 alkylphenols in crude oils and waters. Anal Chem. 68 (20): 3697

24. Bowler BFJ, Larter SR, Clegg H, Wilkes H, Horsfield B, Li M: Dimethylcarbazoles in crude oils: comment on "Liquid Chromatographic Separation Schemes for Pyrrole and Pyridine Nitrogen Aromatic Heterocyclic Fractions from Crude Oils Suitable for Rapid Characterization of Geochemical Samples". Anal Chem. 1997, 69: 3128

25. Bennett B, Larter SR: Quantitative Separation of Aliphatic and Aromatic Hydrocarbons Using Silver Ion-silica Solid Phase Extraction. Anal Chem. 2000, 72: 1039

26. Bennett B, Larter SR: The isolation, occurrence and origin of fluorenones in crude oils and rock extracts. Org Geochem. 2000, 31: 117

27. Chen M, Larter SR, Petch S, Bowler B, Aplin AC: Organic Geochemistry: Developments and Applications to Energy, Climate, Environment and Human History. Edited by: Grimalt JO, Dorronsoro C. 1995, AIGOA Danostia–San Sebastian, Spain, 288-

28. Huang Y, Ringrose PS, Sorbie KS: The effects of heterogeneity and wettability on oil recovery from laminated sedimentary structures. Society of Petroleum Engineers, Texas. 1995, 30781-

29. Van Duin ACT, Larter SR: Application of molecular dynamics calculations in the prediction of dynamical molecular properties. Org Geochem. 1998, 29: 1043

30. Bennett B, Larter SR: Partition behaviour of alkylphenols in crude oil–brine systems under subsurface conditions. Geochim Cosmochim Acta. 1997, 61: 4393

31. Li M, Larter SR, Frolov YB: Adsorptive interactions between nitrogen compounds and organic and/or mineral phases in subsurface rocks. J High Resolut Chromatogr. 1994, 17: 230

32. Chen M: 1995, PhD Thesis, University of Newcastle

33. Bennett B, Chen M, Larter SR: Benzocarbazoles: fractionations in clastic and carbonate systems. Abstract 19, 219th ACS National Meeting. 2000, American Chemical Society, San Francisco

34. Yang Y, Larter SR, Aplin AC, Carruthers D: Theoretical study of natural tracers to determine migration routes and distances: models and applications. Abstracts AAPG Annual Meeting. 1999, American Association of Petroleum Geologists, San Antonio, A156-

35. Li M, Osadetz KG, Fowler MG, Snowdon LR, Stasiuk LD, Yao H, Hwang RJ, Jenden P, Grant B, Idiz E: Case studies on secondary oil migration in the Williston Basin. Proceedings of the 8th International Williston Basin Symposium. Edited by: Christopher J, Gilboy CF, Paterson DF. 1998, Geol Survey Publication 13, Saskatchewan, 247-253.

36. Larter SR, Aplin AC, Bennett B, Bowler B, Noke K, Harrison E, Telnaes N, Wilhelms A, Eglinton T, Whelan J: An assessment of vertical migration mechanisms in mudstone rich basins using organic geochemistry. Extended abstracts 18th Intl. Meeting on Organic Geochemistry 1997. 1998, 13-

37. Li M, Lin R, Wang P, Liao Y, Snowdon LR, Li P: Organic geochemistry of oils and condensates from the Kekeya Field, Tarim basin. Org Geochem. 1999, 30: 15

CHAPTER 9

Neutron production in the lunar subsurface from alpha particles in galactic cosmic rays

Shuya Ota1,2,3, Shingo Kobayashi4, Lembit Sihver3,5,6, Naoyuki Yamashita1, and Nobuyuki Hasebe1

1Research Institute for Science and Engineering, Waseda University, 3-4-1 Okubo, Shinjuku-ku, Tokyo, Japan

2National Institute of Radiological Sciences, 4-9-1 Anagawa, Inage,Chiba, Japan

3Chalmers University of Technology, Goteborg, SE-412 96 Gothenburg, Sweden

4Japan Aerospace Exploration Agency, 3-1-1 Yoshinodai, Sagamihara, Kanagawa, Japan

5Texas A&M University, Texas 77843-3133, USA

6Roanoke College, Salem, Virginia 24153, USA

ABSTRACT

The neutron production from alpha particles in galactic cosmic rays (GCR) in the lunar subsurface has not been estimated with reliable precision despite its importance for lunar nuclear spectroscopy and space dosimetry. Here, we report our estimation of neutron production from GCR nuclei (protons and alpha particles) with the Particle and Heavy Ion Transport code System (PHITS), which includes several heavy ion interaction models. PHITS simulations of the equilibrium neutron density profiles in the lunar subsurface are compared with experimental data obtained in the Apollo 17 Lunar Neutron Probe Experiment. Our calculations successfully reproduced the data within an experimental error of 15%. Our estimation of neutron production from GCR nuclei, estimated by scaling that from protons by a factor of 1.27, is in good agreement within an error of 1% with the calculations using two different alpha particle interaction models in PHITS during a period of average activity of the solar cycle. However, we show that the factor depends on the incident GCR spectrum model used in the simulation. Therefore, we conclude that the use of heavy ion interaction models is important for estimating neutron production in the lunar subsurface.

KEYWORDS

Neutron; alpha particle; lunar and planetary spectroscopy; Moon; gamma ray spectrometer; SELENE PHITS; heavy ion transportation

1. INTRODUCTION

The lunar surface is quite distinct from Earth's surface. Since the Moon has no atmosphere and its magnetic field is very weak, the lunar surface is directly exposed to galactic cosmic rays (GCR). Secondary products, including neutrons and gamma rays, are continuously produced by the nuclear interactions between the GCR nuclei and the materials in the lunar subsurface. In the field of lunar science, it is very important to know the production rates of these secondary products because the neutrons and gamma rays emitted from the lunar surface can be used to estimate the elemental abundance of the lunar subsurface material (Evans *et al.*, 1993; Feldman *et al.*, 1993). The composition of the Moon's surface has been investigated using neutron and gamma ray spectroscopes following the launch of the Lunar Prospector (e.g., Gasnault *et al.*, 2000; Lawrence *et al.*, 2004). Above all, the recent successful mission of the Japanese lunar orbiter SELENE (KAGUYA) equipped with a Ge gamma ray spectrometer is expected to report drastically improved global mapping data of the lunar surface composition (Hasebe *et al.*, 2008, 2009). In order to be able to derive the absolute abundance of elements by nuclear spectroscopy, lunar scientists need a good simulation code for transporting the GCR nuclei and secondary products in the lunar material.

From the view of not only lunar science but also human activities on the Moon, the simulation of particle production and transportation in matter is essential. Future human activities on the lunar surface will lead to further progress in space science. To this end, the building of a lunar base has been seriously discussed for a long period of time. On the lunar surface, human beings would be exposed to not only to both primary GCR and their secondary radiation from the lunar surface but also to high energetic solar energetic particles (SEPs). Consequently, the shielding of human beings from intense radiation exposure is one of major problems to be solved to secure the safety of future human activities on the lunar surface. The unshielded dose equivalent rate on the Moon is currently estimated to be 300–400 mSv/year (Adams *et al.*,2007), depending on the solar cycle, but it has recently been estimated to be as high > 800 mSv/year during the solar minimum of the solar cycle (Hayatsu *et al.*, 2008, 2009); in comparison, the unshielded dose equivalent rate is about 2.4 mSv/year on the Earth (UNSCEAR report,1988). This variability in dose estimation mainly results from uncertainties in the observational data of GCR spectra, composition of the lunar soil, and the particle and heavy ion transportation code. As the qualities and quantities of the observational data are progressively being improved by the launching of continuous lunar missions, such as SELENE, Chang'e, and Chandrayaan-1, transport simulations with an increased accuracy is required.

Many transport codes have been developed for the research of fundamental high-energy particles, and these are used for radioprotection in space and for the treatment planning systems (TPSs), such as heavy ion cancer therapy. These codes typically include MCNP (Briesmeister,1997), FLUKA (Fasso *et al.*, 1993), Geant4 (Agostinelli *et al.*, 2003), and PHITS (Particle and Heavy Ion Transport code System) (Niita *et al.*,1995, 2006). These transport codes are continuously being improved, such as the development of new and better models for nuclear and electromagnetic interactions and the extension of the energy region for the nuclear data libraries, etc. A careful comparison of these codes is found in Sihver *et al.* (2008a). Since each code has its own specific advantages and disadvantages, it is essential to use the best transport code for the purpose being defined, taking the respective advantages/disadvantages into account. Remarkable progress in the calculation of heavy ion reactions and transportation has been made during the last decade (see, for example, Sihver *et al.*, 2007, 2008a, b, 2009; Sihver,2008). In recent years, PHITS has received considerable attention due to its applications in heavy ion transport calculations. Several interaction models for calculating heavy ion reactions, such as JQMD and JAMQMD, are incorporated into PHITS. Consequently, it is possible to transport heavy ions with a considerably high reliability.

By taking advantage of the heavy ion transport models in PHITS, we have estimated neutron production by GCR protons and heavy components. Accurate calculation of the heavy ion reactions and transportation are important when estimating the neutron and gamma ray production from GCR. Although the amount of heavy nuclei ($Z \geq 2$) in the GCR flux is small relative to that of protons, the neutron and gamma ray production rates by heavy ions are much larger than those by the protons. Therefore, neutron production from heavy ion reactions should be taken into account. The neutron production from alpha particles is of particular importance because alpha particles are more abundant than other heavy nuclei ($Z > 2$). The neutrons produced by the protons and alpha particles form the major part of the total neutron density profile produced by all the nuclides in the GCR. We have therefore focused on the production of neutrons from alpha particles in the GCR. To date, neutron production from alpha particles has been mainly estimated by scaling the calculation of protons (Dagge *et al.*, 1991; Masarik and Reedy, 1996). The primary aim of our study was to compare, in detail, the estimation of neutron production from alpha particles by the proton scaling method with that by alpha particle interaction model.

We first calculated the neutron density as a function of depth (neutron density profile) produced by the interaction of GCR particles with lunar material. We then compared our calculated values with the experimental values obtained in the Apollo 17 Lunar Neutron Probe Experiment (LNPE) (Woolum *et al.*, 1973, 1975; Woolum and Burnett, 1974a, b). Since neutron production affects gamma ray production, our calculation is a major contribution in furthering neutron and gamma ray spectroscopy and space dosimetry on the Moon.

2. SIMULATION PROCEDURE

2.1 PHITS

The simulations described in this paper were performed with the three-dimensional Monte Carlo Particle and Heavy Ion Transport code System (PHITS), which was developed by Niita *et al.* (2006) and Iwase *et al.* (2002). PHITS ver. 2.13 was used throughout the work reported here. PHITS has several models for calculating nuclear reactions and the transportation of energetic particles, such as protons, neutrons, heavy ions, and some exotic particles > 10 MeV/n (heavy ions) and > 1 MeV (otherwise) in matter. In addition, PHITS can deal with the complex motion of low-energy neutrons < 20 MeV in a similar manner as the MCNP4C code. PHITS has been successfully used for many radiation transport analyses of the space radiation environment both inside and outside a spacecraft (the reader is referred to Sato *et al.*,2006, 2008a; Sihver *et al.*, 2009; Gustafsson *et al.*, submitted; Sihver *et al.*, submitted). Thus, we concluded it is possible to calculate the neutron production in the lunar subsurface accurately using the PHITS.

2.2 Geometry and material

The simulation procedure for neutron production in the lunar subsurface basically follows the conditions under which the LNPE was performed from Extra-Vehicular Activity (EVA) 1 to EVA 3 in the Apollo 17 mission in 1973. The LNPE measured equilibrium neutrons within the first 400 g/cm^2 using two kinds of track detectors, namely, muscovite mica fission detectors with ^{235}U targets and cellulose triacetate (Triafol TN) plastic detectors with ^{10}B targets (Woolum *et al.*, 1973). A total of eight ^{235}U targets and 23 ^{10}B targets were attached to a 2-m-long rod. The rod was then inserted into the drill hole created in the lunar subsurface. The neutron capture rates of ^{235}U and ^{10}B targets were obtained from the track densities of fission fragments emitted from ^{235}U induced by neutrons with about 5 meV to 5 keV energy and the alpha and some Li recoils emitted via the $^{10}B(n,\alpha)Li$ reaction induced by neutrons with about 5 meV to 500 eV energy (Woolum *et al.*, 1975). Finally, the ^{235}U fission rate and neutron density as a function of depth were measured by converting the ^{235}U and ^{10}B neutron capture rates, respectively, using the neutron capture cross sections. The instrument and its measurement method are described in more detail in Woolum *et al.* (1973). Data analysis and the estimation of experimental error can be found in Woolum and Burnett (1974a) and Woolum *et al.* (1975). In this study, our calculation was compared with the neutron density profile obtained with Triafol TN detectors. The data labeled "Woolum" in this paper were taken from Woolum *et al.* (1975) and are the same as those used by McKinney *et al.* (2006). An overall experimental error of 15% estimated by Woolum and Burnett (1974b), which includes errors of track measurements (7–9%) and several correction effects for conversion to neutron densities, was used in the data (Woolum *et al.*, 1975).

The geometry of the target and beam source in our simulation is modeled after the relationship between the lunar body and the GCR. First of all, a sphere with a radius of 1738 km, which is the same size as the moon, was designated the target. Then, the three concentric spheres were configured inside the lunar sphere to reproduce differences in material and the density dependent on the depth from the surface. The materials and densities in the lunar body used as input are summarized in Table 1. These data were obtained by McKinney *et al.* (2006) based on data obtained from the Soil Mechanics experiment in the Apollo 17 mission carried out by Carrier (1974) and Mitchell *et al.* (1974). The temperature of the material was set at 250 K, as suggested by Woolum and Burnett (1974a). The actual temperature may vary under 333 K, which the LNPE temperature indicators suggested as a maximum, because the LNPE was carried out for 49 h. However, based on the results of our preliminary tests, differences in material temperature primarily affect the shape of the low-energy neutron spectrum but have small effects on the neutron density. Similar results have also been reported by other investigators (Woolum *et* al., 1975; Dagge *et al.*, 1991; Masarik and Reedy, 1996). Therefore, exact information on the temperature is not important. As the beam source, a shell with a radius of 1738 + 0.001 km was configured outside the lunar sphere. Protons and alpha particles were used as the source particles from the GCR. The energy spectra of the proton and alpha particles were generated with the analytical functions described in the following section. The particles are created within PHITS as an external isotropic source on the shell and then originate randomly and inwardly from the shell. The space between the lunar surface and the shell with the source particles was set as a vacuum.

Table 1. Abundances (wt%) and densities of the lunar subsurface used in our calculations. These data were taken from the Lunar Neutron Probe Experiment borehole analysis by McKinney *et al.* (2006). *Read 4.174E-1 as 4.174×10^{-1}.

Depth (cm) (Density (g/cm^3))	0–22 (1.76)	22–71 (2.11)	71–224 (1.78)	>224 (1.79)
Element	Abundance (wt%)			
O	4.174E-01*	4.156E-01	4.230E-01	4.264E-01
Na	2.923E-03	3.136E-03	3.075E-03	3.455E-03
Mg	6.162E-02	6.026E-02	6.156E-02	6.091E-02
Al	6.061E-02	5.977E-02	7.384E-02	7.550E-02
Si	1.903E-01	1.896E-01	1.967E-01	2.022E-01
K	7.262E-04	7.896E-04	9.201E-04	1.632E-03
Ca	7.541E-02	7.668E-02	8.020E-02	7.707E-02
Ti	5.144E-02	4.905E-02	3.380E-02	3.198E-02
Cr	2.872E-03	3.090E-03	2.641E-03	2.545E-03
Mn	1.764E-03	1.778E-03	1.519E-03	1.458E-03
Fe	1.350E-01	1.403E-01	1.228E-01	1.169E-01
Sm	8.334E-06	7.747E-06	7.343E-06	1.051E-05
Eu	1.816E-06	1.782E-06	1.545E-06	1.700E-06
Gd	1.100E-05	1.058E-05	9.818E-06	1.355E-05
Th	9.449E-07	8.022E-07	1.382E-06	3.006E-06

2.3 Beam source

A form of analytical functions for the energy spectra of proton and alpha particles in GCR has been widely accepted as the means to vary the parameters (Castagnoli and Lal, 1980; Dagge *et al.*, 1991; Masarik and Reedy, 1996; McKinney *et al.*, 2006). The 4π GCR flux J(particle/(MeV/n)/cm^2/s) is expressed by

$$J(E,\phi) = C\frac{E\left(E+2m_{\mathrm{p}}c^2\right)(E+\chi+\phi)^{-\gamma}}{(E+\phi)\left(E+2m_{\mathrm{p}}c^2+\phi\right)}, \qquad (1)$$

where C (1/cm²/s/MeV) is a normalization factor, E (MeV/n) is the kinetic energy of GCR nuclei, φ (MV) is the solar modulation parameter to characterize the effect of solar activity, mp (MeV/c²) is the mass of the proton, c is the velocity of light, $\chi = a\exp(-bE)$, and the remaining parameters, such as a, b, and γ, are provided in Table 2. These parameters have been described in detail for protons (Castagnoli and Lal,1980) and alpha particles (Lal, 1985). When the LNPE was carried out, φ corresponds to 550±50 MV. The determination of φ was studied by McKinney *et al.* (2006) in detail. Therefore, we used the energy spectra for proton and alpha particle at φ = 550±50 MV. As can be seen in Table 2, the variability in total flux at φ = 500 and 600 MV ranges over 10% from 1 at φ = 550 MV, respectively, because GCR particles with energies of < 10 GeV/n are extremely affected by the solar activity. Therefore, neutron density also depends on the value of φ. In our calculation, the energy cut-off of the generated source particles was set at a minimum of 10 MeV/n and a maximum of 20 GeV/n.

Table 2. Parameters of proton and alpha particle energy spectra, and total fluxes (particle/cm²/s) used in our calculations. The total fluxes were obtained by integration of the spectra ranging from 10 MeV/n to 20 GeV/n. *Units are particle/cm²/s.

Particle	C (1/cm²/s/MeV)	a (MeV)	b (1/MeV)	γ	Flux* ($\phi = 500$)	Flux* ($\phi = 550$)	Flux* ($\phi = 600$)
Proton	1.24E+06	780	2.50E-04	2.65	3.16	2.88	2.63
Alpha	2.26E+05	660	1.40E-04	2.77	0.233	0.21	0.19

The beam was projected inwardly from the shell with the angular probability distribution of the cosine, i.e., $p(\cos\theta) = \cos\theta$. Here, θ is the angle between the particle incident direction and normal to the shell surface. The cosine angular distribution generated from the shell makes the density of the particles within the sphere uniform. As such, it has been ensured that the lunar surface is bombarded with the GCR nuclei isotropically.

In the simulations described in this paper, typically 100,000 and 25,000 source particles were used for proton and alpha particles, respectively, and the statistical error was estimated to be < 1%.

2.4 Nuclear interaction model

The different theoretical models used for the proton-induced and alpha-induced nuclear interactions in our study are summarized in Table 3. Below 3.5 GeV, PHITS uses the Bertini intranuclear cascade model as a default option for the nucleon-induced interaction. There are three different optional parameterizations of the nucleon-nucleon interaction cross section data used in the Bertini model, namely, “Cugnon Free”, “Cugnon Old”, and “Cugnon New”, respectively. PHITS uses the Cugnon Old parameterization step in the Bertini model as the default. Although the Bertini model is a default model as far as these cross-section data exist (1 MeV < *E* < 3.5 GeV), it is no longer applicable > 3.5 GeV, when the default model switches to the JAM (Jet AA Microscopic Transport Model), which is an original hadronic cascade model (Nara*et al.*, 2000; Niita *et al.*, 2001, 2006). However, it is possible to select other models, such as JAM, JQMD (JAERI Quantum Molecular Dynamics), and JAMQMD for the full energy range (1 MeV < *E* < 200 GeV) (Niita *et al.*, 1995). For the nucleus-induced interaction, PHITS uses JQMD < 3.5 GeV/n and JAMQMD > 3.5 GeV/n as a default. Similarly to the proton interaction model, it is possible to use JQMD and JAMQMD for the full energy range (10 MeV/n < *E* < 100 GeV/n). JQMD describes the behavior of the nucleon-nucleus and nucleus-nucleus reactions by solving the equation of motion for each nucleon-nucleon interaction. JAMQMD is basically the same as JQMD, with the exception that the nucleon-nucleon interaction cross section is calculated using the JAM model. In this work, the calculations for both protons and alpha particles were verified as shown below.

Table 3. Interaction model options in the PHITS (* denotes the default model).

Particle	Model options (*E* denotes the kinetic energy of particle)
Proton	Bertini Free (1 MeV < *E* < 3.5 GeV) + *JAM (*E* > 3.5 GeV)
	*Bertini Cugnon Old (1 MeV < *E* < 3.5 GeV) + *JAM (*E* > 3.5 GeV)
	Bertini Cugnon New (1 MeV < *E* < 3.5 GeV) + *JAM (*E* > 3.5 GeV)
	Isober (1 MeV < *E* < 1 GeV) + *Bertini Cugnon Old (1 GeV < *E* < 3.5 GeV) + *JAM (*E* > 3.5 GeV)
	JAM (*E* > 1 MeV)
	JQMD (*E* > 1 MeV)
	JAMQMD (*E* > 1 MeV)
Alpha	JQMD (*E* > 10 MeV/n)
	JAMQMD (*E* > 10 MeV/n)

2.5 Simulated geometry

In the target sphere, 55 concentric spheres were configured in order to calculate the neutron fluxes at various depths. These spheres were set to depths of 1–10 cm, and then the neutron flux crossing the spherical surface was calculated at each depth. In this way, neutron fluxes were obtained at 55 points from the surface down to the first 3 m, corresponding to about 500 g/cm^2 in the lunar subsurface. The neutron density (ρ_{neutron}) at a given depth (d) was obtained by conversion from the neutron flux (F_{neutron})at d with energies (E) <500 eV using the neutron velocity (v). This conversion is given by

$$\rho_{\text{neutron}}(d) = \int_0^{500\,\text{eV}} \frac{F_{\text{neutron}}(d, E)}{v(E)} dE. \tag{2}$$

3. NEUTRON DENSITY PROFILE ESTIMATED BY GCR PROTONS USING A SCALE FACTOR

The first step was to calculate the neutron density profile using the conventional estimation method, which is scaling of neutron production from the GCR protons (Dagge *et al.*, 1991; Masarik and Reedy, 1996). Since many transportation codes can not transport particles heavier than protons, a scale factor to estimate the neutron contribution from the alpha particles has been used for many years. This scale factor corresponds to the average ratio of alpha particle to the proton in terms of neutron production efficiency in the full energy range of GCR spectra. When we define the scale factor as x at $\varphi = 550$ MV, the neutron production from protons and alpha particles will increase by a factor of $(2.88 + 0.210 \times x)/2.88$ compared to that from only protons (note 2.88 and 0.210 are the total fluxes of protons and alpha particles (particle/cm^2/s), respectively). For example, Reedy and Arnold (1972) estimated the scale factor to be 2.8 for energies > 1 GeV/n, Masarik and Reedy (1996) invoked the scale factor of 3.0 (Reedy, 2009) using the LCS (Prael and Lichtenstein, 1989), and Dagge *et al.* (1991) reported 3.5 to be the optimum factor using the HERMES code (Prael and Lichtenstein, 1989; Prael, 1993), which supports the alpha particle irradiation. By indirect means, Yamashita *et al.* (2006) measured the factor to be 3.5 from the ratio of the efficiencies of the alpha particle and proton to produce gamma rays. On the other hand, McKinney *et al.* (2006) suggested a factor of 3.8 using a preliminary version of MCNPX 2.6.0, which supports the alpha particle transportation. In this section, we calculated the neutron production from proton and alpha particles in GCR using the scale factor of 3.5.

The calculated neutron density profiles are shown in Figs. 1 and 2. The results calculated by all patterns of modelsinTable3at φ = 550 MV are shown in Fig. 1(a) and (b). The trend that peaks of neutron density profiles exist around 150–180 g/cm^2 is similar to the experimental results. How-ever, the greatest difference in the absolute values is that of 30% by JAMQMD around 200 g/cm^2 and 50% by Bertini Cugnon Old around 350 g/cm^2. Compared to the LNPE data, it is evident that JAM provides the best agreement. As mentioned above, the solar modulation parameter φ also has a large influence on neutron production. Therefore, we calculated the neutron density profile at φ = 500, 550, and 600 MV using the JAM in Fig. 2. The agreement between the calculation at φ = 600 MV and experimental values is much improved. Thus, our calculation results successfully reproduced the LNPE data within the measurement error of 15%.

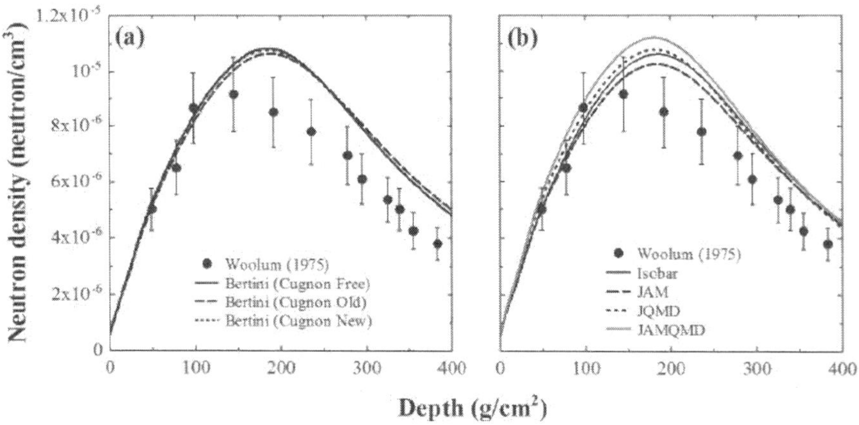

Figure 1. Comparison of calculated neutron density profiles with the experimental results reported by Woolum *et al.* (1975). The calculations were performed using seven different kinds of proton interaction models; (a) Cugnon Free, Cugnon Old, Cugnon New, (b) Isobar, JAM, JQMD, JAMQMD. In both figures, neutron production from alpha particles in the GCR was estimated using a scale factor of 3.5; this factor was based on that for protons in the GCR.

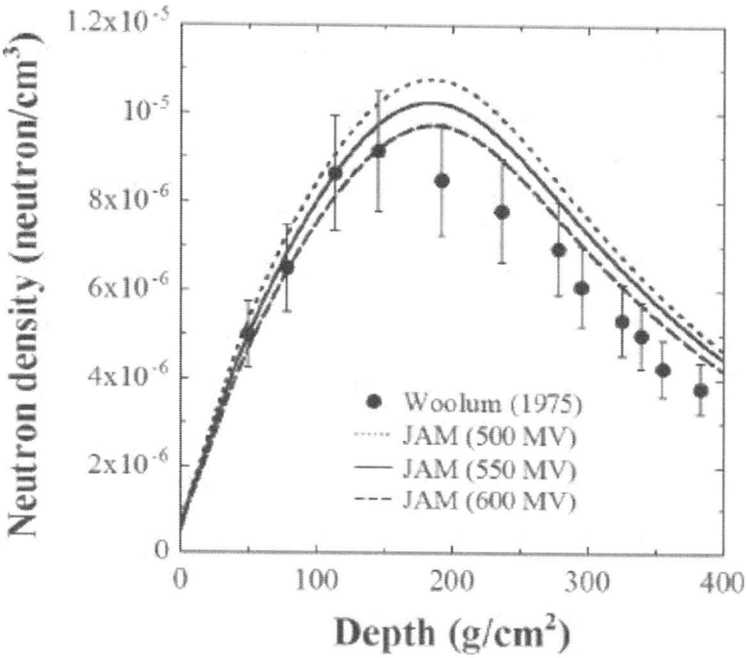

Figure 2. Comparison of calculated neutron density profiles with the experimental results resported by Woolum *et al.* (1975). The calculations were performed using the JAM interaction model for the three different solar modulation parameters. Neutron production from alpha particles in the GCR was estimated using a scale factor of 3.5; this factor was based on that for protons in the GCR.

4. NEUTRON PRODUCTION FROM THE GCR ALPHA PARTICLES

In this section, we report our estimation of neutron production from the GCR alpha particles using two different heavy ion interaction models: JQMD and JAMQMD (see Table 3). The calculated results of neutron density profile using JQMD and JAMQMD for alpha particles are shown in Fig. 3. In these calculations, JAM was used for the proton interaction. The results by JQMD and JAMQMD showed a good agreement with each other, although the results from both JQMD and JAMQMD are a few percentage points higher than that achieved by the proton scaling method. This difference suggests that a scale factor of 3.5 is a little low to estimate correctly the contribution from the GCR alpha particles. To study the contribution of alpha particles in more detail, we changed the scale factor to 3.6, 3.8, and 4.0, respectively. Comparisons of the neutron density profile by the proton scaling method with those using JQMD and JAMQMD are shown in Fig. 4. As can be seen in Fig. 4, a scale factor of 3.8 provides a better agreement with the calculations performed with both the JQMD and JAMQMD. The standard deviations of such distributions are about 1%, suggesting that no special difference was found between the proton scaling method and the alpha particle interaction models. We conclude that 3.81 and 3.72 are the best factors by which to obtain the best fits to the results by JQMD and JAMQMD, respectively, in the more detailed comparison. More detailed differences from the view of thermal ($E \leq 1$ eV), epithermal (1 eV $< E \leq 1$ MeV), and fast neutrons ($E > 1$ MeV) using the scale factor of 3.77 are shown in Fig. 5(a) (3.77 is the mean of 3.81 and 3.72). Detailed comparisons are made in Fig. 5(b). In this figure, one can see that there is an agreement between the proton scaling method and the alpha interaction models within an error of 1% < 250 g/cm^2 for all energy ranges. On the other hand, the discrepancies increase up to a few percent points > 250 g/cm^2. However, these discrepancies result in only a small specific difference in the actual application method, e.g., neutron and gamma ray spectroscopy and space dosimetry.

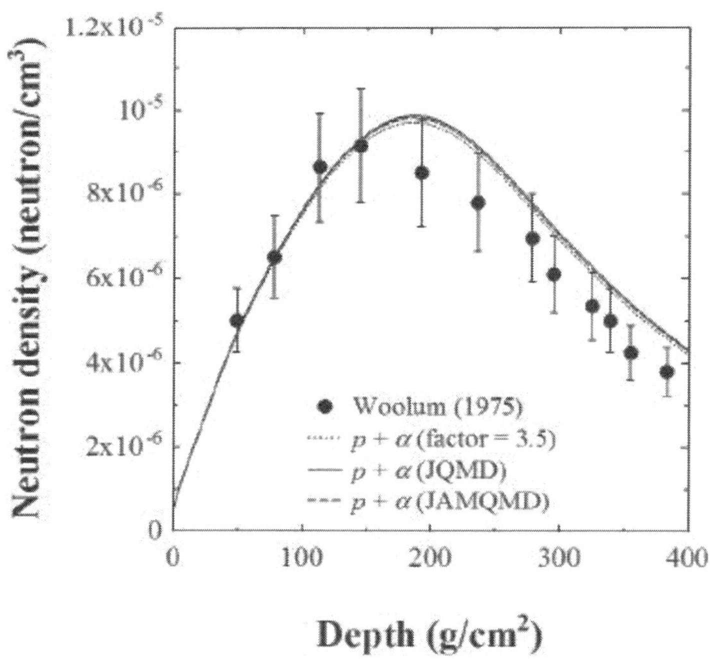

Figure 3. Comparison of calculated neutron density profiles with the experimental results reported by Woolum *et al.* (1975). The calculations were performed using three modes at $\varphi = 600$ MV. In the calculation using the scale factor, the neutron production from alpha particles in the GCR was estimated using a scale factor of 3.5; this factor was based on that for protons in the GCR.

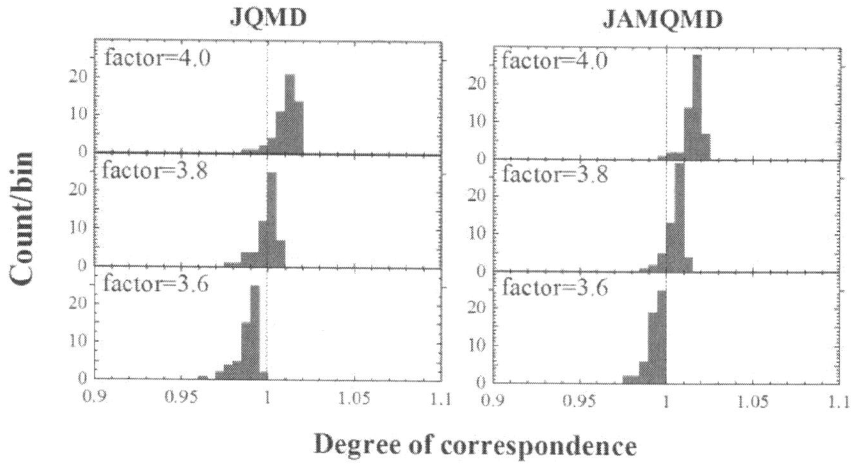

Figure 4. Degree of correspondence of neutron density profiles calculated at 55 points from the lunar surface to the first 3 m at $\varphi = 600$ MV by the proton scaling method with those calculated by the two different alpha particle interaction models of JQMD and JAMQMD. The degree of correspondence was defined as the ratio of the result obtained by the alpha particle interaction model to the proton scaling method.

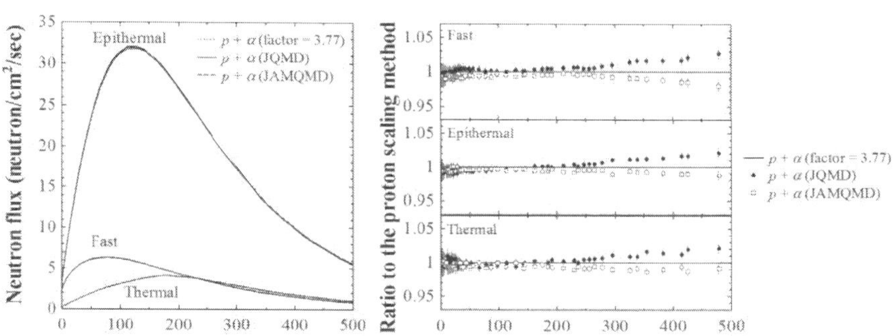

Figure 5 .(a) Distributions of neutron fluxes in the lunar subsurface calculated by the proton scaling method and alpha particle interaction models. Note that all of the patterns of calculations overlap in the individual distributions of fast, epithermal, and thermal neutrons, respectively. (b) Comparisons of neutron fluxes calculated at 55 points from the lunar surface to the first 3 m by the alpha particle interaction models of JQMD and JAMQMD with that by the proton scaling method.

5. DISCUSSION

5.1 Correlation between the scale factor and solar cycle

We have determined the scale factor for estimating the neutron production from alpha particles in the lunar subsurface at φ = 600 MV, or around a period of average activity of the solar cycle. In this subsection, the validity of the scale factor is verified at the solar minimum (φ = 300 MV) and solar maximum (φ = 1000 MV) of solar cycle activity. The calculated results at φ = 300 and 1000 MV are shown in Fig. 6. We found that a scale factor of 3.77 reproduced neutron production from alpha particles for the cases of both φ = 300 and 1000 MV within an accuracy of 1% (standard deviation). Therefore, the scale factor can be applied throughout the entire period of solar cycle. When we use our estimated scale factor of 3.77, neutron production from the GCR particle (proton + alpha particles) increases by a factor of 1.29, 1.27, and 1.26 for φ = 300, 600, and 1000 MV, respectively, compared to the one from only protons.

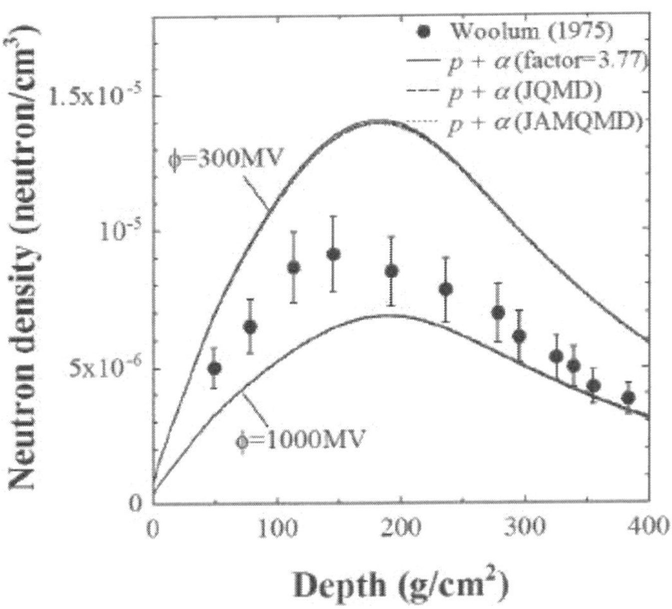

Figure 6. Comparisons of calculated neutron density profiles by the proton scaling method and the alpha particle interaction models (JQMD and JAMQMD) in the periods of minimum and maximum activities of the solar cycle.

5.2 Reliability of transport models

As shown in Fig. 3, our calculation successfully reproduced the LNPE neutron density profile data within the measurement error of 15%. However, there is still room for improvement of both the simulations and the evaluation of the measurements. For example, Woolum *et al.*(1975) suggested that the LNPE data may have 10–30% systematic errors that result from the different observer-dependent criteria for nuclear track recognition. We therefore discuss possible options for improvement in this and the next subsection. We also discuss the reliability of our calculation from the points of uncertainty of the transport model and the GCR spectra.

In our calculations, the neutron density profile is directly dependent on the correlation of the neutron production cross section and the energy of the GCR particles. Therefore, we calculated the excitation functions of the total neutron production cross sections for protons and alpha particles on the lunar material (aluminum as an example). A thin cubic target made of aluminum, $4 \times 4 \times 4\ \mu m^3$, was established, and the source particles of protons and alpha particles were then generated at the point of 1 cm from the target. The space surrounding the target was a vacuum, and the target was bombarded with pencil and monochromatic beams that were perpendicular to the target surface. Thus, the total neutron production cross section for the incident particles on the Al target was determined from the numbers of source particles and from the neutrons produced. The geometry of this simulation can be found in more detail in Mancusi *et al.* (2007). The cross sections were calculated from 10 MeV/n to 10 GeV/n.

The excitation functions of the total neutron production cross sections produced from protons and alpha particles reacting with aluminum targets are shown in Fig. 7(a) and (b), respectively. Appreciably differences can be observed for the production of both protons and alpha particles. However, the difference in the neutron density profiles between the alpha particle interaction models of JQMD and JAMQMD is extremely small, as shown in the Fig. 3, suggesting that the accuracy of those heavy ion reaction models does not greatly influence neutron production. On the other hand, as can be seen in Fig. 1, the differences in the calculated neutron density profiles between the proton interaction models of Cugnon Free, Cugnon Old, Cugnon New, Isobar, JAM, JQMD, and JAMQMD fluctuated within 20%, which implies that these fluctuations are caused by differences in the calculated energy-dependent cross sections in Fig. 7(a). Consequently, it is important to use an accurate proton interaction model—and not the alpha particle interaction model—to be able to calculate the cross sections at all energy ranges. When the proton interaction models that differ from JAM are used, the calculated results do not agree with the LNPE data. However, all of the calculation patterns agree with the LNPE data within 30% systematic error. Benchmarking and improvement of these models are therefore important. At the present time, we do not have any means to determine the best model for our calculation. In the near future, therefore, our aim must be to carefully benchmark the different models available in PHITS against the measurement of neutron production double differential cross sections for nuclear spallation reactions induced by 0.8, 1.5 and 3.0 GeV protons, respectively (Ishibashi *et al.*,1997). In addition, the Japanese Evaluated Nuclear Data Library High Energy File (JENDL-HE) has recently become available (Fukahori *et al.*, 2002; Watanabe *et al.*, 2005). It has been frequently used to reproduce various experimental values (Sato and Niita, 2006; Sato *et al.*,2008b). We are in progress of incorporating the JENDL-HE into our calculation. These results will be reported in future publications.

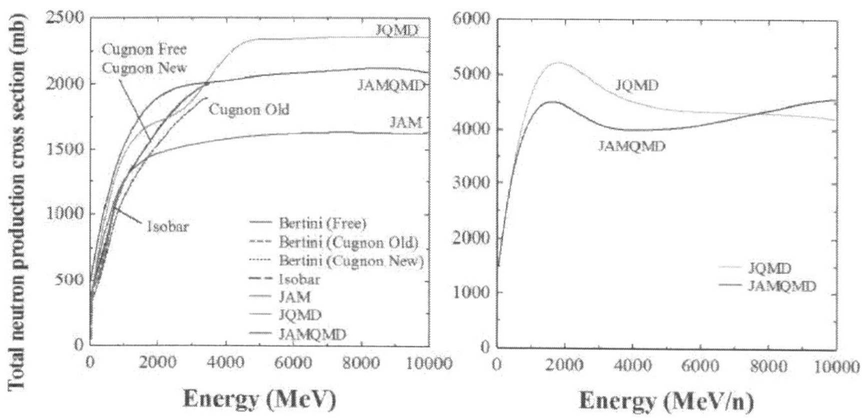

Figure 7. (a) Excitation functions of the total neutron production cross section for protons on the Al target calculated using seven different interaction models. (b) Excitation functions of the total neutron production cross section for alpha particles on the Al target using JQMD and JAMQMD.

5.3 Reliability of the GCR spectra

We emphasize the importance of accurate GCR spectral data for the precise calculation of the neutron production in the lunar subsurface. Since Simpson's first report on the GCR spectra (Simpson, 1983), many researchers have made calculations based on his spectra. However, some recent measurements of GCR spectra performed within the framework of the AMS, CAPRISE98, and BESS balloon experiments reveal a number of important discrepancies from Simpson's data (Alcaraz *et al.*, 2000; Boezio *et al.*, 2003; Shikaze *et al.*, 2007; Hayatsu *et al.*, 2008). For example, according to Hayatsu *et al.* (2008), the total fluxes of protons and alpha particles calculated from the GCR spectra (10 MeV/n-20 GeV/n) by BESS at φ = 600 MV are 3.37 and 0.346 particle/cm^2/s, which are 28 and 82%, respectively, higher than the fluxes used in this paper (see Table 2). Figure 8 shows the GCR proton and alpha particle spectra calculated in our work and by Hayatsu *et al.* (2008) The neutron density profiles using the GCR spectra given by Hayatsu *et al.* (2008) are shown in Fig. 9(a). These are apparently higher than the results from our primary calculations. However, we should note that the results using the GCR spectra by Hayatsu *et al.* also give good agreement with the LNPE experimental values, within the systematic error of 30%, as shown in Fig. 9(b). Thus, the uncertainties in the GCR spectra may cause overestimations or underestimations of neutron production. Therefore, further studies of GCR spectra are needed for a precise estimation of radiation environment on the Moon.

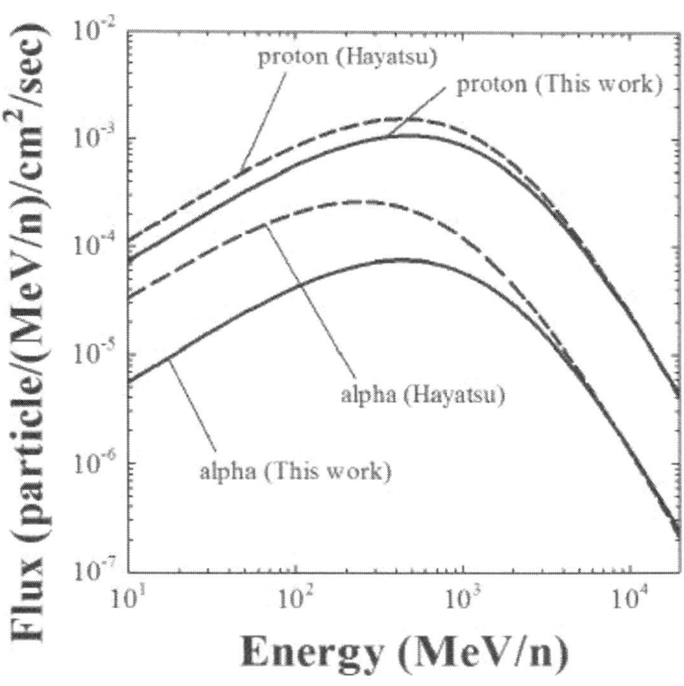

Figure 8 Comparison of the GCR proton and alpha particle spectra used in this work and given by Hayatsu *et al.* (2008).

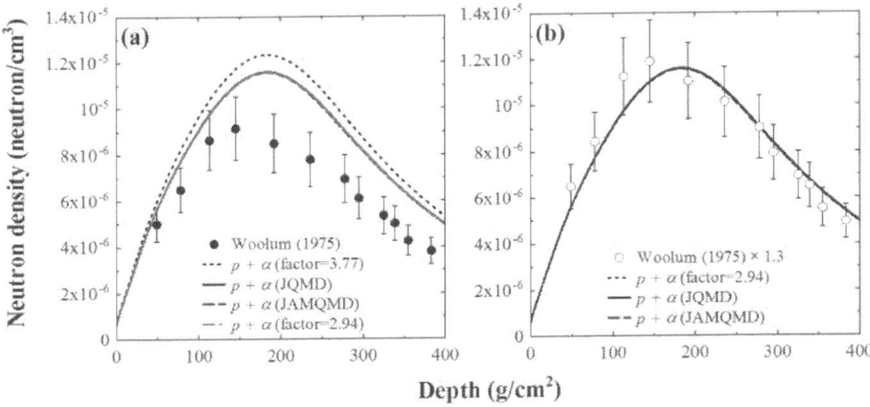

Figure 9. (a) Neutron density profiles in the case of the GCR spectra reported by the BESS balloon experiment. (b) Comparison of neutron density profiles calculated using the GCR spectra by BESS with the experimental data reported by Woolum et al. (1975). The experimental data are multiplied by a factor of 1.3 to take into account the systematic error of 30%.

Another interesting topic is how the difference in the scale factor is dependent on the GCR spectra. In Fig. 9(a), the estimation using the scale factor of 3.77 is described. However, this scale factor does not seem to reproduce neutron production from alpha particles. The best fit to the calculations by the alpha particle interaction models is achieved with a scale factor of 2.94. This is in agreement with a scale factor of 3.0 that was previously estimated by Masarik and Reedy (1996) and Reedy (2009). Thus, we can safely state that variations in the scale factor are largely dependent on the GCR spectra, thereby demonstrating the need for more accurate data on the GCR spectra. On the other hand, it is not necessary to use a scale factor when using a heavy ion interaction model. Therefore, the use of the heavy ion interaction model in the radiation environment on the Moon will give more accurate results when calculating the neutron production in the lunar subsurface.

5.4 Applications and future prospects

The suitability of PHITS for the simulation of neutron spectroscopy was verified by the calculations reported here. In this subsection, we discuss future prospects foradditional calculations in the application fields of lunar science and space dosimetry using the PHITS.

For neutron spectroscopy, the leakage neutron spectrum from the lunar surface is essential. Therefore, we calculated the leakage neutron spectra from the lunar surface at $\varphi = 600$ MV by the proton scaling method and by the alpha particle interaction models in Fig. 10(a). As can be seen, there is a good agreement between all patterns of calculation results. Detailed comparisons are shown in Fig. 10(b), and the total neutron fluxes from the surface are shown in Table 4. It is obvious that no specific differences exist between the proton scaling method and the alpha particle interaction models.

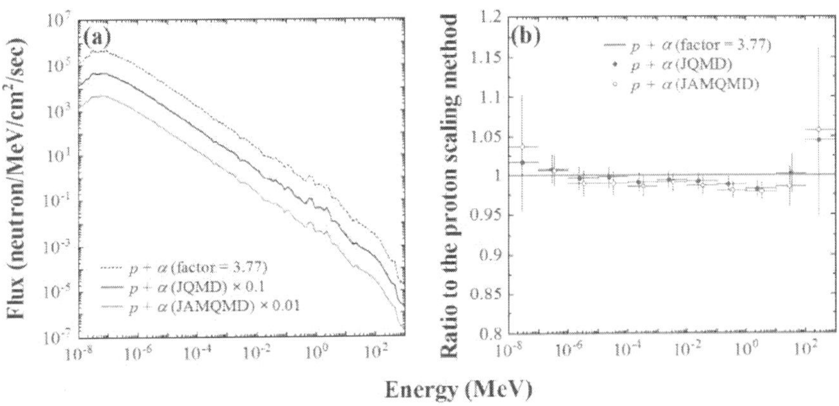

Figure 10. (a) Leakage neutron spectra from the lunar surface calculated by the proton scaling method and the alpha particle interaction models (JQMD and JAMQMD). Note that the results by JQMD and JAMQMD are multiplied by a factor of 0.1 and 0.01, respectively. (b) Comparison of the leakage neutron spectra in various energy ranges by the alpha particle interaction models with that by the proton scaling method. Longitudinal error bars are the statistical error.

Table 4. Total leakage fluxes using different calculation modes. Percentages in parentheses indicate deviations from the proton scaling method.

Calculation mode	Total leakage flux (neutrons/cm^2/s)
Proton + Alpha (factor = 3.77)	6.21
Proton + Alpha (JQMD)	6.17 (−0.6%)
Proton + Alpha (JAMQMD)	6.12 (−1.5%)

Also, the consideration that the contribution of GCR nuclei is heavier than that of the alpha particles may be important. While GCR nuclei comprise only about 1% of GCR particles, they contain more than 10% of the nucleons in the GCR (Shapiro and Silberberg, 1970). However, it is more difficult to obtain accurate spectra of these particles than it is to obtain accurate spectra for the protons and alpha particles. Therefore, further detailed observations of GCR heavy components ($Z > 2$) over a wide energy range are needed. According to our preliminary calculations, if we assume (1) the same parameters and shape of the spectrum as those for the alpha particles used in this work and (2) total fluxes as those given by Simpson (1983), the neutron production in the lunar subsurface by those particles is 2% of that by protons and alpha particles. We can therefore conclude that neutron production from the GCR heavy components ($Z > 2$) is negligibly small.

Heavy ion interaction plays an essential role in terms of space dosimetry. According to Hayatsu *et al.* (2008), who considered only GCR as incident particles, secondary neutron and GCR heavy components ($Z \geq 2$) contribute about 9 and 84%, respectively, to the ambient dose equivalent on the lunar surface. Therefore, it is of importance that the transport and interactions of heavy ions are preformed with accuracy.

Finally, taking into account the availabilities in these applications fields, we conclude that PHITS is a suitable transport code to estimate the radiation environment, including the neutron components, in the lunar environment.

6. SUMMARY

Galactic cosmic rays and solar energetic particles, which are main contributors to the neutron production in the lunar subsurface, include not only protons but also heavy components, such as alpha particles. It is important to be able to precisely estimate the neutron production from such components for the lunar spectroscopy and space dosimetry. Therefore, we calculated neutron production from the interactions of GCR protons and alpha particles with the lunar subsurface using the Particle and Heavy Ion Transport code System.

Neutron density profiles calculated by PHITS successfully reproduced the experimental data from the Apollo 17 Lunar Neutron Probe Experiment. In future studies, further detailed simulations of the emission and transport of radiation produced by cosmic radiation, including secondary products such as neutrons and gamma rays, are essential to determine the relationship between the chemical abundances and the intensities of gamma ray lines for planetary nuclear spectroscopy. This can also be applied to space dosimetry for the design of shielding and protection for a lunar base.

In this work, we found that the best scale factor to estimate neutron production from GCR alpha particles impinging on the lunar surface was 3.77±0.05 using the conventional GCR spectra given by Castagnoli and Lal (1980) and Lal (1985). No major difference was found between neutron production from alpha particles calculated using the proton scaling method and that calculated using the alpha particle interaction models. The main reason for this lack of difference is that the flux of alpha particles is relatively small so the difference in neutron production by alpha particles from that by protons is not a large contributory factor. However, the factor cannot be determined in advance, as it depends on the GCR spectra. Actually, when using the GCR spectra based on the latest observation data by BESS, the best scale factor was 2.94. This is almost in agreement with the estimation by Masarik and Reedy (1996) and Reedy (2009). Thus, the variation in the scale factor is largely dependent on the GCR spectra. This clearly shows the need for more accurate data on the GCR spectra. On the other hand, it is not necessary to use a scale factor when using a heavy ion interaction model. Therefore, we conclude that the use of heavy ion interaction models, such as JQMD and JAMQMD, are important for estimating neutron production from the interactions of GCR with the lunar subsurface.

For future work in this area to be more accurate, two main uncertainties should be decreased. The uncertainties in the GCR spectra need to be decreased because these may cause overestimations or underestimations in the calculations. Improved modeling of GCR spectra has been recently reported by some investigators (Alcaraz *et al.*, 2000; Boezio *et al.*, 2003; Shikaze *et al.*, 2007). Further detailed study on GCR spectra is necessary. On the other hand, it is also necessary to evaluate the accuracy of the different proton interaction models available in PHITS for neutron production. In the near future, we therefore aim to carefully benchmark the different models available in PHITS against the measurement of neutron-production double-differential cross sections for nuclear spallation reactions induced by 0.8, 1.5, and 3.0 GeV protons (Ishibashi *et al*, 1997). We will also compare the calculated results using the Japanese Evaluated Nuclear Data Library High Energy File (JENDL-HE). Those results will be reported in a future publication.

ACKNOWLEDGMENTS

We would like to express our thanks to the developers of the PHITS code for their kind support in its use. We also would like to thank Dr. S. Kunieda

(JAEA, Japan) and Mr. H. Kitamura (NIRS, Japan) for their valuable discussions on the JENDL High Energy Nuclear Data Library and the design of the geometries used in this simulation, respectively. Moreover, we would like to express our thanks to Dr. R. C. Reedy (Planetary Science Institute, USA) and Dr. T. Sato (JAEA, Japan) for their very helpful comments. S. Ota is supported by the Practical Training Program for Doctoral Students at Waseda University. S. Kobayashi is supported by Research Fellowships of the Japan Society for the Promotion of Science (JSPS) for Young Scientist. L. Sihver would also like to acknowledge, with much gratitude, the support from JSPS, which made it possible for him to spend two very fruitful and interesting months as a short-term JSPS Fellow at NIRS, in 2009. This paper is a part of the outcome of research performed under a Waseda University Grant for Special Research Projects (Project number: 2009A-956).

REFERENCES

1. Adams, J. H., M. Bhattacharya, Z. W. Lin, G. Pendleton, and J. W. Watts, The ionizing radiation environment on the moon, Adv. Space Res., **40**, 238–341, 2007.

2. Agostinelli, S. et al., Geant4—a simulation toolkit, Nucl. Instr. Meth. A, **506**, 250–303, 2003.

3. Alcaraz, J. et al, Cosmic protons, Phys. Lett. B, **490**, 27–35, 2000.

4. Boezio, M. et al., The cosmic-ray proton and helium spectra measured with the CAPRICE98 balloon experiment, Astropart. Phys, **19**, 583–604, 2003.

5. Briesmeister, J. F., MCNP-A general Monte Carlo N-particle transport code, version 4B, Rep. LA-12625-M, Los Alamos Natl. Lab., Los Alamos, N. M., 1997.

6. Carrier, W. D., Apollo drill core relationships, Moon, **10**, 183–194, 1974.

7. Castagnoli, G. C. and D. Lal, Solar modulation effects in terrestrial production of carbon-14, Radiocarbon, **22** (2), 133–158, 1980.

8. Dagge, G., P. Dragovitsch, D. Filges, and J. Bruckner, Monte Carlo simulation of Martian gamma-ray spectra induced by galactic cosmic rays, Proc. Lunar Planet. Sci. Conf, **21**, 425–435, 1991.

9. Evans, L. G., R. C. Reedy, and J. I. Trombka, Introduction to planetary remote sensing gamma ray spectroscopy, in Remote Geochemical

Analysis: Elemental and Mineralogical Composition, 167–198, Cambridge Univ. Press, New York, 1993.

10. Fasso, A., A. Ferrari, J. Ranft, and P. R. Sala, Proceedings ofthe IV International Conference on Calorimetry in High Energy Physics, La Biodola, Italy, 1993.

11. Feldman, W. C., W. V. Boynton, and D. M. Drake, Planetary neutron spectroscopy from orbit, in Remote Geochemical Analysis: Elemental and Mineralogical Composition, 213–234, Cambridge Univ. Press, New York, 1993.

12. Fukahori, T., Y. Watanabe, N. Yoshizawa, F. Maekawa, S. Meigo, C. Konno, N. Yamano, A. Yu. Konobeyev, and S. Chiba, JENDL high energy file, J. Nucl. Sci. Technol., Suppl., **2**, 25–30, 2002.

13. Gasnault, O., C. d'Uston, W. C. Feldman, and S. Maurice, Lunar fast neutron leakage flux calculation and its elemental abundance dependence, J. Geophys. Res., **E105**, 4263–4271, 2000.

14. Gustafsson, K., L. Sihver, D. Mancusi, and T. Sato, PHITS simulations of the Matroshka experiment, Adv. Space Res., (submitted).

15. Hasebe, N., E. Shibamura, T. Miyachi, T. Takashima, M. Kobayashi, O. Okudaira, N. Yamashita, S. Kobayashi, T. Ishizaki, K. Sakurai, M. Miyajima, M. Fujii, K. Narasaki, S. Takai, K. Tsurumi, H. Kaneko, M. Nakazawa, K. Mori, O. Gasnault, S. Maurice, C. d'Uston, R. C. Reedy, and M. Grande, Gamma-ray spectrometer (GRS) for lunar polar orbiter SELENE, Earth Planets Space, **60**, 299–312, 2008.

16. Hasebe, N., E. Shibamura, T. Miyzchi, T. Takashima, M. Kobayashi, O. Okudaira, N. Yamashita, S. Kobayashi, M. Hareyama, Y. Karouji, S. Kodaira, K. Sakurai, K. Iwabuchi, K. Hayatsu, S. Nemoto, M. Ebihara, T. Hihara, C. d'Uston, S. Maurice, O. Gasnault, B. Diez, and R. C. Reedy, First results of High Performance Germanium Gamma-Ray Spectrometer onboard Lunar Orbiter SELENE (KAGUYA), J. Phys. Soc. Jpn. Suppl., **78**, 18–25,2009.

17. Hayatsu, K., M. Hareyama, S. Kobayashi, N. Yamashita, M. Miyajima, K. Sakurai, and N. Hasebe, Radiation doses for human exposed to galactic cosmic rays and their secondary products on the lunar surface, Biol. Sci. Space, **22** (2), 59–66, 2008.

18. Hayatsu, K., M. Hareyama, S. Kobayashi, N. Yamashita, K. Sakurai, and N. Hasebe, HZE particle and neutron dosages from cosmic rays on the lunar surface, J. Phys. Soc. Jpn. Suppl, **78**, 149–152, 2009.

19. Ishibashi, K. et al., Measurement of neutron-production doubledifferential cross sections for nuclear spallation reaction induced by 0.8, 1.5 and 3.0 GeV protons, J. Nucl. Sci. Tech, **34** (6), 529–537, 1997.

20. Iwase, H., K. Niita, and T. Nakamura, Development of general purpose particle and heavy ion transport monte carlo code, J. Nucl. Sci. Tech., **39**, 1142, 2002.

21. Lal, D., Theoretically expected variations in the terrestrial cosmic ray production rates of isotopes, in the Last Millenia, Proc. Int. Sch. Phys. Enrico Fermi, **95**, 216–233, 1985.

22. Lawrence, D. J., S. Maurice, and W. C. Feldman, Gamma-ray measurements from Lunar Prospector: Time series data reduction for the Gamma-Ray Spectrometer, J. Geophys. Res., **109**, E07S05, 2004.

23. Mancusi, D., L. Sihver, K. Gustafasson, C. L. Tessa, S. B. Guetersloh, C. J. Zeitlin, J. Miller, L. H. Heilbronn, K. Niita, T. Sato, H. Nakashima, T. Murakami, and Y. Iwata, PHITS—benchmark of partial charge-changing cross sections for intermediate mass systems,Nucl. Instr. Meth. B, **254**, 30–38, 2007.

24. Masarik, J. and R. C. Reedy, Gamma-ray production and transport in Mars, J. Geophys. Res, **101**, 18,891–18,912, 1996.

25. McKinney, G. W., D. L. Lawrence, T. H. Prettyman, R. C. Elphic, W. C. Feldman, and J. J. Hagerty, MCNPX benchmark for cosmic ray interactions with the Moon, J. Geophys. Res., **111**, E06004, 2006.

26. Mitchell, J. K., W. D. Carrier III, N. C. Costes, W. N. Houston, R. F. Scott, and H. J. Hovland, Apollo 17 preliminary report, NASA Spec. Publ., SP-330, 8–1, 1974.

27. Nara, Y., N. Otuka, A. Ohnishi, K. Niita, and S. Chiba, Relativistic nuclear collision at AGS energies fromp + BetoAu + Au with hadronic cascade model, Phys. Rev. C, **61**, 024901, 2000.

28. Niita, K., S. Chiba, T. Maruyama, H. Takada, T. Fukahori, Y. Nakahara, and A. Iwamoto, Analysis of the (N, xN') reactions by quantum molecular dynamics plus statistical decay model, Phys. Rev. C, **52**, 5, 1995.

29. Niita, K., S. Meigo, H. Takada, and Y. Ikeda, High energy particle transport code NMTC/JAM, Nucl. Instr. Meth. B, **184**, 406, 2001.

30. Niita, K., T. Sato, H. Iwase, H. Nose, H. Nakashima, and L. Sihver, PHITS—a particle and heavy ion transport code system, Radiat. Measure., **41**, 1080–1090, 2006.

31. Prael, R. E., Model cross-section calculations using LAHET, in Nuclear Data Evaluation Methodology, edited by C. L. Dunford, 525–534, World Sci., River Edge, N. J., 1993.

32. Prael, R. E. and H. Lichtenstein, User guide to LCS. The LAHET Code System, Rep. LA-UR-89-3014, 76, Los Alamos Natl. Lab., Los Alamos, N. M., 1989.

33. Reedy, R. C., private communication, 2009.

34. Reedy, R. C. and J. R. Arnold, Interaction of solar and galactic cosmic ray particles with the moon, J. Geophys. Res., **77**, 537–555, 1972.

35. Sato, T. and K. Niita, Analytical functions to predict cosmic-ray neutron spectra in the atmosphere, Radiat. Res., **166**, 544–555, 2006.

36. Sato, T., K. Niita, H. Iwase, H. Nakashima, Y. Yamaguchi, and L. Sihver, Applicability of particle and heavy ion transport code PHITS to the shielding design of spacecrafts, Radiat. Measure., 41,1142–1146,2006.

37. Sato, T., L. Sihver, K. Gustafsson, D. Mancusi, and Niita, Shielding design of spacecrafts using PHITS, Am. Nucl. Soc. Trans., **99**, 592, 2008a.

38. Sato, T., H. Yasuda, K. Niita, A. Endo, and L. Sihver, Development of PARMA: PHITS based analytical radiation model in the atmosphere,Radiat. Res., **170**, 244–259, 2008b.

39. Shapiro, M. M. and R. Silberberg, Heavy cosmic ray nuclei, Ann. Rev. Nucl. Part. Sci., **20**, 323, 1970.

40. Shikaze, Y. et al., Measurements of 0.2-20 GeV/n cosmic-ray proton and helium spectra from 1997 through 2002 with the BESS spectrometer, Astropart. Phys., **28**, 154–167, 2007.

41. Sihver, L., Transport calculations and accelerator experiments needed for radiation risk assessment in space, Z. Med. Phys., **18**, 253–264, 2008.

42. Sihver, L. et al., Recent developments and benchmarking of the PHITS code, Adv. Space Res., **40**, 1320–1331, 2007.

43. Sihver, L. et al., Benchmarking of calculated projectile fragmentation cross-sections using the 3-D, MC codes PHITS, FLUKA, HETC-HEDS, MCNPXHI, and NUCFRG2, Acta Astronaut., **63**, 865–877, 2008a.

44. Sihver, L., D. Matthiä, T. Koi, and D. Mancusi, Dose calculations at high altitudes and in deep space with Geant4 using BIC and JQMD models for nucleus-nucleus reactions, New J. Phys., **10**, 105019, 2008b.

45. Sihver, L., T. Sato, K. Gustafsson, V. A. Shurshakov, and G. Reitz, Simulations of the MTR-R and MTR Experiments at ISS, and Shielding Properties Using PHITS, Proc. 2009 IEEE Aerospace Conf., Big Sky, MT, 2009 (peer-reviewed).

46. Sihver, L., K. Gustafsson, D. Mancusi, T. Sato, K. Niita, H. Iwase, Y. Iwamoto, N. Matsuda, H. Nakashima, and Y. Sakamoto, An update of recent PHITS code, Adv. Space Res., (submitted).

47. Simpson, J. A., Elemental and isotopic composition of the galactic cosmic rays, Ann. Rev. Nucl. Part. Sci., **33**, 323–381, 1983.

48. United Nations Scientific Committee on the Efffects of Atomic Radiation (UNSCEAR) report, http://www.unscear.org/docs/ reports/ 1988/1988c_unscear.pdf, 1988.

49. Watanabe, Y. et al., T, Nuclear data evaluations for JENDL high-energy file, Proc. Int. Conf. Nucl. Data. Sci. Tech, 326–331, 2005.

50. Woolum, D. S., D. S. Burnett, and C. A. Bauman, Apollo 17 preliminary science report, NASA, Spec. Publ, SP-330, 18–1, 1973.

51. Woolum, D. S. and D. S. Burnett, In-situ measurement of the rate of 235U fission induced by lunar neutrons, Earth Planet. Sci. Lett, **21**, 153–163, 1974a.

52. Woolum, D. S. and D. S. Burnett, Lunar neutron capture rates and surface mixing of the regolith, Abstr. Lunar Planet. Sci. Conf., **5**, 848–850, 1974b.

53. Woolum, D. S., D. S. Burnett, M. Furst, and J. R. Weiss, Measurement of the lunar neutron density profile, Moon, **12**, 231–250, 1975.

54. Yamashita, N., N. Hasebe, T. Miyachi, M. Miyajima, K. Sakurai, M. Kobayashi, O. Okudaira, H. Yamamoto, E. Shibamura, Y. Uchihori, and M. Kase, Energy spectra of prompt gamma rays from Al and Fe thick targets irradiated by helium and proton beams: Concerning planetary gamma-ray spectroscopy, J. Phys. Soc. Jpn., **75**, 054201, 2006.

CHAPTER 10

2D structural and thermal models in southeastern Norway based on the recently drilled Årvollskogen borehole and 2D density, magnetic and thermal modelling

Yuriy Petrovich Maystrenko*, Harald Kristian Elvebakk, Guri Venvik Ganerød, Ole Lutro, Odleiv Olesen and Jan Steinar Rønning

ABSTRACT

Background

We have made an attempt to understand the main mechanism which controls the conductive heat transfer in the Årvollskogen borehole. This has been done in order to determine the 2D subsurface temperature distribution within the deep-seated crystalline rocks and, therefore, to estimate the geothermal potential in the Moss area near Oslo.

Methods

An integrated 2D density, magnetic and conductive thermal analysis has been performed in order to recognise the major structural features and thermal pattern of the crystalline crust.

Results

Based on 2D density and magnetic modelling, a 2D structural model has been constructed for the Moss area. This 2D model has been used during the 2D thermal modelling. The results of the 2D thermal modelling demonstrate that a significant decrease of the Earth's surface temperatures during the last glaciations still affects the subsurface thermal field of the study area in terms of reduced temperatures within the uppermost crystalline crust. The modelled temperatures are characterised by almost horizontal isotherms without considerable vertical disturbances, reflecting the predominance of subhorizontal layering within the crystalline crust of the Moss area.

Conclusion

The 2D density and magnetic modelling, with consideration of all available geological and structural data, allows us to reveal the deep structure of the crystalline crust within the Moss area. According to the results of the 2D thermal modelling, the predicted temperatures within the upper crystalline crust are in the range of expected values for this part of Fennoscandia.

KEYWORDS

Density; Magnetic and thermal modelling; Deep structure; Subsurface temperature; Conductive heat transfer; Geothermal potential; Norway

1. BACKGROUND

The ecological aspects of energy production permanently increase the importance of environment-friendly sources of energy, one of which is a geothermal energy. During the last decades, the Geological Survey of Norway (NGU) has studied the crustal structure and the subsurface thermal regime of Norway in terms of both direct measurements and modelling approaches (Ebbing et al. [2007], Olesen et al. [2007], [2013], Pascal et al. [2010a], [2010b], Slagstad [2008], Slagstad et al. [2008a], [2008b], [2009]). As a result, in the spring of 2012, NGU has been asked to join Statkraft Varme AS in a research project with the goal to evaluate the geothermal potential in the town of Moss, which is located near Oslo in southeastern Norway.

An integrated lithosphere-scale, 2D density, magnetic and conductive thermal analysis has been carried out in order to understand some of the structural features and thermal pattern of the crystalline crust within the Moss area. There, the Årvollskogen borehole was drilled in the town of Moss in 2012 (10.7°E, 59.4°N; Figure 1). The 2D density, magnetic and conductive thermal analysis has been carried out along a N-S-trending cross-section, called Line 1 (for location, see Figure 2). The Årvollskogen borehole has revealed the configuration of the uppermost part of the crystalline crust in the Moss area. Thermal well logging (Figure 3) was performed in this borehole at the beginning of January 2013, allowing us to constrain the thermal regime of the subsurface within the uppermost crystalline crust of Line 1. In addition, there is the Fredrikstad borehole at 28 km southeast of the Årvollskogen borehole (Figures 2, 4, 5 and 6). The Fredrikstad borehole shows similar measured temperatures (Figure 3), implying that thermal field along Line 1 can be rather stable without any significant perturbations. The cross-section had purposely not been chosen to run in the geologically more representative E-W direction across the major fault of the Oslo Graben due to a lack of detailed geological and gravity data offshore within the Oslofjord (Figures 4 and 5).

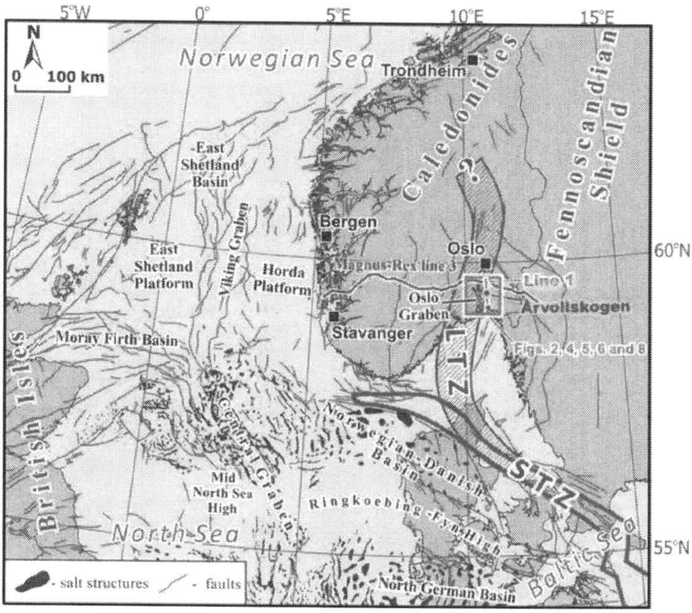

Figure 1. Tectonic setting within northwestern Europe. This map has been compiled after Ziegler ([1990]), Vejbæk and Britze ([1994]), Lokhorst ([1998]), Pharaoh ([1999]), Baldschuhn et al. ([2001]), Sigmond ([2002]) and Maystrenko et al. (Maystrenko et al. [2012]). The location of the Årvollskogen borehole, Line 1 and the Magnus-Rex deep seismic line 3 from Stratford et al. [2009] (magenta line) are also shown. LTZ - Lithospheric Transition Zone and STZ - Sorgenfrei-Tornquist Zone.

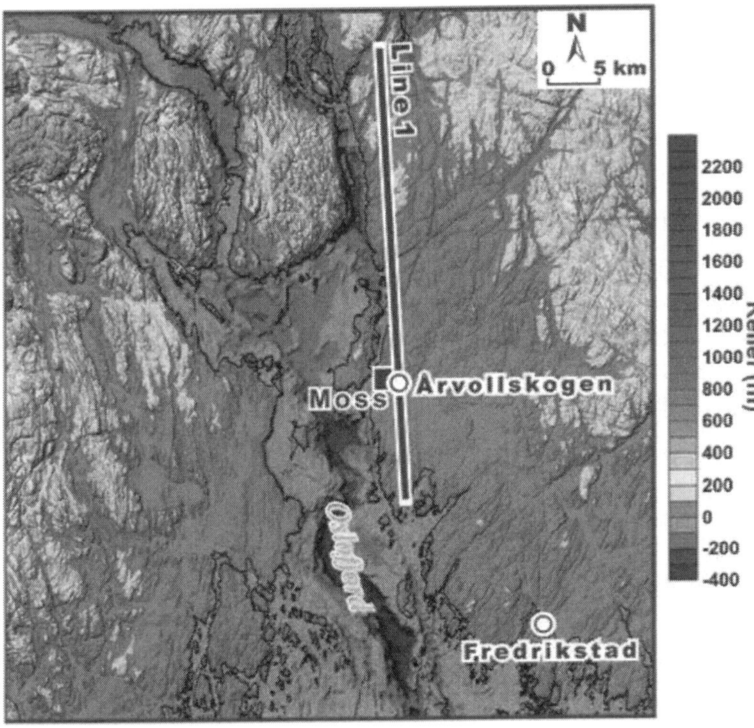

Figure 2. Bathymetry and topography within the study area (Norwegian Mapping Authority).

A total of 108 core samples from the 811 m-deep Årvollskogen borehole have been investigated in the laboratory of NGU to obtain values of density and thermal conductivity for the upper-crustal rocks. In addition, the NGU petrophysical database, called "Petbase" (Olesen et al. [1993]), provides additional measured values of densities, magnetic susceptibilities and thermal properties of crystalline rocks which are exposed at the Earth's surface within the study area.

The 2D density and magnetic modelling approach has been used to obtain a 2D density and magnetic model along Line 1 in order to understand the configuration of the deep structure within the study area. Furthermore, this gravity- and magnetic-consistent, 2D structural model has been used to evaluate the present-day conductive thermal field beneath the study area. Therefore, the main results of our study are related to the 2D configuration of the structural and thermal pattern within the Moss area based on 2D density, magnetic and thermal modelling.

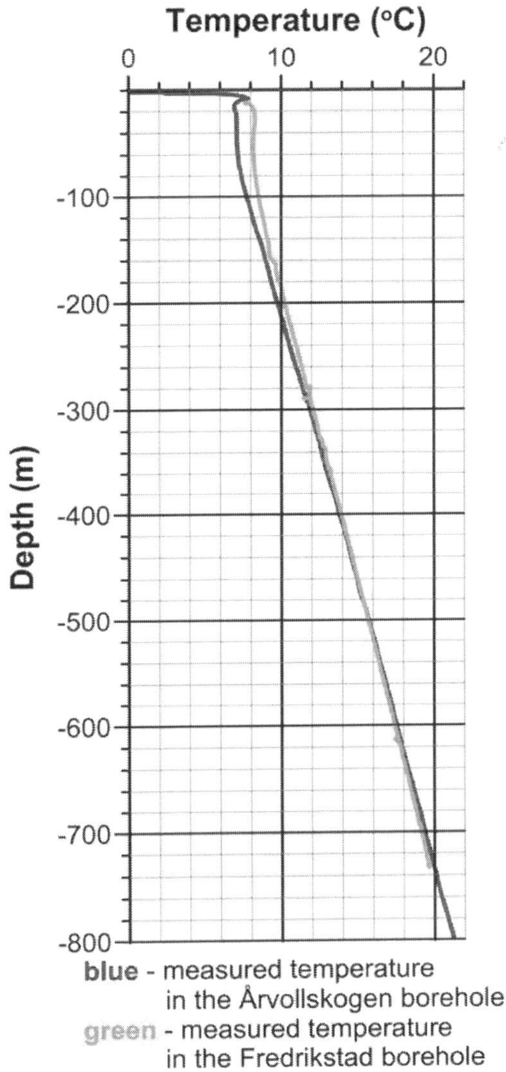

Figure 3. Measured temperatures in the Årvollskogen (blue line) and Fredrikstad (green line) boreholes.

1.1 Geological setting

The topography (Figure 2) in the vicinity of the Årvollskogen borehole, which is located at 36 m.a.s.l., is mainly characterised by comparatively smoothed landforms. Changes in relief are notable only on the western side of the Oslofjord where the relief is 200–400 m above mean sea level. The bathymetry demonstrates that depths to the sea bottom reach more than 400 m within the axial parts of the Oslofjord (Figure 2).

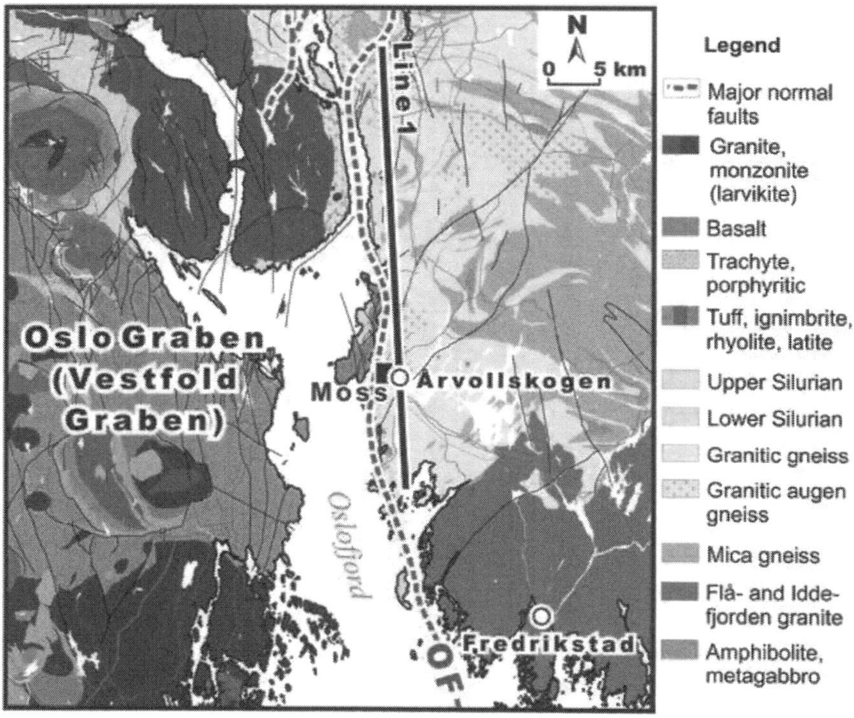

Figure 4. Geological map of the study area (after Lutro and Nordgulen[2008]**).** OF - Oslofjord Fault.

The major tectonic element of the Moss area is the Late Carboniferous-Early Permian Oslo Graben (Figure 1) which contains large volumes of plutonic and volcanic rocks (Ro et al. [1990], Heeremans and Faleide [2004]). These magmatic rocks are well mapped at the surface within the Oslo Graben, whereas some local magmatic intrusions are present within the flanks of this graben structure (Figure 4; Lutro and Nordgulen [2008]).

At the large scale, the Oslo Graben is situated above the lithospheric transition zone (Figure 1) which is characterised by the steep gradient of the eastward-dipping lithosphere-asthenosphere boundary (Medhus [2009], Gradmann et al. [2013]). Tectonically, the Oslo Graben was affected by a major extensional event when a large part of Europe was subjected to a regional rifting event in the Late Carboniferous-Early Permian times (e.g. Heeremans and Faleide [2004]). Localisation of the most prominent extensional deformation structures occurred along the major boundary fault which is located about 4 km west of the Årvollskogen borehole. Consequently, this NNE-SSW-striking graben is structurally bounded by extensional boundary faults. The average amount of vertical displacement is more than 500 m along these boundary faults (e.g. Larsen et al. [2008]).

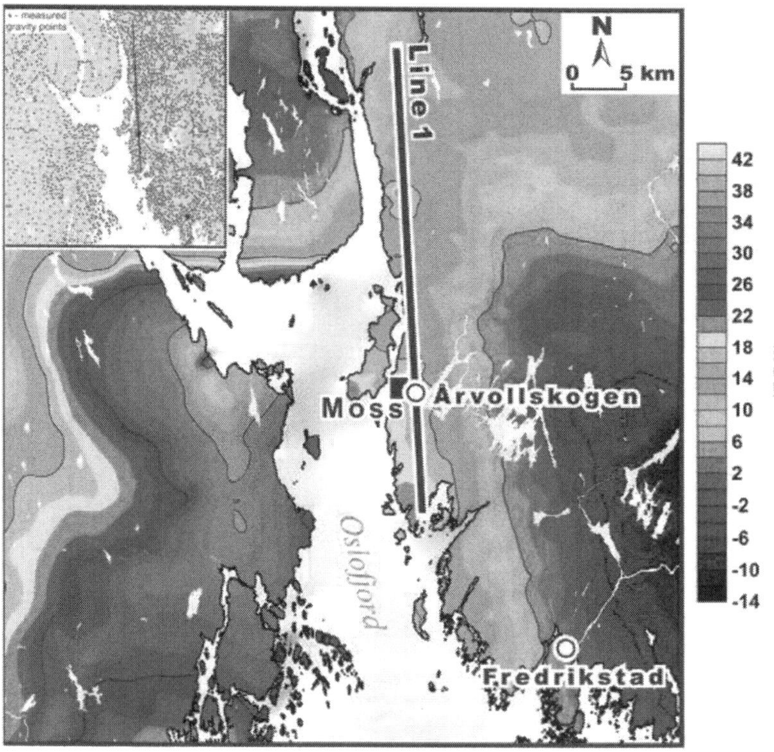

Figure 5 Bouguer gravity anomalies over the study area**.** Gravity data are from the NGU database.

The bedrock geology of the Moss region (Lutro and Nordgulen, [2008]) is shown in Figure 4. The Årvollskogen borehole is located within the eastern flank of the Vestfold Graben, which is the southern segment of the Oslo Graben. Meso- and Neoproterozoic granites and different kinds of gneisses are exposed in the Moss area (Figure 4). There are also Palaeo- and Mesoproterozoic mafic and metamorphic rocks mostly represented by gabbroic intrusions and amphibolites, as well as Late Carboniferous-Early Permian and Early Triassic intrusions. To the west, the Oslofjord Fault separates the Palaeo-Mesoproterozoic crystalline rocks of the eastern flank from the Vestfold Graben, which is filled with sedimentary rocks and the relatively thick, Upper Carboniferous-Lower Permian, synrift volcanic rocks and intrusions, including alkaline basalts, tholeiites, rhomb-porphyry lavas, larvikites and syenites (Figure 4; Neumann et al. [2004], Larsen et al. [2008]).

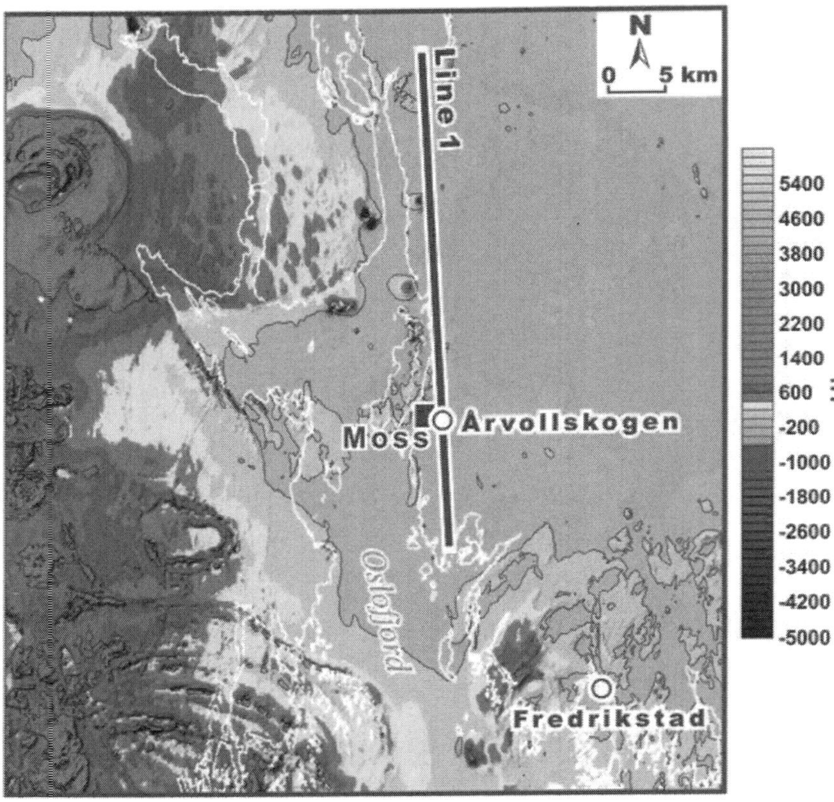

Figure 6. Observed magnetic field over the study area. Magnetic data are from the NGU GEOS fixed-wing survey (Fugro Airborne Surveys [2003]).

The drillcore of the Årvollskogen borehole is very heterogeneous with alternations of rocks of different lithologies. For that reason, the borehole section has been subdivided based only on the dominant lithologies at different depth intervals (Figure 7). The upper interval of 70–350 m of the drillcore is characterised by amphibolites and subordinate metagabbro, whereas below that down to slightly more than 755 m of the drillcore, rocks are mainly represented by granitic to quartz-dioritic mica gneisses and mica gneisses with interlayers of granitic pegmatites. Amphibolites with minor metagabbro constitute the deepest c. 50 m of the drillcore.

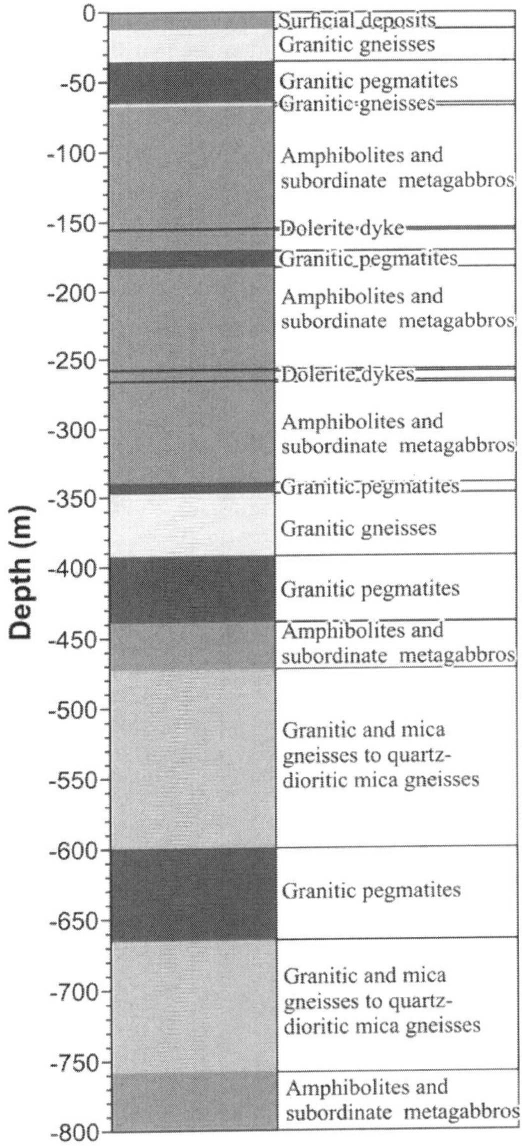

Figure 7. Lithological subdivision of the Årvollskogen borehole. Dominant lithologies are indicated by different colors (modified after Rønning et al. [2013]).

2. METHODS

2.1 Observed gravity and magnetic anomalies

A detailed Bouguer gravity anomaly map from the NGU database (Gellein et al. [1993], Olesen et al. [2010]) has been used in our study (Figure 5). This map of the Moss area has been compiled by use of more than 5000 measured gravity points with a station spacing over land areas between 500 and 3000 m.

The most prominent feature of the observed gravity field over the investigated area (Figure 5) is the presence of a wide positive gravity anomaly over the western side of the Oslofjord. This NE-SW-trending anomaly is located over the axial part of the Oslo Graben. The observed gravity field along Line 1 is characterised by a NW-SE-trending positive anomaly in the south, which is possibly an eastern continuation of the previously mentioned regional gravity anomaly. In addition, a short-wavelength positive gravity anomaly is observed along the northern part of Line 1 (Figure 5).

The observed magnetic field over the study area (Figure 6) has been extracted from the NGU GEOS fixed-wing survey which was carried out in 2003 in the Oslo region by Fugro Airborne Surveys (Pty) Limited (Fugro Airborne Surveys [2003]). The total magnetic field is referred to the International Geomagnetic Reference Field (IGRF).

The most pronounced feature of the observed magnetic field is related to the high-amplitude magnetic anomalies over the western part of the map shown in Figure 6. These prominent positive magnetic anomalies are spatially coincident with the distribution of the Late Carboniferous-Early Permian, extrusive and intrusive/igneous rocks (cf. Figures 4 and 6). Along Line 1, several short-wavelength magnetic anomalies are present, indicating a locally inhomogeneous composition of the crystalline crust within the study area.

2.2 Structural data

Configuration of the deep crystalline crust within the Moss area is based on the results of forward modelling ray-tracing solutions along Line 3 of the Magnus-Rex deep seismic refraction experiment across southern Norway (see Magnus-Rex Line 3 in Figure 1; Stratford et al. [2009], Stratford and Thybo [2011]). Along the Magnus-Rex Line 3, the structure of the crust is characterised by almost horizontal layering with a local uplift of the major crustal boundaries beneath the Oslo Graben (Stratford et al. [2009], Stratford and Thybo [2011]).

In contrast, the structure of the upper crystalline crust is based on the surface geology (Figure 4) extrapolated to great depth and also borehole data.

The Moho topography has been adopted from Stratford et al. ([2009]) who have recently constructed a Moho map beneath southern Norway and southwestern Sweden based on all available deep seismic lines. The extracted part of this regional-scale map is shown in detail in Figure 8a, and reveals that the Moho topography is relatively flat along Line 1.

The lithosphere-asthenosphere boundary beneath the study area (Figure 8b) has been derived from a 3D integrated geophysical modelling study by Gradmann et al. ([2013]), who used the lithospheric geometry from published data sets (Calcagnile [1982], Artemieva [2006], Ebbing et al. [2012]) with adjustments applied to the depth of the lithosphere in order to match seismic velocity constraints. According to the map in Figure 8b, the depth to the lithosphere-asthenosphere boundary is characterised by a steep gradient in the vicinity of Line 1.

2.3 Densities, magnetic and thermal properties

The adjusted values of the density, magnetic susceptibility, thermal conductivity, specific heat capacity and radiogenic heat production are shown in Table 1.

The upper-crustal densities and magnetic susceptibilities are generally taken from NGU's petrophysical database, "Petbase". The average constant values of upper-crustal densities and magnetic susceptibilities for each layer have been assigned according to the location of the sampling points. In addition, densities within the upper crust have been obtained by laboratory measurements of drillcore samples at NGU. The assigned upper-crustal densities vary in the range 2640–2695 kg/m^3. The density for the upper-middle crystalline crust is 2740 kg/m^3 on average. The middle-lower crystalline crust is characterised by an average density of 2830 kg/m^3. A high-velocity lower crustal layer is characterised by velocities which exceed 6700–6800 m/s (Stratford et al. [2009], Stratford and Thybo [2011]). As a result, the average density of this high-velocity layer has been taken to be 3056 kg/m^3. In addition, a density of 2940 kg/m^3 has been chosen for the high-density middle crust. Besides, a marked velocity change from 7100–7400 m/s within the lowest crust to more than 8000–8300 m/s within the uppermost mantle occurs at the Moho (Stratford et al. [2009], Stratford and Thybo [2011]). This prominent change in P-velocities indicates a distinct increase of densities within the uppermost mantle compared to the lower crust. Considering the thermal state of the lithospheric mantle, an average density of the lithospheric mantle has been taken to be around 3222 kg/m^3. The assigned densities of the crystalline rocks (Table 1) are generally consistent with previous gravity studies within this region (Ebbing et al. [2007], Maystrenko and Scheck-Wenderoth [2013]) and are in the range of the empirical velocity-density relationship according to Barton ([1986]).

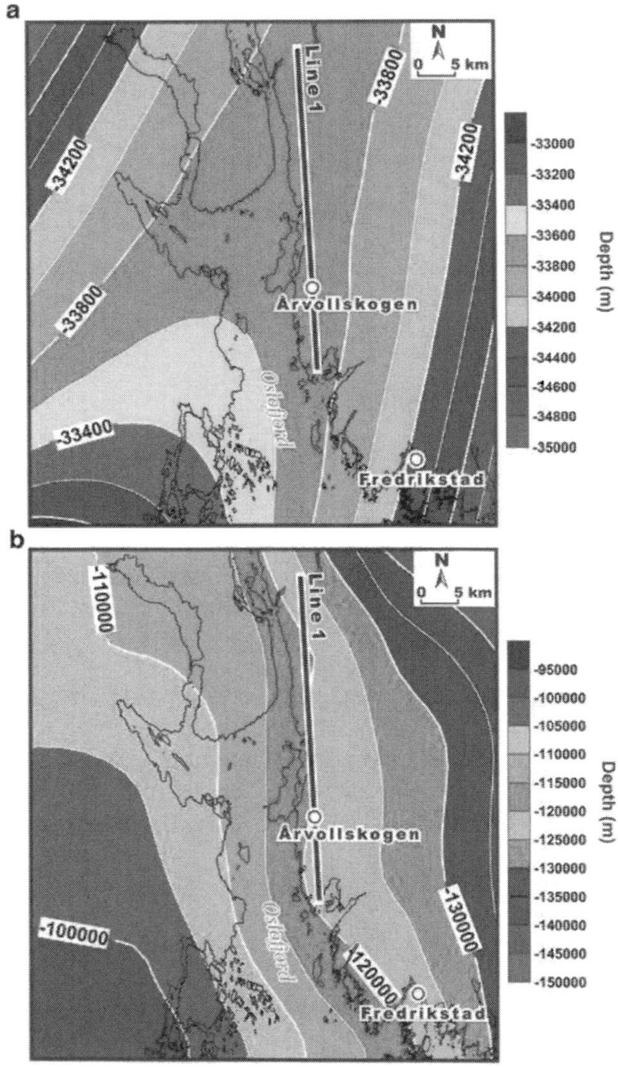

Figure 8. Main deep surfaces. (a) Moho topography and **(b)** depth to the lithosphere-asthenosphere boundary. Data extracted from the regional compilations by Stratford et al. ([2009]) for the Moho and by Gradmann et al. ([2013]) for the lithosphere-asthenosphere boundary. Note that the contour intervals are different.

Table 1. Densities, magnetic susceptibilities and thermal properties (specific heat capacity, thermal conductivity and radiogenic heat production) of the layers within the Moss area used during the 2D density, magnetic and thermal modelling (k_r is the thermal conductivity at room temperature)

Layer of the 2D structural model	Density ρ (kg/m³)	Magnetic susceptibility χ_m (SI)	Thermal conductivity scale value k_r (W/mK)	Specific heat capacity C_p (J/kgK)	Radiogenic heat production S (μW/m³)
Gabbro	2950	0.03-0.09	2.90	780	0.4
Pyroxenite	3100	0.07	4.00	780	0.4
Amphibolite and metagabbro	2850-3050	0.001-0.06	2.90	790	1.2
Granitic to quartzdioritic gneiss	2654-2683	0.00003-0.0095	3.00	800	4.4
Granitic augen gneiss	2640-2695	0.0001-0.01	3.05	800	2.7
Mica gneiss	2645-2670	0.0007-0.008	3.05	800	2.1
Upper/middle crystalline crust	2740	0.001	3.10	930	1.6
Lower/middle crystalline crust	2830	0.001	3.20	1050	0.7
High-density crust	2940	0.003	3.10	1050	0.3
High-density lower crustal layer	3056	0.006	2.85	1100	0.25
Lithospheric upper mantle	3222	-	4.79	1200	0.03

Magnetic susceptibilities for the middle-lower crystalline crust are adopted from the estimations published by Fichler et al. ([2011]) and Ebbing et al. ([2007]).

Thermal properties (specific heat capacity and thermal conductivity) within the upper crust of the study area have been obtained by laboratory measurements of drillcore samples at NGU. The distribution of the measured thermal conductivities on core samples with depth is shown in Figure 9. Moving average values of 20 and 100 m have been applied to all samples with the exception of two samples with a thermal conductivity higher than 4 W/mK. According to the averaging, most of the measured values are around 3 W/mK in the upper part of the borehole and slightly higher (around 3.3 W/mK on average) within the lower part, reflecting the predominance of amphibolites and metagabbros at a depth interval of 100–350 m and gneisses with pegmatites at depths between 500 and 750 m. In particular, the measured thermal conductivity of amphibolites varies from 2.3 to 3.94 W/mK, whereas the thermal conductivity of gneisses and pegmatites are in the ranges of 2.26-3.92 and 2.02-4.87 W/mK, respectively. Moreover, four measurements show that thermal conductivity varies from 3.05 to 3.83 W/mK in the case of dolerites (diabases). Only three samples of breccia have been measured and gave values in range of 3.2-3.96 W/mK.

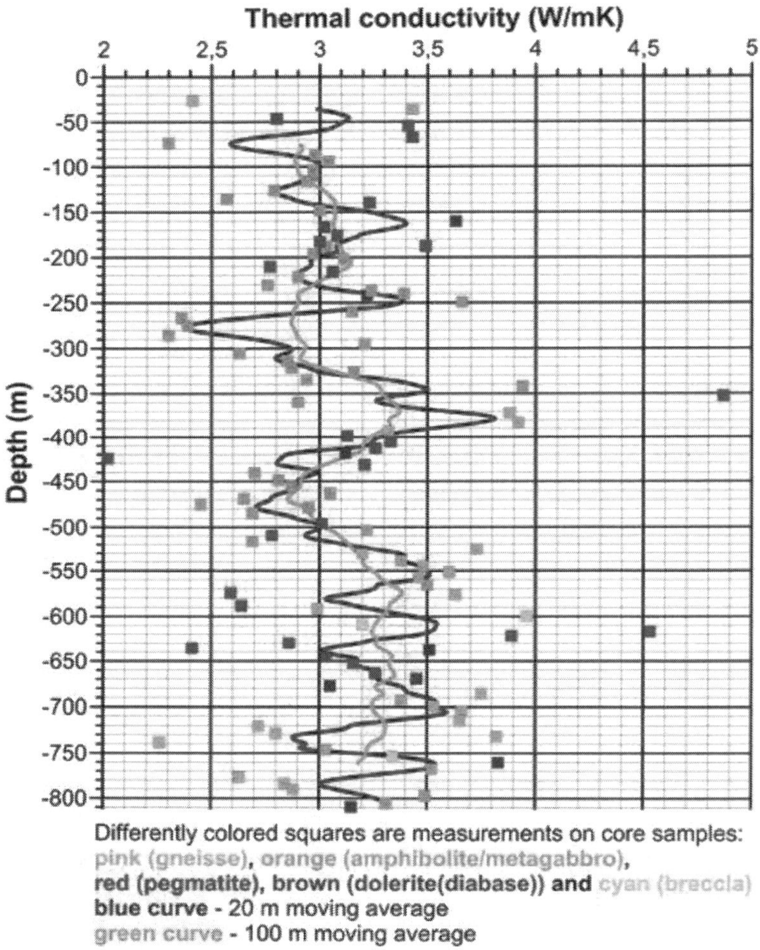

Figure 9. Distribution of thermal conductivities with depth in the Årvollskogen borehole. Measurements on core samples and averaged values are shown. Two non-typical values of thermal conductivities higher than 4 W/mK have been excluded from the averaging calculations.

In this study, the thermal conductivities of rocks have been set to be updated during the modelling in response to changes in calculated temperatures. This has been done in order to monitor changes in the thermal conductivities due to increasing temperature with depth.

Temperature-dependent values of the thermal conductivities for the uppermost crust have been calculated according to the empirical equations (1) and (2) from Sass et al. ([1992]):

$$k(T) = k_o / \left(1.007 + T\left(0.0036 - 0.0072/k_o\right)\right) \quad (1)$$

$$k_o = k_r\left(1.007 + 25\left(0.0037 - 0.0074/k_r\right)\right) \quad (2)$$

where k(T) is the thermal conductivity [W/mK] at temperature T in [°C], k(0) is the thermal conductivity [W/mK] at 0°C, k_r is the thermal conductivity [W/mK] at room temperature of 25°C and T is the temperature [°C].

The empirical equations (3) and (4) from Vosteen and Schellschmidt ([2003]) have been used to calculate the temperature-dependent thermal conductivities for the rest of the crystalline crust where temperatures are higher than 300°C:

$$k(T) = k_o / \left(0.99 + T\left(a - b/k_o\right)\right) \quad (3)$$

$$k_o = 0.53k_r + 1/2\left(1.13(k_r)^2 - 0.42k(25)\right)^{1/2} \quad (4)$$

where k(T) is the thermal conductivity [W/mK] at temperature T [K], k_o is the thermal conductivity [W/mK] at 0°C, k_r is the thermal conductivity [W/mK] at room temperature of 25°C, T is the temperature [K], a and b are constants which vary within the following range: $a = 0.0030 \pm 0.0015$ and $b = 0.0042 \pm 0.0006$.

To define the temperature- and pressure-dependent thermal conductivities within the lithospheric mantle, the empirical equations (5) and (6) from Hofmeister ([1999]) have been used:

$$k(T,P) = k_r(298/T)^a \exp[-(4\gamma + 1/3)\,\alpha(T-298)](1 + K'_oP/K_o) + k_{rad} \tag{5}$$

$$k_{rad} = 4.7\left(0.01753 - 0.00010365T + 2.2451T^2/10^7 - 3.407T^3/10^{11}\right) \tag{6}$$

where k(T,P) is thermal conductivity [W/mK] at temperature T [K] and under pressure [Pa], k_r is the thermal conductivity [W/mK] at room temperature, T is the temperature [K], γ is Grueneisen parameter (γ =1 to 1.4), a is the phonon fitting parameter (a =0.25 to 0.45), α(T-298) is the volume coefficient of thermal expansion as a function of temperature, K_o is the bulk modulus [Pa] (K_o = 261 GPa), K'_o is the pressure derivative of the bulk modulus (K'_o = 5) and k_{rad} is the radiative component of the thermal conductivity, enhanced according to van den Berg et al. ([2001]).

The assigned radiogenic heat production of the upper-crustal layers is based on the average values which are calculated from airborne gamma spectrometry surveys (heat generation map of the Oslo Region in Pascal et al. [2010]) and/or obtained from average heat-production rates for geological units in Norway according to rock sample measurements (Slagstad [2008], Slagstad et al. [2009]). In addition, the results of gamma spectrometry logging in the Årvollskogen borehole have been used to obtain values of the radiogenic heat production derived from uranium (U), thorium (Th) and potassium (K) concentrations. The relationship between radiogenic heat production and concentrations of the radiogenic elements (7) from Rybach ([1988]) has been used to calculate the radiogenic heat production of upper-crustal rocks.

$$S = \rho\,(9.52C_U + 2.56C_{Th} + 3.48C_K) * 10^{-5} \tag{7}$$

where S is the radiogenic heat production [$\mu W/m^3$], ρ is the density [kg/m^3], C_U and C_{Th} are the concentrations of U and Th in ppm, and C_K is the concentration of K in wt.%.

The obtained radiogenic heat production in the Årvollskogen borehole varies mainly from 0.5 to more than 8 $\mu W/m^3$ (Figure 10). These values clearly correlate with the predominant lithology. There is a clear difference in the calculated radiogenic heat production between the upper part of the borehole's section where amphibolites and metagabbro predominate and the lower part of the borehole where granitic to quartzdioritic gneisses are mainly present. Amphibolites are characterised by 1.2-1.5 $\mu W/m^3$ on average, whereas the radiogenic heat production of granitic to quartzdioritic gneisses is much higher, reaching more than 8 $\mu W/m^3$ locally (Figure 10). An average value of 4.4 $\mu W/m^3$ has been assigned to the upper-crustal granitic gneisses, assuming that radiogenic heat production decreases with depth. Local peaks with very high values of the calculated radiogenic heat production are mostly related to the intervals where pegmatites are particularly abundant.

The values of thermal properties for the lithospheric mantle, and the middle and lower crystalline crust are inferred from the results of measurements of rock samples with similar lithological composition (Čermak and Rybach [1982], Clauser [2011]). In addition, the assigned thermal properties of the crystalline crust and the lithospheric mantle have been compared with published values from Wollenberg and Smith ([1987]), Hofmeister ([1999]), Artemieva et al. ([2006]) and Scheck-Wenderoth and Maystrenko ([2008], [2013]). It should be mentioned that the derived density, magnetic susceptibility, specific heat capacity and radiogenic heat production have been set to be constant for each layer (Table 1).

2.4 Methodology

The 2D density, magnetic and thermal modelling have been applied to understand the present-day deep structure and thermal state of the crystalline crust within the Moss area.

The 2D density and magnetic modelling have been carried out by use of the commercial software package Geosoft Oasis montaj. During this modelling, the GM-SYS Profile Modelling module has been used to model the structure of the crystalline crust along a 2D profile (Line 1 in Figures 2, 4, 5 and 6) by interactive changes of the geometries, densities and magnetic properties of the layers, in order to obtain a fit between the observed gravity and magnetic anomalies and the modelled gravity and magnetic responses of the 2D structural model. The advantage of the GM-SYS Profile Modelling is that a great number of 2D layers/bodies with any polygonal shape can be used to simulate the deep structure within the study area. This means that any geometrical complexity of the geological structure can be accurately represented and examined during the 2D density and magnetic modelling.

Therefore, GM-SYS Profile Modelling allows us to create a complex 2D structural model which is consistent with both the observed gravity and magnetic fields.

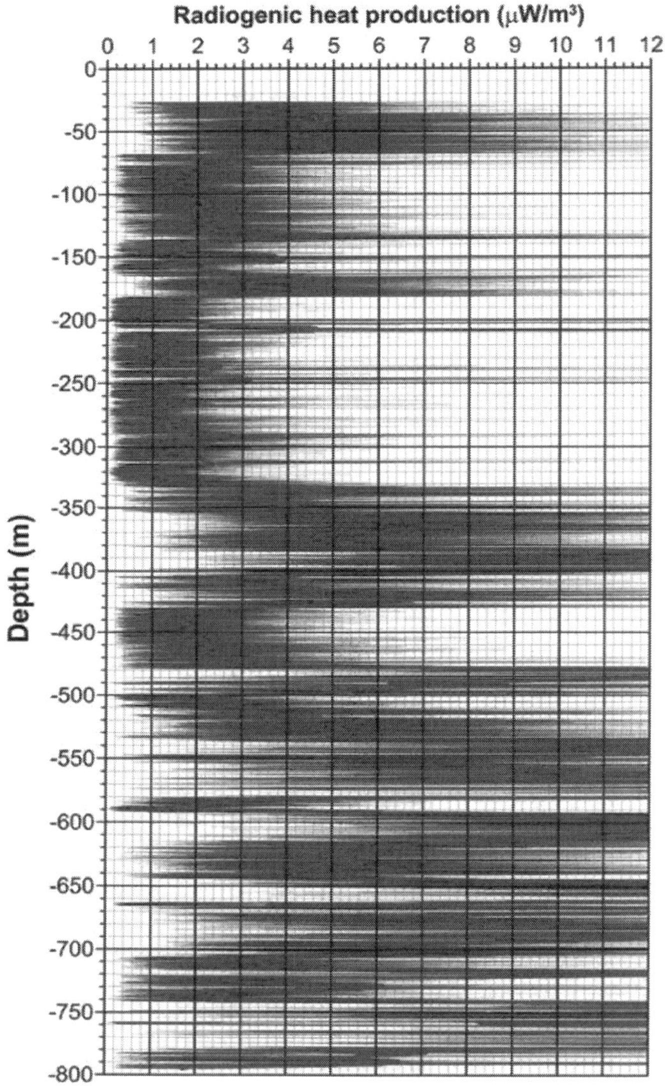

Figure 10. The radiogenic heat production of rocks in the Årvollskogen borehole based on a gamma spectrometry logging. Radiogenic heat production has been calculated according to Rybach ([1988]). The blue plot is the radiogenic heat production obtained from the gamma spectrometry logging, and the red one is calculated by use of running-mean averages at depth intervals of 5 m.

A 2D temperature distribution has been calculated with the help of the commercial software package COMSOL Multiphysics, which is a finite-element analysis software for a variety of physical processes.

During the 2D thermal modelling, the Heat Transfer Module has been used to simulate the stationary and time-dependent heat transfer in solids by heat conduction, which is assumed to be the dominant mechanism of heat transfer at the regional scale within the study area. Therefore, these simulations have been carried out based on physical principles of the conductive 2D thermal field by solving the heat equation (8):

$$\rho C\ (\delta T/\delta t) = \nabla \cdot (k \nabla T) + Q \tag{8}$$

where ρ is the density [kg/m^3], C is the heat capacity [J/kgK], T is the temperature [K], k is the thermal conductivity [W/mK], ∇T is the temperature gradient [K/m], t is the time [s], Q is the heat source (radioactive heat production) [W/m^3], δT is the change in temperature per time interval δt, and $\nabla \cdot$ is the operator giving the spatial variation in temperature.

The heat flow q [W/m^2] has been calculated according to Fourier's law of heat conduction:

$$q = -k \nabla T \tag{9}$$

where k is the thermal conductivity [W/mK] and ∇T is the temperature gradient [K/m].

The 2D thermal modelling has been performed by means of a finite-element method in 2D which is a suitable approach for a relatively complex geological structure. Free triangular elements with different size have been used for the mesh. The mesh has been set to be more detailed within the upper part of the model compared to the deeper part. The reason for this is related to the fact that the resolved structure is relatively detailed within the upper crust.

The solution of the heat equation (8) is sensitive to the thermal boundary conditions. The lateral boundaries are closed to heat transfer, assuming that the temperature gradient is zero across the thermally insulated lateral boundaries. The base of the lithosphere (Figure 8b) has been chosen as a lower thermal boundary, which corresponds to the 1300°C isotherm (Turcotte and Schubert [2002]). For the upper boundary, the annual average air temperature and time-dependent palaeotemperatures at the Earth's surface (Figure 2) have been applied.

2.5 Palaeoclimatic corrections

The 2D thermal modelling has been carried out considering the palaeoclimatic changes of the surface temperature during the last 228,000 years before present (BP). During this time interval, the Moss area was affected by glaciations during the Saalian glacial/Eemian interglacial period (220,000-110,000 years BP) and the Weichselian glacial period (~110,000–10,000 years BP), as well as by the Holocene interglacial period (10,000 years BP to present day). The present-day temperature at the Earth's surface has been taken from the database at the Norwegian Meteorological Institute (Tveito et al. [2000]) and is 6.2°C.

The palaeotemperature at 8000 years BP has been assumed to be about 1°C below the present-day average air temperature. The difference between the present-day temperature and the temperature at 8000 years BP has been inferred from the reconstructed, area-average, mean annual temperature anomalies for northwestern Europe (Davis et al. [2003]).

To reconstruct the older palaeoclimatic history, a recently constructed model of temporal ice-cover variations within the Scandinavian peninsula during the Weichselian glacial period (Olsen [2006], Slagstad et al. [2009], Olsen et al. [2013]) has been used. Figure 11 shows the limits of the ice sheet through time (10,500-110,000 years BP; reproduced from Slagstad et al. [2009]) over southern Scandinavia with the location of the Årvollskogen borehole. At times when the study area was glaciated, a temperature of −0.5°C is assumed at the Earth's surface beneath the ice sheet. This is similar to the assumption used by Slagstad et al. ([2009]) for palaeoclimatic corrections of heat-flow measurements from several sites in Norway. A near-melting point temperature of around 0°C is in agreement with published estimates of the subglacial thermal regime beneath the large polar ice sheets in Antarctica (Pattyn [2010]), which can be taken as comparable analogues of ice sheets developed during the Quaternary glacial cycles in Europe. In contrast, a different scenario has been used for time intervals when the Moss area was free of ice or close to the limit of the ice sheet (Figure 11). Mean annual temperatures of −5°C within the Moss area during the Younger Dryas (Renssen and Isarin [1998]) have been taken as representative temperatures at the Earth's surface for 12,000 years BP and for the time when the borehole site was ice free or located close to the edge of the ice cap.

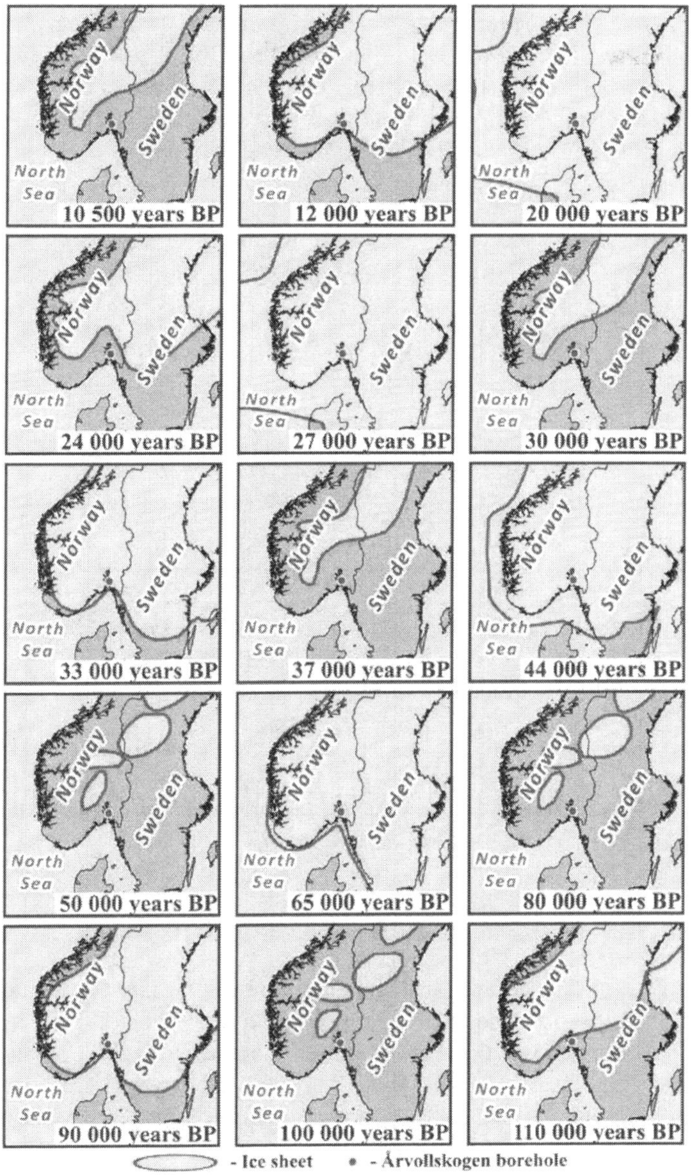

Figure 11. Ice cover during the Weichselian glaciation in southern Scandinavia with location of the Årvollskogen borehole. Limits of ice cover at different time steps are simplified after Olsen ([2006]), Slagstad et al. ([2009]) and Olsen et al. ([2013]). The location of the Årvollskogen borehole is shown by the red circle.

The Weichselian glacial (~110,000–10,000 years BP) and Holocene interglacial (10,000 years BP to present day) palaeoclimatic settings were also applied for the Saalian glacial/Eemian interglacial period (220,000-110,000 years BP), assuming that climatic conditions were similar during the Weichselian glacial/Holocene interglacial and the Saalian glacial/Eemian interglacial periods (Andersen and Borns [1994], Slagstad et al. [2009]). The summarised changes of palaeotemperatures during the Weichselian glacial/Holocene interglacial and Saalian glacial/Eemian interglacial periods are shown in Figure 12 and Table 2. This scenario has been applied during the time-dependent 2D conductive thermal modelling. It should be mentioned that the temperature used for the ice-free time steps is derived from the results of modelling of mean annual temperatures for the Younger Dryas (Renssen and Isarin [1998]) and can theoretically vary in the range of ±8°C (Figure 12 and Table 2) during the Weichselian and Saalian glacial periods.

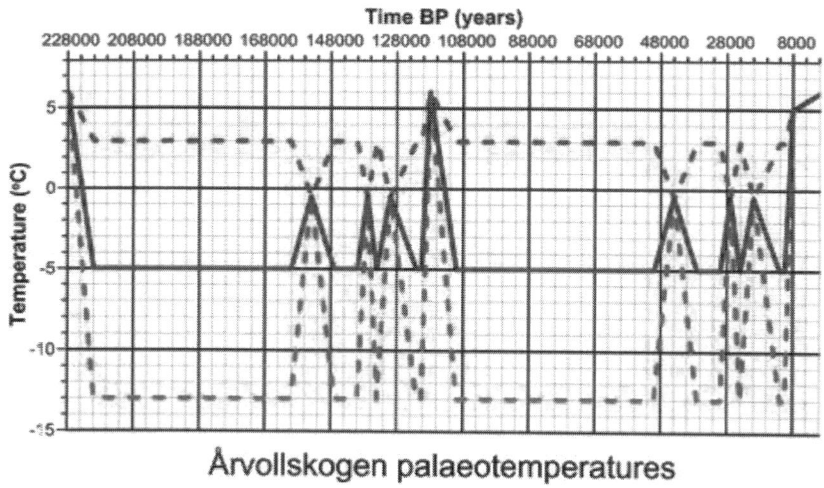

Figure 12. Time-dependent surface temperatures (palaeotemperatures) for the Årvollskogen borehole. Temperatures within the ice-free areas are according to the simulated mean annual temperatures for the Younger Dryas (Renssen and Isarin [1998]). Possible deviations of the used palaeotemperatures (blue line) within the ice-free areas are shown by red (+8°C) and magenta (−8°C) dashed lines.

Table 2. Time-dependent surface temperatures (palaeotemperatures) for the Årvollskogen borehole

Time BP (years)	Average palaeo-temperature (° C)	Deviations of palaeotemperature in the range ±8°C (° C)
0	6	6
8000	5	5
10,500	−5	−13 - 6
12,000	−5	−13 - 6
20,000	−0.5	−0.5
24,000	−5	−13 - 6
27,000	−0.5	−0.5
30,000	−5	−13 - 6
33,000	−5	−13 - 6
37,000	−5	−13 - 6
44,000	−0.5	−0.5
50,000	−5	−13 - 6
65,000	−5	−13 - 6
80,000	−5	−13 - 6
90,000	−5	−13 - 6
100,000	−5	−13 - 6
110,000	−5	−13 - 6
118,000	6	6
120,500	−5	−13 - 6
122,000	−5	−13 - 6
130,000	−0.5	−0.5
134,000	−5	−13 - 6
137,000	−0.5	−0.5
140,000	−5	−13 - 6
143,000	−5	−13 - 6
147,000	−5	−13 - 6
154,000	−0.5	−0.5
160,000	−5	−13 - 6
175,000	−5	−13 - 6
190,000	−5	−13 - 6
200,000	−5	−13 - 6
210,000	−5	−13 - 6
220,000	−5	−13 - 6
228,000	6	6

3. RESULTS AND DISCUSSION

3.1 Results of 2D crustal (2D density and magnetic) modelling

The 2D density and magnetic analyses have been applied along Line 1 (for location see Figures 2, 4, 5 and 6) to reveal the major structural features of the crystalline crust in the vicinity of the Årvollskogen borehole. Results of the 2D density and magnetic modelling demonstrate that the upper crystalline crust is not so homogeneous laterally (Figure 13). In contrast, layers of the middle-lower crust are almost horizontal, reflecting the subhorizontal distribution of the velocities along the Magnus-Rex deep seismic Line 3 (Stratford et al. [2009], Stratford and Thybo [2011]), which is situated slightly to the north of the Moss area (Figure 1).

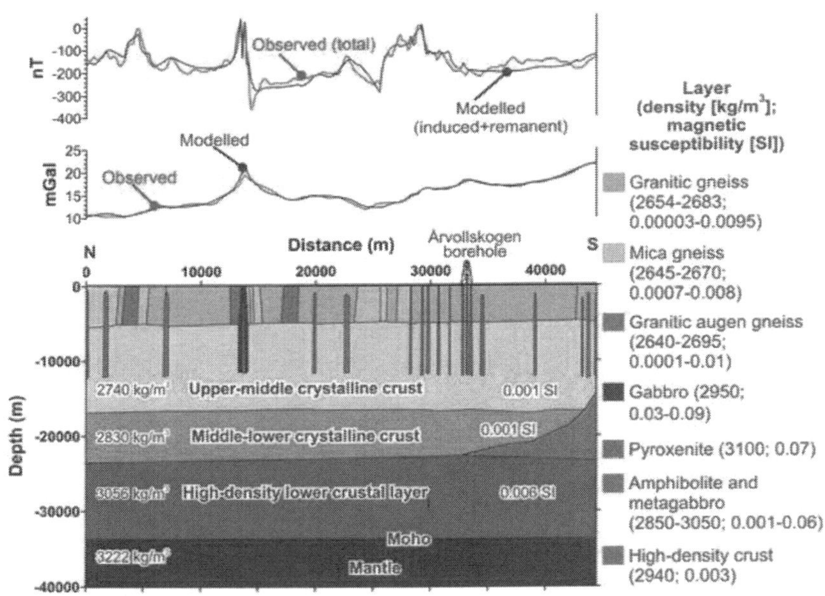

Figure 13. Results of the 2D gravity and magnetic modelling along Line 1. For location of the line see Figures 1, 2, 4, 5 and 6.

On the regional scale, there is a gradual increase in the observed gravity within the southern part of Line 1 (Figure 13). According to the results of the 2D density modelling, this regional-scale positive gravity anomaly is related to the presence of the high-density zone within the middle crust (the violet body in Figure 13) which can be at least partially associated with the basin-scale positive gravity anomaly over the axial part of the Oslo Graben (Figure 5). On the other hand, this regional gravity anomaly does not really correlate with the series of high-amplitude positive magnetic anomalies over the western part of the Oslo region (c.f. Figures 4 and 6) and therefore, to some extent, these anomalies could have a different origin.

An obvious correlation between the high-amplitude magnetic anomalies and areas underline by the Upper Carboniferous-Lower Permian magmatic rocks (c.f. Figures 4 and 6) indicates an upper-crustal origin for the positive magnetic anomalies over the western part of the magnetic map in Figure 6. On the other hand, the long-wavelength positive gravity anomaly over the axial part of the Oslo Graben could be mostly associated with the deep-seated high-density body. The presence of this body is supported by zones of increased seismic velocities within the middle crust of the Oslo Graben according to the Magnus-Rex deep seismic experiment (Stratford et al. [2009]; Stratford and Thybo [2011]). Therefore, the modelled, middle-lower, crustal high-density zone within the southern part of Line 1 (Figure 13) could represent the southeastern extension of the middle-lower crustal high-velocity zone beneath the central part of the Oslo Graben.

The long-wavelength component of the observed gravity anomalies is complicated by a short-wavelength positive gravity anomaly over Line 1 between 10,000 and 20,000 metres (Figure 13). In contrast to the previously described case, there is a clear correlation between this local-scale gravity anomaly and the observed magnetic anomalies (Figure 13, c.f. Figures 4 and 6). Moreover, these short-wavelength gravity and magnetic anomalies are situated over the magmatic intrusive rocks (gabbro and pyroxenite) according to the geological data (Figure 4; Lutro and Nordgulen [2008]). As a result, the magmatic intrusion has been included into the upper and middle crystalline crust during the 2D crustal modelling in order to fit the observed and modelled gravity anomalies. According to the geological data, this vertically elongated intrusive body (the brown body in Line 1 in Figure 13) may have a Late Carboniferous, Permian or Early Triassic age.

In addition to the above-mentioned Upper Palaeozoic-Lower Mesozoic (possibly Early Triassic) intrusion, several smaller intrusions (orange bodies) have been included into the upper-middle crust based on the geological map (Lutro and Nordgulen [2008]) and the potential field data. However, these intrusions are shown rather schematically because they are not so strongly pronounced by the observed gravity and magnetic fields as compared to the previously described intrusion. This is most likely due to the fact that these smaller intrusions are much older and are mainly represented by the Palaeo-Mesoproterozoic (meta)gabbros and metamorphosed intrusions (amphibolites). The Årvollskogen borehole is actually located within the area where the Palaeo-Mesoproterozoic gabbros and amphibolites are exposed at the surface and are also encountered in our borehole. In this particular case, the thickness of the drilled amphibolites is around 300 m within the upper part of the Årvollskogen borehole. The amphibolites and gabbros are definitely characterised by different thermal properties compared to the surrounding granitic gneisses and, for that reason, these rocks have been included into the 2D crustal model in order to enhance the accuracy of the 2D conductive thermal modelling. On the other hand, the detailed geometry of these intrusions is somewhat uncertain.

For the final fit of the observed and modelled gravity and magnetics, the upper crust has been subdivided into several blocks with slightly different densities and varying magnetic susceptibilities. This detailed structure of the upper crystalline crust has been modelled to reproduce the short-wavelength components of the observed gravity and magnetic fields, indicating the presence of upper-crustal blocks with varying densities and magnetic properties. The latter, together with the geological data (Lutro and Nordgulen [2008]), demonstrate that the lithological composition of the upper-crustal layer varies spatially and, therefore, the thermal properties of the upper-crustal rocks can vary as well.

3.2 Results of 2D conductive thermal modelling

The final step of our integrated study concerns 2D thermal simulations, which have been performed to understand the thermal pattern of the subsurface beneath the Moss area. The obtained 2D crustal model along Line 1 (Figure 13), consistent with the gravity and the magnetics, has been used as a structural skeleton during the 2D conductive thermal modelling. A number of testing models have been generated and validated during the 2D thermal modelling in order to simulate a fit between the measured and the modelled temperatures in the Årvollskogen borehole.

The starting point of the 2D thermal modelling was an initial thermal model that has been calculated without consideration of the palaeoclimatic corrections. The annual average air temperature during the period 1961–1990 (Tveito et al. [2000]) near the location of the Årvollskogen borehole has been used for the upper thermal boundary at the Earth's surface (Figure 4). In particular, a constant temperature of 6.2°C has been applied during the initial 2D thermal modelling at the surface within the Moss area. The lithosphere-asthenosphere boundary (Figure 8b; Gradmann et al. [2013]) has been taken as a lower thermal boundary, which corresponds approximately to the 1300°C isotherm (e.g., Turcotte and Schubert [2002]).

The calculated distribution of temperatures with depth in the Årvollskogen borehole is shown in Figure 14. The modelled temperatures have been carefully compared with the measured temperatures in the Årvollskogen borehole. The result of the comparison demonstrates (Figure 14) that, as expected, there is a relatively big misfit between the measured temperature-depth plot and the results of the modelling. This misfit is more than 4°C at a depth of 800 m. Within the uppermost 10–20 m of the borehole, the large misfit is caused by the seasonal changes of present-day surface temperatures.

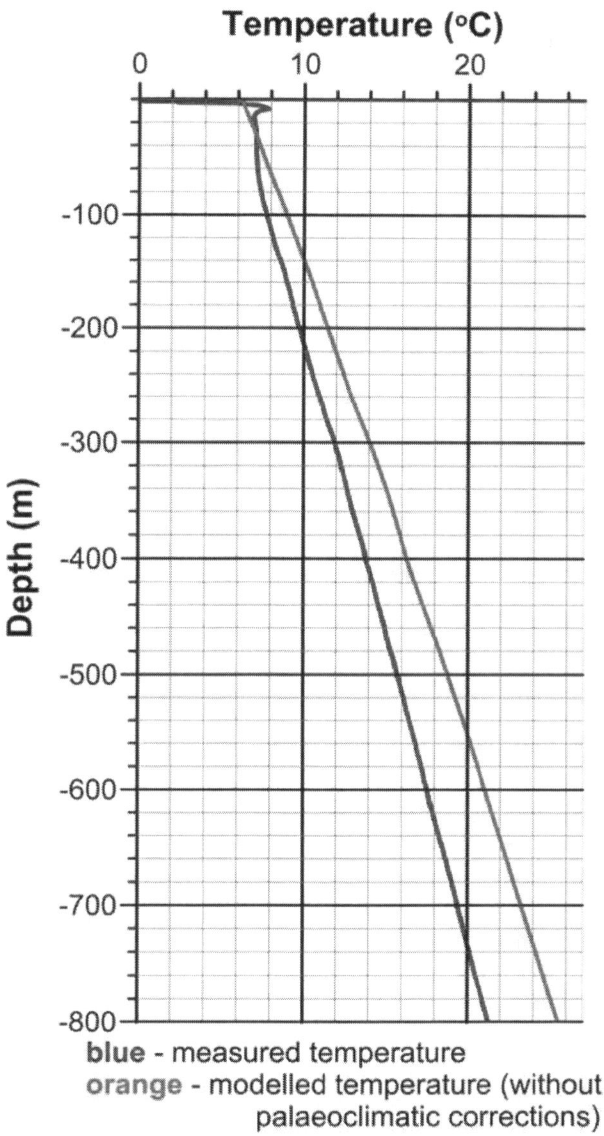

Figure 14. The Årvollskogen borehole: measured versus modelled temperatures. Modelled temperatures (orange line) are calculated without palaeoclimatic corrections.

According to Kukkonen ([1989]), Popov et al. ([1999]), Slagstad et al. ([2009]), Kukkonen et al. ([2010]) and others, the significant changes in the shallow temperatures can be related to palaeoclimatic effects and/or advective cooling due to groundwater flow. In the case of the Moss area, drilling and logging in the Årvollskogen borehole indicate the presence of a number of fracture zones. On the other hand, the continuous character of the measured temperature-depth plot is characterised by an absence of large peaks related to strong groundwater flow (Figure 14). Therefore, the influence of only the palaeoclimatic changes has been considered in the present study, following the modified assumptions suggested by Slagstad et al. ([2009]). The correction made to heat flow for palaeoclimate can reach up to 25 mW/m^2 in the Årvollskogen borehole, and is generally around 20 mW/m^2 according to the European-scale estimation by Majorowicz and Wybraniec ([2010]) within this part of Scandinavia. Palaeoclimatic corrections of heat flow are also relatively high in the British Isles, ranging from 13.4 to 28.2 mW/m^2 (Westaway and Younger, [2013]). Therefore, palaeoclimatic corrections can significantly change the uppermost thermal pattern within the study area.

The 2D thermal modelling has been performed with taking into account a fall in the air and ground temperatures during the Eemian and Weichselian glaciations (Figure 12 and Table 2). The results of the 2D thermal modelling, taking into account the palaeoclimatic corrections, are displayed in Figure 15, showing the 2D distribution of the modelled temperatures, heat flow and thermal gradients. The calculated distribution of temperatures along Line 1 within the Moss area is characterised by almost horizontal modelled isotherms (Figure 15a), reflecting the predominance of subhorizontal layering within the crystalline crust (Figure 13). The modelled heat flow (Figure 15b) and thermal gradient (Figure 15c) along Line 1 are characterised by a subhorizontal pattern without any considerable vertical disturbances. The modelled heat flow locally reaches more than 70 mW/m^2 within the uppermost crust. On the other hand, deep heat flow at the Moho level is only around 33 mW/m^2. Therefore, there is, as expected, a general decrease of heat-flow values with depth. This kind of heat-flow pattern is a result of the fact that the crystalline crust is characterised by an increased content of radiogenic elements compared to the lithospheric mantle (Table 1) and, for that reason, the crystalline crust acts as an additional heat source, contributing essentially to a deep heat flow from the Earth's interior. Furthermore, the calculated heat flow and thermal gradient are characterised by the presence of a clearly recognisable zone with reduced values within the uppermost part of the 2D model in Figures 15b, c. This effect is related to an influence of the palaeoclimatic changes of the surface temperature during the last glaciations (Figure 12). The calculated thermal gradient is generally in agreement with the measured thermal gradient in the Årvollskogen borehole which is on average 19.3°C/km.

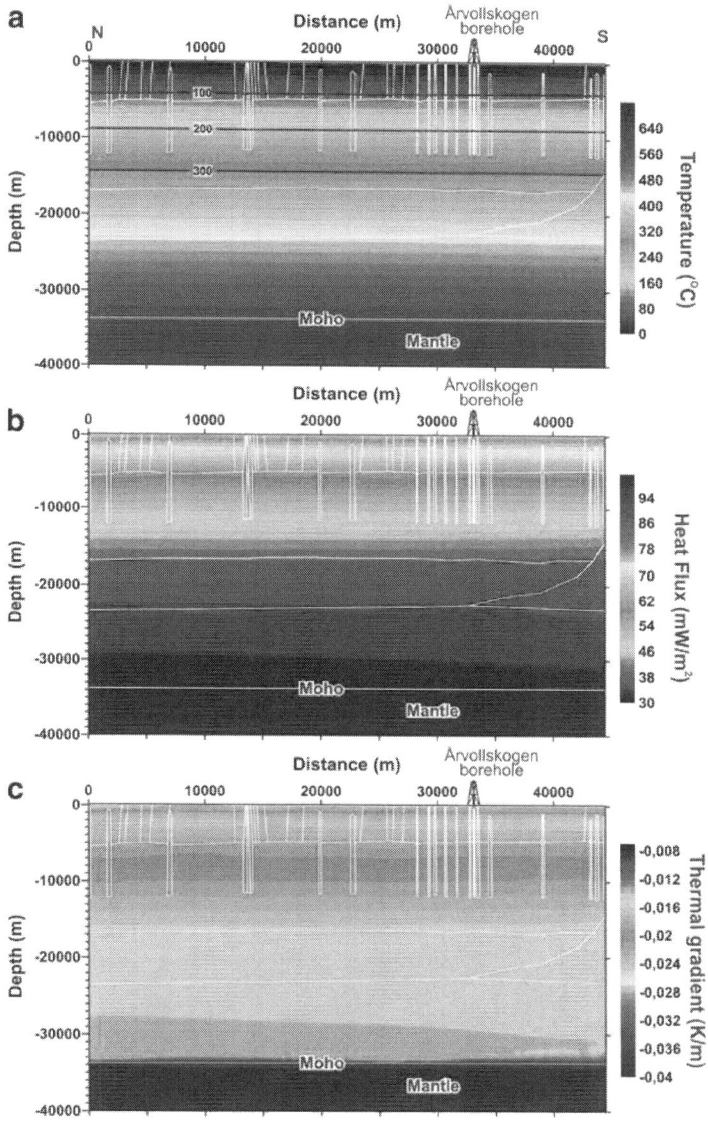

Figure 15. Results of the 2D thermal modelling along Line 1 within the study area. Modelled temperature **(a)**, heat flow **(b)** and thermal gradient **(c)** are calculated considering the palaeoclimatic influence according to the scenario in Figure 12.

The comparison of modelled and measured temperature-depth plots (Figure 16) demonstrates a very good fit in the case when the palaeoclimatic perturbations (Figure 12) have been considered during the 2D thermal simulations. Results of the 2D thermal modelling for greater depths (Figure 17) demonstrate that the modelled temperature is around 120°C at a depth of 5000 m, around 141°C at a depth of 6000 m and 100°C isotherm has been modelled at around a depth of 4130 m.

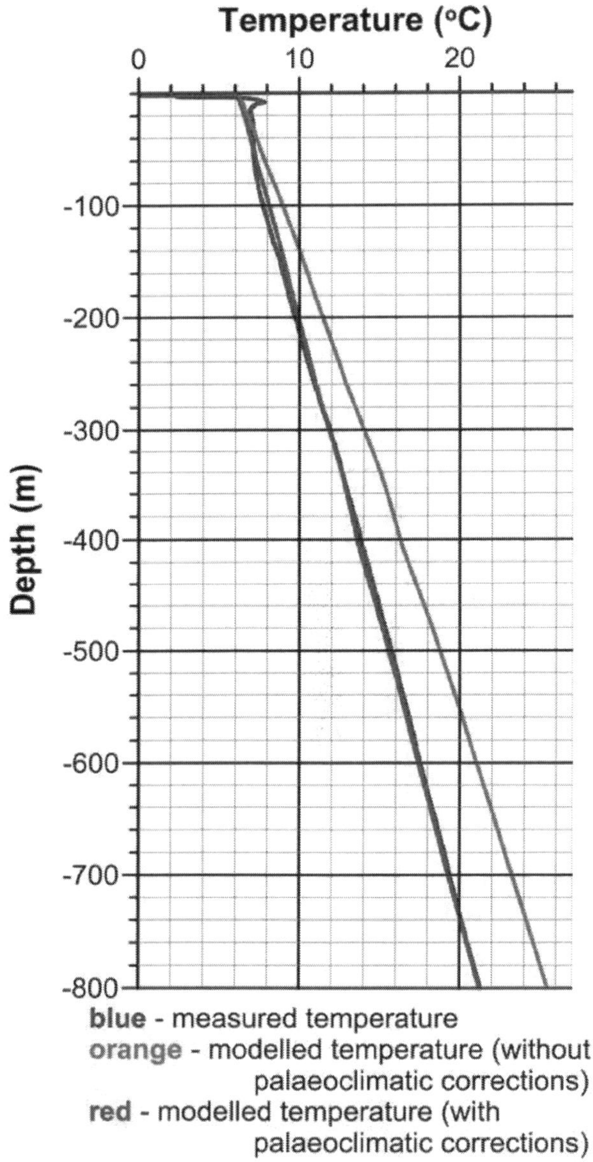

Figure 16. The Årvollskogen borehole: measured versus modelled temperatures. The orange line is assuming steady-state conditions and the red one is with taking into account palaeoclimatic corrections.

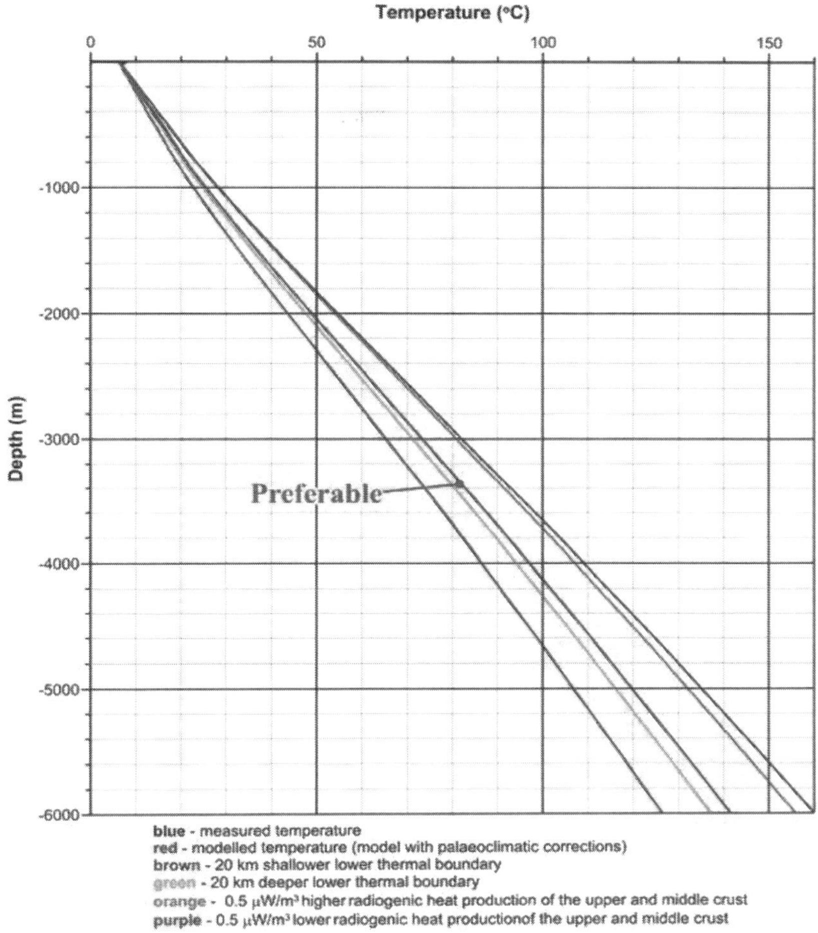

Figure 17. The Årvollskogen borehole: measured versus modelled temperatures with the results of the sensitivity analyses. The preferable modelled temperature-depth plot is shown by the red colour. The brown line is according to a 20 km shallower lower thermal boundary whereas the green line represents a 20 km deeper lower thermal boundary. The orange line corresponds to the model with a 0.5 $\mu W/m^3$ higher radiogenic heat production of the upper and middle crust, and the purple line shows a 0.5 $\mu W/m^3$ lower radiogenic heat production of the upper and middle crust.

4. DISCUSSION

It is obvious that various density, magnetic and structural models can independently reproduce the observed gravity or magnetic fields. Therefore, we have to consider the inherent non-uniqueness of the methodology.

During this investigation, most of the limitations of the approach have been drastically reduced by use of additional constraining structural data. Sensitivity analysis demonstrates that the modelled gravity response is mainly sensitive to modifications of the layer's geometry at the interfaces where a strong density contrast is present, such as internal crustal boundaries, the Moho discontinuity and the lithosphere-asthenosphere boundary. For instance, at a depth of 4500–5500 m, the modelled lithological change in the crust is supported by seismic data (Stratford and Thybo [2011]). At this depth there is a P-wave velocity change from 5560–6030 m/s in the upper layer to more than 6100 m/s below this layer. This difference in seismic velocity is reflected by 2D density modelling which indicates changes of average densities from 2645–2695 kg/m^3 to 2740 kg/m^3 within this part of the upper crust (Figure 13). Furthermore, the depth position of the Moho is well constrained by deep seismic data over large parts of the modelled area (Figure 1, 6a; Stratford et al. [2009], Stratford and Thybo [2011]). The uncertainties in geometry, densities and magnetic properties actually increase with depth as a result of the decreasing resolution of the input data. This point is especially relevant for the depth to the lithosphere-asthenosphere boundary which is the deepest boundary of the model and at the same time the lower thermal boundary. To remove all uncertainties, additional data would be required which we do not have at the moment. Alternatively, the influence of the probable uncertainties in the depth of the lithosphere-asthenosphere boundary can be examined. The modelled temperatures for two possible cases with a 20,000 m shallower and a 20,000 m deeper base of the lithosphere have been calculated in addition to our preferred configuration of the lower thermal boundary in Figure 8b. Furthermore, 2D thermal models with reduced and increased values of the radiogenic heat production within the upper and middle crust have also been evaluated. The reduction and increase of the radiogenic heat production have been chosen to be equal to 0.5 μW/m^3, representing an influence of possible variations of the radiogenic heat production within the upper and middle crust of the Moss area at the regional scale. The position of the lower thermal boundary controls the distance to a deep heat from the Earth's interior, which is one of the main heat sources in the 2D thermal model, and the radiogenic heat production of the upper and middle crust represents the largest source of heat in the model in addition to the internal heat of the Earth. Therefore, the influence of two major factors, such as (1) the vertical position of the lower thermal boundary and (2) the radiogenic heat production within the middle-upper crystalline crust, which control the pattern and magnitude of heat flow, have been examined, allowing us to estimate the first-order uncertainty of the modelled temperature. The results of the sensitivity analysis (Figure 17) indicate that the differences in temperatures strongly increase with depth, reaching deviations of the modelled temperature of about ±12% down to a depth of 6000 m.

5. CONCLUSIONS

The regional structure of the upper crust is relatively complex, reflecting the long-lasting tectonic evolution since Palaeoproterozoic time when the majority of the gneissic crystalline rocks of the Moss area were formed. The results of the 2D density and magnetic modelling indicate that the crystalline crust is characterised by subhorizontal layering which is strongly complicated by the presence of the magmatic intrusions within the upper part of the crystalline crust (Figure 13).

Based on the gravity- and magnetic-consistent structural background, the modelled temperature in the subsurface of the Moss area reflects the interaction between the thermal properties of the rocks and the deep heat flow from the base of the lithosphere. At great depths, the limited information that we have about the thermal properties of rocks implies that the chosen values of the radioactive heat production, thermal conductivity and specific heat capacity can vary within a reasonable range and, therefore, the modelled temperature-depth profile can deviate from the obtained values (Figure 15a). The thermal effects of changes of the thermal properties and the depth position of the lithosphere-asthenosphere boundary have been examined, showing that the deviations of the modelled temperature are around ±12% at a depth of 6000 m. It should be mentioned that if more than one parameter (radiogenic heat production, depth to the lithosphere-asthenosphere boundary, thermal conductivity, etc.) does not fit the predicted range of values, the differences can increase to more than 12% in total. On the other hand, the relative stability of the results is supported by a coincidence between the calculated and the measured temperatures in the Årvollskogen borehole when the palaeoclimatic corrections are considered (Figure 16).

The effect of the palaeoclimatic changes on the surface temperature is relatively great within the Moss area, which was affected by two separate glaciations (Saalian and Weichselian glacial periods) during the last 228,000 years.

In summary, it can be stated that the modelled temperatures in the Moss borehole are within a reasonable range of values which would be expected within the eastern flank of the Oslo Graben, according to available information about the structure of the study area.

NOMENCLATURE

NGU: Geological Survey of Norway (Norges geologiske undersøkelse).

ACKNOWLEDGEMENTS

The authors are grateful to Statkraft Varme AS (and Torbjørn Mehli in particular) for the initiation and support of this study within the framework of the project "Evaluation of the deep geothermal potential in Moss area, Østfold county". We would also like to thank Statkraft Varme AS for permission to publish the obtained results, and colleagues in the NGU laboratory for performing the analyses. We appreciate the helpful reviews and suggestions provided by two anonymous reviewers. Special thanks to David Roberts for improvement of English.

COMPETING INTERESTS

The authors declare that they have no competing interests.

AUTHORS' CONTRIBUTIONS

MYP performed the 2D density, magnetic and thermal modelling, made the scientific analysis and interpretation of the obtained results and wrote the greater part of the text, HKE carried out the data acquisition (thermal, natural gamma and gamma spectrometry logging in the borehole), GVG performed the lithological characterization of the drillcore and sampling. The NGU laboratory was responsible for measurements of densities, thermal conductivities and specific heat capacities, OL assisted with the geology of the study area and described the lithologies in the borehole, OO participated in the 2D modelling and was deeply involved in all discussions, JSR supervised the project, participated in text writing and discussed the results. All authors participated in the discussing the results and in approval of the manuscript.

REFERENCES

1. Andersen BG, Borns JHW: The Ice Age World. Scandinavian University Press (Universitetsforlaget AS), Oslo; 1994.

2. Artemieva IM: Global 1° × 1° thermal model TC1 for the continental lithosphere: Implications for lithosphere secular evolution. Tectonophysics 2006, 416: 245–277.

3. Artemieva IM, Thybo H, Kaban MK: Deep Europe Today: Geophysical Synthesis of the Upper Mantle Structure and Lithospheric Processes Over 3.5 Ga. In European Lithosphere Dynamics, Geol Soc London Mem. Edited by: Gee DG, Stephenson RA. the Geological Society Publishing House, Bath; 2006.

4. Baldschuhn R, Binot F, Fleig S, Kockel F: Geotektonischer Atlas von Nordwest-Deutschland und dem deutschen Nordsee-Sektor - Strukturen, Struckurenwicklung, Paläogeographie. Geologisches Jahrbuch A 2001, 153: 1–88. 3 CD-Rs 3 CD-Rs

5. Barton PJ: The relationship between seismic velocity and density in the continental crust - a useful constraint? Geophys J R Astron Soc 1986, 87: 195–208.

6. Calcagnile G: The lithosphere-asthenosphere system in Fennoscandia. Tectonophysics 1982, 90: 19–35.

7. Čermak V, Rybach L: Thermal Properties: Thermal Conductivity and Specific Heat of Minerals and Rocks. In Landolt-Börnstein, New Series, Geophysics and space research Volume 1: 305–343, subvolume A. Edited by: Angenheister G. Springer, Berlin; 1982.

8. Clauser C: Thermal Storage and Transport Properties of Rocks. In Heat capacity and latent heat, Encyclopedia of Solid Earth's Geophysics 2nd edition. Edited by: Gupta H. 2011, 1423–1431.

9. Davis BAS, Brewer S, Stevenson AC, Guiot J, Contributors D: The temperature of Europe during the Holocene reconstructed from pollen data. Quat Sci Rev 2003, 22(15–17):1701–1716. 10.1016/S0277-3791(03)00173-2View Article

10. Ebbing J, Skilbrei JR, Olesen O: Insights into the magmatic architecture of the Oslo Graben by petrophysically constrained analysis of the gravity and magnetic field. J Geophys Res 2007., 112:

11. Ebbing J, England R, Korja T, Lauritsen T, Olesen O, Stratford W, Weidle C: Structure of the Scandes lithosphere from surface to depth. Tectonophysics 2012, 536–537: 1–24.

12. Fichler C, Odinsen T, Rueslåtten H, Olesen O, Vindstad JE, Wienecke S: Crustal inhomogeneities in the Northern North Sea from potential field modeling: Inherited structure and serpentinites? Tectonophysics 2011, 510: 172–185.

13. Fugro Airborne Surveys:Logistics report, fixed-wing borne magnetic, radiometric and VLF-EM survey in the Oslo region, southern Norway. Report FCR 2241. 2003.

14. Gellein J, Olesen O, Reitan M, Sæther PO: Brukerbeskrivelse - database og prosessering av gravimetriske data. NGU Internrapport1993, 93: 019.

15. Gradmann S, Ebbing J, Fullea J: Integrated geophysical modelling of boundary zone in lithospheric mantle under Norway and Sweden. Geophys J Int 2013, 194(3):1358–1373.

16. Heeremans M, Faleide JI: Permo-Carboniferous rifting in the Skagerrak, Kattegat and the North Sea: evidence from seismic and borehole data. In Permo-Carboniferous Magmatism and Rifting in Europe. Edited by: Wilson M, Neumann E-R, Davies G, Timmerman MJ, Heeremans M, Larsen BT. Geological Society London Special Publications, London, United Kingdom; 2004:157–176.

17. Hofmeister AM: Mantle values of thermal conductivity and the geotherm from phonon lifetimes. Science 1999, 283: 1699–1706.

18. Kukkonen IT: Terrestrial heat flow and radiogenic heat production in Finland, the central Baltic Shield. Tectonophysics 1989, 164:219–230.

19. Kukkonen IT, Rath V, Kivekäs L, Šafanda J, Čermak V: Geothermal studies of the Outokumpu Deep Drill Hole, Finland: Vertical variation in heat flow and palaeoclimatic implications. Phys Earth Planet Inter 2010, 188: 9–25.

20. Larsen BT, Olaussen S, Sundvoll B, Heeremans M: The Permo-Carboniferous Oslo Rift through six stages and 65 million years. Episodes 2008, 31(1):52–58.

21. Lokhorst A: The Northwest European GasAtlas. Netherlands Institute of Applied Geoscience TNO, Haarlem, the Netherlands; 1998.

22. Lutro O, Nordgulen Ø: Oslofeltet, berggrunnskart M 1: 250,000. Norges geologiske undersøkelse. 2008.

23. Majorowicz J, Wybraniec S (2010) New terrestrial heat flow map of Europe after regional palaeoclimatic correction application. Int J Earth Sci doi:10.1007/s00531–010–0526–1 Majorowicz J, Wybraniec S (2010) New terrestrial heat flow map of Europe after regional palaeoclimatic correction application. Int J Earth Sci doi: 10.1007/s00531-010-0526-1

24. Maystrenko YP, Scheck-Wenderoth M: 3D lithosphere-scale density model of the Central European Basin System and adjacent areas. Tectonophysics 2013, 601: 53–77.

25. Maystrenko YP, Bayer U, Scheck-Wenderoth M: Regional-scale structural role of Permian salt within the Central European Basin

System. In Salt Tectonics, Sediments and Prospectivity. Edited by: Alsop GI, Archer SG, Hartley AJ, Grant NT, Hodgkinson R. Geological Society London Special Publications, London, United Kingdom; 2012:409–430.

26. Medhus AB, Balling N, Jacobsen BH, Kind R, England RW: Deep-structural differences in southwestern Scandinavia revealed by P-wave travel time residuals. Nor J Geol 2009, 89: 203–214.

27. Neumann E-R, Wilson M, Heeremans M, Spencer EA, Obst K, Timmerman MJ, Kirstein L: Carboniferous-Permian rifting and magmatism in southern Scandinavia, the North Sea and northern Germany: a review. In Permo-Carboniferous Magmatism and Rifting in Europe. Edited by: Wilson M, Neumann E-R, Davies G, Timmerman MJ, Heeremans M, Larsen BT. Geological Society London Special Publications, London, United Kingdom; 2004:1–10.

28. Olesen O, Reitan M, Sæther PO: Petrofysisk database PETBASE 3.0, Brukerbeskrivelse. 1993.

29. Olesen O, Balling N, Barrère C, Breiner N, Davidsen B, Ebbing J, Elvebakk H, Gernigon L, Koziel J, Lutro O, Midttømme K, Nordgulen Ø, Olsen L, Osmundsen PT, Pascal C, Ramstad RK, Rønning JS, Skilbrei JR, Slagstad T, Wissing B: KONTIKI final report, continental crust and heat generation in 3D. NGU Report 2007, 2007: 042.

30. Olesen O, Brönner M, Ebbing J, Gellein J, Gernigon L, Koziel J, Lauritsen T, Myklebust R, Sand M, Solheim D, Usov S: New aeromagnetic and gravity compilations from Norway and adjacent areas – methods and applications. Petroleum Geology Conference Series 2010, 7: 559–586.

31. Olsen L: Fennoscandia M-LWeichselian glacial variations - much more than just local ice-margin fluctuations. In The 27th Nordic Geological Wintermeeting. Bulletin of the Geological Society of Finland, Special Issue. Edited by: Peltonen P, Pasanen A. Geological Society of Finland, Oulu, Finland; 2006:118.

32. Olsen L, Sveian H, Ottesen D, Rise L: Quaternary glacial, interglacial and interstadial deposits of Norway and adjacent onshore and offshore areas. In Quaternary Geology of Norway. Edited by: Olsen L, Fredin O, Olesen O. Geological Survey of Norway Special Publication, Trondheim, Norway; 2013:79–144.

33. Pascal C, Aarseth M, Ebbing J, Lauritsen T, Lutro O, Olesen O: Quantitative assessment of geothermal resources in the Oslo and Bergen regions. NGU Report 2010, 2010: 054.

34. Pascal C, Barrère C, Davidsen B, Ebbing J, Elvebakk H, Mesli M, Roberts D, Slagstad T, Wissing B: HeatBear Final Report 2010, Basement Heat Generation and Heat Flow in the western Barents Sea - Importance for hydrocarbon systems. NGU Report 2010,2010: 030.

35. Pattyn F: Antarctic subglacial conditions inferred from a hybrid ice sheet/ice stream model. Earth's Planet Sci Lett 2010, 295: 451–461.

36. Pharaoh TC: Palaeozoic terranes and their lithosphere boundaries within the Trans-European Suture Zone (TESZ): a review. Tectonophysics 1999, 314: 17–41.

37. Popov YA, Pevzner SL, Pimenov VP, Romushkevich RA: New geothermal data from the Kola superdeep well SG-3. Can Geotech J1999, 306(3–4):345–366.

38. Renssen H, Isarin RFB: Surface temperature in NW Europe during the Younger Dryas: AGCM simulation compared with temperature reconstructions. Clim Dyn 1998, 14: 33–44.

39. Ro HE, Larsson FR, Kinck JJ, Husebye ES: The Oslo Rift-its evolution on the basis of geological and geophysical observations. Tectonophysics 1990, 178: 11–28.

40. Rønning JS, Dalsegg E, Elvebakk H, Ganerød GV, Lutro O, Maystrenko Y, Olesen O: Evaluation of the deep geothermal potential in Moss area, Østfold county. NGU Report 2013, 2013: 008.

41. Rybach L: Determination of heat production rate. In Handbook of Terrestrial Heat-Flow Determination. Edited by: Hänel R, Rybach L, Stegena L. Kluwer Academic Publishers, Dordrecht; 1988:125–142.

42. Sass JH, Lachenbruch AH, Moses TH Jr, Morgan T: Heat flow from a scientific research well at Cajon Pass, California. J Geophys Res 1992, 97(B4):5017–5030.

43. Scheck-Wenderoth M, Maystrenko Y: How warm are passive continental margins? A 3D lithosphere-scale study from the Norwegian margin. Geology 2008, 36(5):419–422.

44. Scheck-Wenderoth M, Maystrenko YP: Deep control on shallow heat in sedimentary basins. Energy Procedia 2013, 40: 266–275.

45. Sigmond EMO: Geological Map, Land and Sea Areas of Northern Europe, Scale 1: 4 Million, Geological Survey of Norway. 2002.

46. Slagstad T: Radiogenic heat production of Archaean to Permian geological provinces in Norway. Nor J Geol 2008, 88: 149–166.

47. Slagstad T, Barrère C, Davidsen B, Ramstad RK: Petrophysical and thermal properties of pre-Devonian basement rocks on the Norwegian continental margin. Geological Survey of Norway, Bulletin 2008, 448: 1–6.

48. Slagstad T, Midttømme K, Ramstad RK, Slagstad D: Factors influencing shallow (<1000 m depth) temperatures and their significance for extraction of ground-source heat. In Geology for Society, Geological Survey of Norway Special Publication Edited by: Slagstad T. 2008, 99–109.

49. Slagstad T, Balling N, Elvebakk H, Midttømme K, Olesen O, Olsen L, Pascal C: Heat-flow measurements in Late Palaeoproterozoic to Permian geological provinces in south and central Norway and a new heat-flow map of Fennoscandia and the Norwegian-Greenland Sea. Tectonophysics 2009, 473: 341–361.

50. Stratford W, Thybo H: Crustal structure and composition of the Oslo Graben, Norway. Earth Planet Sci Lett 2011, 304: 431–442.

51. Stratford W, Thybo H, Faleide JI, Olesen O, Tryggvason A: New Moho Map for onshore southern Norway. Geophys J Int 2009, 178:1755–1765.

52. Turcotte DL, Schubert G: Geodynamics (2nd edition): Cambridge. Press, Cambridge University; 2002.

53. Tveito OE, Førland E, Heino R, Hanssen-Bauer I, Alexandersson H, Dahlström B, Drebs A, Kern-Hansen C, Jónsson T, Vaarby Laursen E, Westman Y: Nordic temperature maps. DNMI-Report 09/00 KLIMA. 2000.

54. van den Berg AP, Yuen DA, Steinbach V: The effects of variable thermal conductivity on mantle heat-transfer. Geophys Res Lett2001, 28: 875–878.

55. Vejbaek OV, Britze P: Geological map of Denmark 1: 750,000. Top pre-Zechstein (two-way traveltime and depth). Sub- and supercrop map. 1994.

56. Vosteen H-D, Schellschmidt R: Influence of temperature on thermal conductivity, thermal capacity, and thermal diffusivity for different types of rock. Phys Chem Earth 2003, 28(9–11):499–509.

57. Westaway R, Younger PL: Accounting for palaeoclimate and topography: a rigorous approach to correction of the British geothermal dataset. Geothermics 2013, 48: 31–51.

58. Wollenberg HA, Smith AR: Radiogenic heat production of crustal rocks: an assessment based on geochemical data. Geophys Res Lett 1987, 14(3):295–298.

59. Ziegler P: Geological atlas of western and central Europe. 2nd edition. Shell Internationale Petroleum Maatschappij BV, the Netherlands, the Hague; 1990.

CHAPTER 11

Magnetotelluric and temperature monitoring after the 2011 sub-Plinian eruptions of Shinmoe-dake volcano

ABSTRACT

Three sub-Plinian eruptions took place on 26–27 January 2011 at Shinmoe-dake volcano in the Kirishima volcanic group, Japan. During this event, GPS and tiltmeters detected syn-eruptive ground subsidence approximately 7 km to the WNW of the volcano. Starting in March 2011, we conducted broad-band magnetotelluric (MT) measurements at a site located 5 km NNW of the volcano, beneath which the Shinmoe-dake magma plumbing system may exist. In addition, temperature monitoring of fumaroles and hot-springs near the MT site was initiated in July 2011. Our MT data record changes in apparent resistivity of approximately ±5%, along with a ±1° phase change in the off-diagonal component of the impedance tensor (Z_{xy} and Z_{yx}). Using 1-D inversion, we infer that these slight changes in resistivity took place at relatively shallow depths of only a few hundred meters, at the transition between a near-surface resistive layer and an underlying conductive layer. Resistivity changes observed since March 2012 are correlated with the observed temperature increases around the MT monitoring site. These observations suggest the existence beneath the MT site of pathways which enable volatile escape.

KEYWORDS

Magnetotellurics; temperature monitoring; resistivity monitoring; structural change

1. INTRODUCTION

The Kirishima volcanic group, situated at the southern end of Kyushu Island, Japan, consists of more than 20 Quaternary volcanic cones that occur in a ~20 × 20 km area (Fig. 1). Historical eruptions have occurred at three volcanoes from this group: Iwo-Yama, Ohachi, and Shinmoe-dake. Since the 1923 eruption of the Ohachi volcano, all other eruptions (i.e., the 1959, 1962, 1991, 2010, and 2011 eruptions) occurred at Shinmoe-dake volcano, with the 2011 eruptions being the largest on record over the past 290 years (i.e., since a similar-scale eruption in 1717). The 2011 event started with a series of small eruptions on 19 and 22 January, followed by three sub-Plinian eruptions on 26–27 January and the subsequent emergence of lava within the crater. Synchronous with these events, global positioning system (GPS) instruments and tiltmeters detected significant ground subsidence in an area approximately 7 km WNW of Shinmoe-dake (Fig. 1), at an estimated source depth of ~10 km (Nakao et al., 2011; Ueda et al., 2013; Kozono et al., 2013). The estimated volume of this deflation is consistent (within one order of magnitude) with the erupted dense rock equivalent (0.021–0.027 km^3; Nakada et al., 2013). These observations suggest that a magma chamber exists approximately 7 km WNW of Shinmoe-dake volcano, connected by a subsurface plumbing system to the Shinmoe-dake crater prior to the 2011 eruptions. In February 2011, the style of eruption of this volcano has switched from sub-Plinian to Vulcanian, during which a 400-kg volcanic bomb was ejected as far as 3.2 km from the vent. Due to the dangerous nature of these eruptions, people were prohibited from entering within a 4-km radius of the Shinmoe-dake crater. This eruptive activity then gradually subsided, and the latest eruption as of this writing (on 7 September 2011) comprised only an ash emission. After the sub-Plinian eruptions, GPS instruments initially detected a continuous inflation event occurring at approximately the same location as the previously documented syn-eruptive deflation event, although this late-stage inflation has subsequently ceased in late 2011 (Nakao et al., 2012).

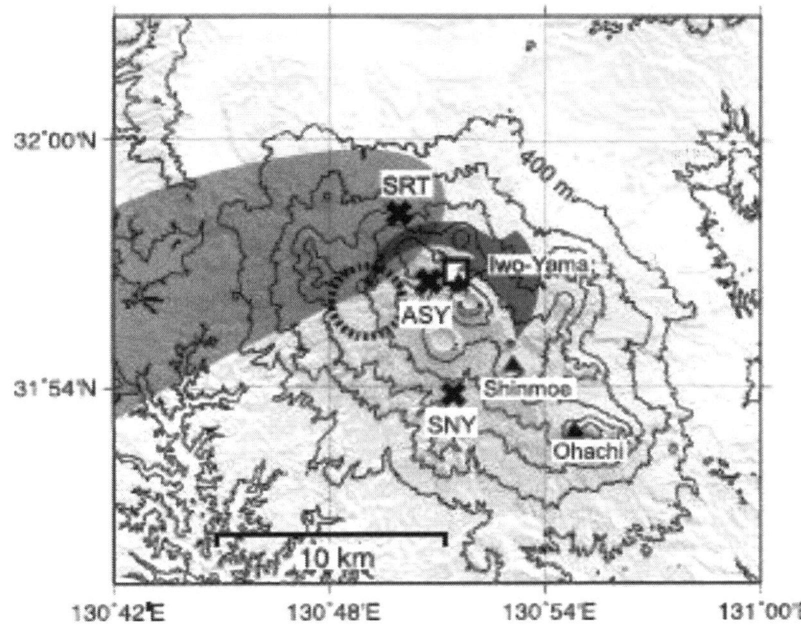

Figure 1. The MT monitoring site at Kirishima volcanic group. The topographic contour interval is 200 m. The square shows the location of the MT site, some 400 m north of Iwo-Yama volcano; crosses with name labels (SRT, ASY, and SNY) indicate the location of temperature measurement sites, and the dashed circle indicates the possible location of the deformation source (Nakao et al., 2011; Ueda et al., 2013; Kozono et al., 2013). The dark shaded area and the arrow indicate the possible location of a deep conductive body and inferred magma movements during the 2011 Shinmoe-dake eruptions, respectively (Aizawa et al., submitted).

Pre-eruptive activity also occurred. For example, GPS data indicate that ground inflation had commenced as early as February 2010 (Nakao et al., 2011). The source of these pre- and post-eruptive inflations is identical to the source region of the syn-eruptive deflation (7 km WNW of Shinmoe-dake and at a depth of 10 km). However, no significant seismic activity or ground deformation occurred between 22 and 26 January, meaning that the timing, magnitude, and location of the three sub-Plinian eruptions on 26–27 January was almost entirely unexpected. Only studies carried out on the erupted material itself suggested the possibility of any impending large magmatic eruptions. In particular, Suzuki et al. (2013) analyzed the volcanic ash erupted on 19 January, and found that it contains 8% juvenile micro-pumice in particle number, which indicates that magma must have reached at a relatively shallow level beneath the Shinmoe-dake crater at that time. On the days of the sub-Plinian eruptions, subtle changes in strain were detected several hours prior to each eruption by a vault-housed extensometer deployed approximately 18 km NW of the Shinmoe-dake crater (Yamazaki et al., submitted).

To understand the current state of Shinmoe-dake volcano and to forecast the next large-scale magmatic eruption, we deployed a number of scientific instruments at distances of 3–10 km from this volcano (Fig. 1) to monitor and measure changes in temperature, and at one station (situated 5 km away) installed equipment to measure electrical resistivity. In this paper, we present the results of that monitoring study, and report on observed temporal variations in electrical resistivity and temperature that occurred after the 2011 sub-Plinian eruptions. Although the imaging of subsurface magma transport is usually quite difficult due to the relatively small volumes of magma involved, some previous seismic studies have indicated that abundant (and observable) structural changes are common in the rocks surrounding the magma conduit. A number of different causes for these structural changes have been proposed, including changes in local stress fields, crack opening/fracture propagation, and/or volatile degassing (Miller and Savage, 2001; Patane et al., 2006; Brenguier et al., 2008).

2. METHODS OF OBSERVATION AND INSTRUMENTAL SETUP

2.1 Magnetotelluric monitoring

The magnetotelluric (MT) method uses the natural electromagnetic field variation to image subsurface resistivity structure, and usually involves measuring two horizontal electric field components (E_x and E_y) and three magnetic field components (B_x, B_y, and B_z) at the Earth's surface, where the subscripts x and y indicate the N-S and E-W directions, respectively. From time series data, the impedance tensor Z and geomagnetic transfer functions T are calculated in the frequency domain as

$$\begin{pmatrix} E_x(f) \\ E_y(f) \end{pmatrix} = \begin{pmatrix} Z_{xx}(f) & Z_{xy}(f) \\ Z_{yx}(f) & Z_{yy}(f) \end{pmatrix} \begin{pmatrix} B_x(f) \\ B_y(f) \end{pmatrix}, \qquad (1)$$

and

$$B_z(f) = (T_x(f)\ T_y(f))\begin{pmatrix} B_x(f) \\ B_y(f) \end{pmatrix}. \quad (2)$$

B_z is the geomagnetic field in the vertical direction. Impedance tensors and geomagnetic transfer functions are complex numbers to calculate mathematically and are generally referred to as MT response functions. Each component of the impedance tensor is usually expressed as an apparent resistivity ρ_a (ohm-m) and phase ϕ (degree), as follows:

$$\rho_{a\,ij}(f) = \frac{1}{5f}|Z_{ij}(f)|^2, \quad (3)$$

And

$$\phi_{ij}(f) = \arctan\left[\frac{\mathrm{Imaginary}(Z_{ij}(f))}{\mathrm{Real}(Z_{ij}(f))}\right], \quad (4)$$

where i,j denotes either x or y. The impedances (Z_{ij}) in Eqs. (3) and (4) are based on units of mV/km and nT, respectively. In the case of homogeneous half-space, Z_{xx} and $Z_{yy} = 0$, $Z_{xy} = Z_{yx}$, and the apparent resistivity is equal to resistivity in the ground.

Using the real part of the geomagnetic transfer function, the induction vector is graphically represented in the xy plane as

$$(-\mathrm{Real}(T_x(f)),\ \ -\mathrm{Real}(T_y(f))). \quad (5)$$

This representation is frequently used to evaluate the dimensionality (D) of the subsurface resistivity structure, because it tends to point to the relatively conductive material in the vicinity of the measurement site. In the 1-D case (the resistivity changes only in the vertical direction) the induction vector becomes (0, 0).

The MT method is useful for investigating static resistivity structures and temporal variations in these structures; however, in practice, this technique is mainly used to investigate static structures. The resistivity structure is deduced by the inversion of impedances and/or geomagnetic transfer functions. The higher-frequency data correspond to shallower resistivity structure, whereas the lower-frequency data correspond to deeper resistivity structure. Because the MT method can be used to conduct deep (a few hundred kilometers) sounding without preparing an artificial electric source, this method has been widely applied in studies aimed at imaging the various scales of resistivity structure. In particular, we have already used the MT method to evaluate the large-scale subsurface resistivity structure present within the seismogenic zone beneath the Japanese Archipelago Island Arc system (Ogawa et al., 2001; Uyeshima et al., 2005; Yoshimura et al., 2009), and to evaluate the smaller-scale (shallow) resistivity structure beneath a number of historically active Japanese volcanoes (Ogawa et al., 1998; Hase et al., 2005; Aizawa et al., 2009b; Kanda et al., 2010).

MT monitoring studies are relatively underused compared with the other techniques that can be employed to survey resistivity structures. Resistivity monitoring generally involves the measuring of DC electric currents, as undertaken near the active craters of Izu-Oshima and Miyakejima volcanoes, Japan (Yukutake et al., 1990; Utada, 2003; Zlotnicki et al., 2003). More recent long-term MT monitoring has been used to determine whether precursory resistivity changes occur prior to earthquakes in the area around Ru-apehu volcano (Hanekop and Simpson, 2006), and in California (Kappler et al., 2010). MT methods have also been used to monitor fluid movements beneath Sakurajima volcano in Japan (Aizawa et al., 2011) and during geothermal exploration in Australia (Peacock et al., 2012).

In the present study, an MT monitoring site was established 5 km NNW of the Shinmoe-dake crater in March of 2011 (Fig. 1). This site is located 400 m NE of the Iwo-Yama volcano, which has a historical record of phreatic eruptions. Prior to 1990, a fumarolic area spanning about 500×500 m also developed on Iwo-Yama, although there is currently no fumarolic activity near this volcano. At the MT site, geomagnetic fields (B_x, B_y, and B_z) and Earth currents (E_x and E_y) were recorded using a Metronix ADU07 system. The sampling frequencies were 32 Hz (00:00–23:50 UT) and 1024 Hz (17:00–19:00 UT). Non-polarized Pb-$PbCl_2$ electrodes were placed at intervals of 40 m in the N-S direction and 35 m in the E-W direction. The contact resistance between the soil and a given electrode was maintained at <6,000 ohm-m, but E_y channel values drifted to >10,000 ohm-m up to December 2011. This drift resulted in drifting of impedance phases to around 80–320 Hz; these values were corrected immediately by changing electrodes in December 2011, and the apparent resistivity across all frequency ranges was not influenced by this changing of electrodes. The electrical power supply was maintained by a series of solar panels (80 W each×six panels), a charging controller, and a number of deep-cycle batteries

(105 Ah each × six batteries). Approximately every two months, we visited the MT observation site to maintain the MT apparatus, replacing the electrodes if necessary and collecting the digital time series data that had been recorded.

2.2 Temperature monitoring

For temperature monitoring in this study, the thermistor sensors and data loggers (TandD TR-5420 and TR-52i, respectively) were deployed at the surface of three separate geothermal zones (labeled SRT, ASY, and SNY in Fig. 1), and the temperature measurements commenced in June 2011. The temperature resolution of the data logger is 0.1°C. Measurement site SRT (Shiratori Jigoku) is situated at an altitude of 820 m, where numerous high-temperature fumaroles and boiling hot-springs exist in an area measuring approximately 200 × 200 m. At this site, the sensor was set into one of the open holes, which was observed to spout out vapor with a hissing sound. Measurement site ASY (Ebino-kogen Asiyu) is situated at an altitude of 1190 m, and is within 1.3 km of the MT monitoring site and Iwo-Yama volcano. At this hydrothermally active site, low-temperature hot-spring water is drawn up from a depth of 60 m, and then used as a footbath for tourists; the ASY site represents an unconfined rather than an artesian well. At a distance of 30 m from this footbath, there exists a small outpouring of hot-spring water and gas bubbles at the surface due to fracturing of the casing pipe, and it is within this surface outpouring that the temperature monitoring sensor was placed at the ASY site. Site SNY (Shinyu) is located at an altitude of 920 m, at a place where a broad (300 × 300 m) zone of fumarolic and hot-spring activity exists. At this site, the sensor was inserted into one of the fumaroles. All temperature monitoring data in this study were recorded every 10 minutes. During our excursions to the MT site about once every two months, we also visited these three temperature-monitoring sites to maintain the apparatus and to collect the digital time series data that had been recorded.

3. DATA

3.1 MT response function

Although there are no electrical power lines near the MT monitoring site, artificial 60-Hz noise (and associated odd-order overtones at 180, 300, and 420 Hz) originating from distant commercial power lines is still observed in the recorded power spectra of all five measured components

(E_x, E_y, B_x, B_y, and B_z). Due to these different frequency noises, the quality of the MT response functions is not good in frequency ranges of >60 Hz. To reduce these noises, we applied a notch filter to the time series. A Fourier transform was applied to the raw data, the result of which was then transformed to a frequency domain. In this step, the amplitudes of the 60 Hz odd-order overtones were reduced to 1/100 of their initial values, with a notch width of 0.1 Hz. Finally, the time series data were recovered by applying an inverse Fourier transform. By applying this comblike notch filter, the quality of MT response functions was significantly improved, especially at frequencies near 60 Hz and at odd-order overtones.

The MT response functions in this study are calculated using the robust estimation code (Chave and Thomson, 2004). At high frequencies (above 0.1 Hz), we also employed remote-reference processing (Gamble et al., 1979) using MT data recorded at Sakurajima volcano (50 km south of Kirishima), where an identical apparatus measures the MT data at that location, based on the same time schedule as our MT apparatus. In the low-frequency range below 0.1 Hz, the geomagnetic data recorded at the Kakioka Geomagnetic Observatory of Japan Meteorological Agency (located 1000 km ENE of Kirishima) were used for remote-reference processing, primarily as the lower-frequency data obtained at proximal stations were of poorer quality than the data obtained at the magnetic observatory. Using these procedures, we obtained the daily MT response functions across a broad frequency range of 320–0.001 Hz.

Figure 2 shows the resultant sounding curves for the MT monitoring site, obtained by stacking the daily impedances, as shown in Fig. 3. The lowest-quality 10% of the data, as defined using the size of error bars on the data, was removed to improve interpretation quality; only the remaining 90% higher-quality data were stacked. Uncertainties within the stacked data were evaluated by an error propagation equation by assuming that the errors in the daily impedance were independent. At frequencies above 1 Hz, the apparent resistivity of Z_{xy} is approximately the same as that of Z_{yx}, but these values gradually diverge from one another as frequencies decrease progressively below 1 Hz. The apparent resistivities of Z_{xx} and Z_{yy} are minor in comparison with the apparent resistivities of Z_{xy} and Z_{yx}, except at lower frequencies where these values converge with those for Z_{xy} and Z_{yx}. These results indicate that the overall resistivity structure can be approximated as 1-D at shallow regions beneath the MT site, but gradually shows 3-D features at deeper levels. The mean values of off-diagonal components (yellow and green circles in Fig. 2) are discussed in Section 4.

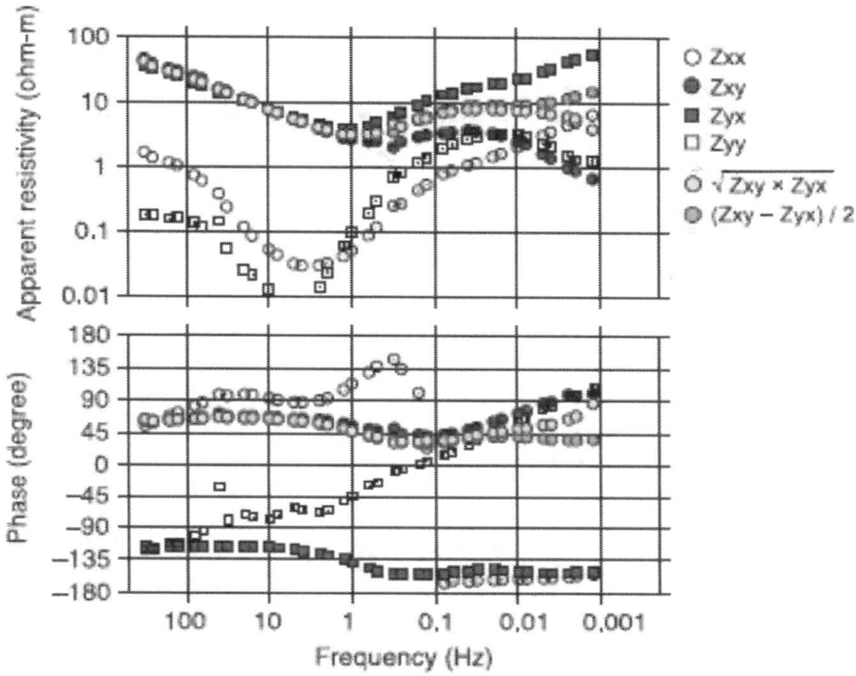

Figure 2. Sounding curves of the MT monitoring site. In the calculation, all data for the observation period are used. Errors, which are shown vertical bars, are smaller than symbols.

Figure 3 shows two representative examples (i.e., for the off-diagonal components Z_{xy} and Z_{yx}) of the observed temporal variations of apparent resistivity and phase measurements at the MT site, each of which displays five selected (i.e., representative) subsets of data that were measured at specific frequencies of 20, 1.25, 0.16, 0.04, and 0.01 Hz. The data for the diagonal components (Z_{xx} and Z_{yy}) are not shown in Fig. 3 because their amplitudes are (for the most part) minor in comparison with those for the off-diagonal data (see Fig. 2), but also because these data-points have relatively large associated error bars. It is clear from Fig. 3 that the quality of the impedance data is better for the higher-frequency datasets than for the lower, and this is because the cumulative number of stacked responses increases with increasing frequency. The poorest-quality data corresponds to about 0.16 Hz, which is known as a dead band where natural electromagnetic signals are relatively weak. Figure 3 also shows that the measurement of Z_{xy} data was poorly conducted in October-December 2011 and in March 2012 due to electrode problems. Neglecting the small-scale (i.e., short-duration) fluctuations in the data, we observe no major changes in apparent resistivity or phase measurements throughout the entire duration of our MT monitoring study. These remarkably stable impedances indicate that the overall resistivity structure beneath the MT site remained stationary throughout this period. However, upon closer inspection, the 20 and 1.25 Hz data show small progressive changes in apparent resistivity and phase measurement.

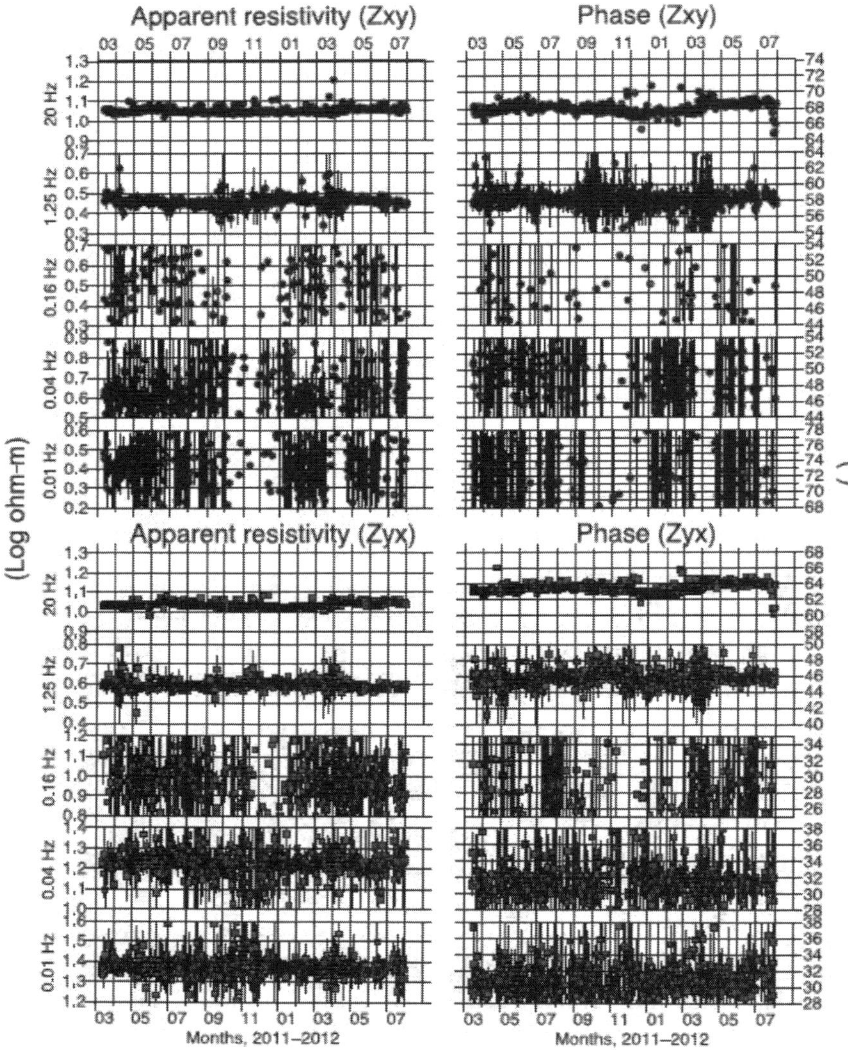

Figure 3. Temporal changes of daily apparent resistivity (left) and phase (right) of off-diagonal components of impedance tensor (Z_{xy}, and Z_{yx}). The error bars are double sided 95 per cent jackknife confidence limits (Chave and Thomson, 2004). Data in the five frequencies (20, 1.25, 0.16, 0.04, and 0.01) are shown.

Figure 4 shows a representative example of the temporal variations of amplitude and orientation observed for the induction vector. These results show that throughout the observation period, these amplitudes remained small and stable, which indicates that the resistivity structure beneath the MT site can be approximated as a first-order 1-D structure during this entire interval of time. We note, however, that in the data from December 2011, the 20 Hz data show a small step-like variation, but the absolute cause or significance of this pattern is unclear.

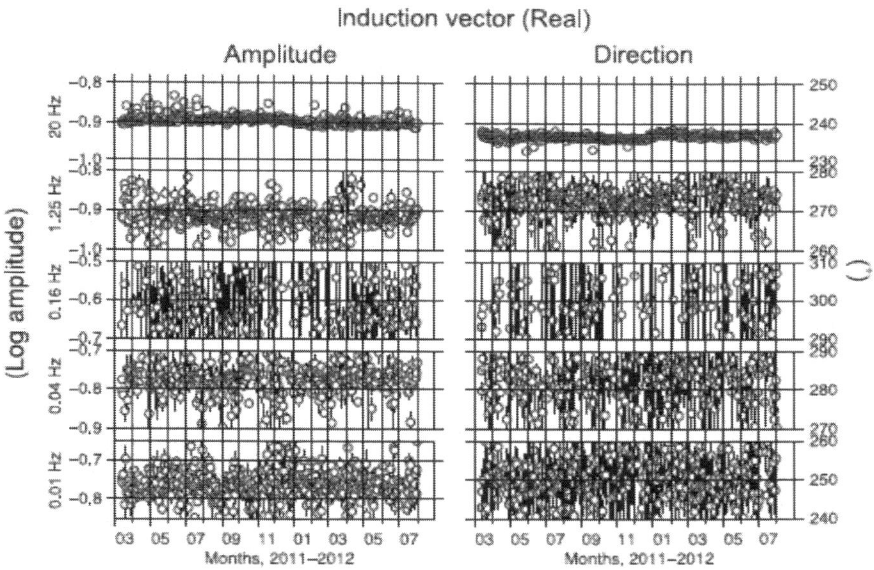

Figure 4. Time change of induction vector (real part). Directions are based on Parkinson criteria, and are shown by counter clockwise from East.

3.2 Temperature

Figure 5 shows the temporal evolution of the temperature data collected during our monitoring study, with each data-point representing a single 1-day average. For comparison, precipitation data and eruption activity for the region are also shown. At the SRT and SNY sites, the instrument sensors were not perfectly inserted into the open fumarole holes from July to November 2011, resulting in a high degree of fluctuation in these data. These sensors were adjusted, placed in a more secure location within the fumarole hole, and rocks were placed to fix the sensor cable on 30 November 2011.

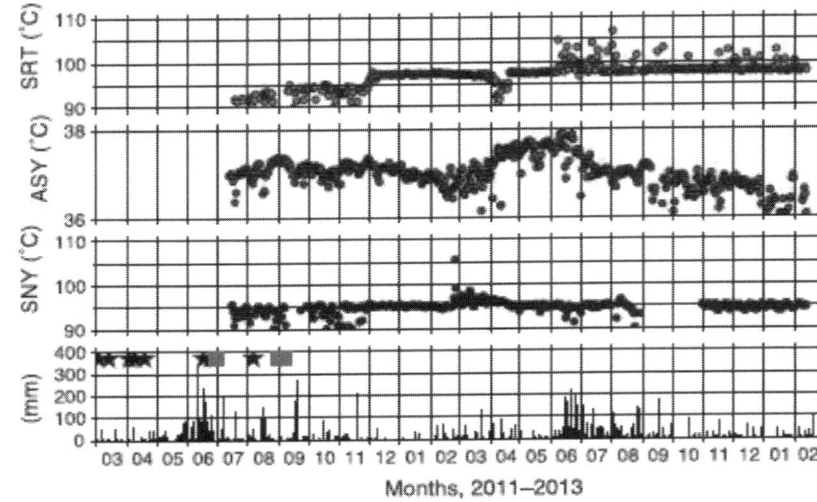

Figure 5. Temperature variation at the monitoring sites. 1 day average values are shown. Sensor positions at SRT and SNY are moved 10 cm on November 30, 2011. For comparison, daily precipitation recorded at a site located 1 km west of ASY is shown at the bottom panel. Stars and grey rectangles show eruptions and continuous ash emission of Shinmoe-dake volcano, respectively.

It is clear that increased rainfall leads to an instantaneous decrease in recorded temperature, based on correlations of daily trends in precipitation and raw temperature data (not shown). However, the one-day time-averaged data in Fig. 5 reveal that the long-term temperature trends actually remain quite stable against a few days of rainfall. For example, the temperature decrease linked with typhoon rainfall in September 2011 soon recovered several days later. The temporal variations in temperature recorded in this study also show significant contrasts between fumarole versus hot-spring settings, whereby the high-temperature fuma-role data (sites SRT and SNY) show relatively punctuated fluctuations, while the low-temperature hot-spring data (site ASY) show more gradual variations.

4. ANALYSIS

4.1 1-D inversion of the MT data

Images of subsurface resistivity structure are estimated by the inversion of MT response functions, and in the present case (i.e., deployment of a single

MT site), 1-D inversion is the most practical approach. Therefore, in this study we used the >0.01 Hz apparent resistivity data (and corresponding phase data) to generate a first-order, 1-D approximation of the subsurface resistivity structure beneath the MT site. The data at frequencies below 0.01 Hz were excluded from the inversion because the apparent resistivity of the diagonal component exceeds that of $Z_{xy}$and shows 3-D features; in addition, geomagnetic transfer functions were not used in the inversion due to the assumption of 1-D structure.

The apparent resistivity and phase values were calculated from the arithmetic average $(Z_{xy} - Z_{yx})/2$, and then used as input for the 1-D inversion. Although the determinant $(\sqrt{Z_{xx} * Z_{yy} - Z_{xy} * Z_{yx}})$ of the impedance tensor is also commonly used in this type of 1-D analysis, this approach was not used during this study, primarily as calculated values had significant scatter at frequencies of <1 Hz, due to relatively poor estimations of Z_{xx} and Z_{yy} values. It should be noted that there is no significant difference between the arithmetic average and determinant values at <1 Hz, primarily as Z_{xx} and Z_{yy} values are very small within this frequency range (Fig. 2). We also applied the Occam approach (Constable et al., 1987), which adapts the smoothest model according to the target root mean square (RMS) misfit, which is usually set at 1.0. The initial inversion model uses a half-space with a resistivity of 100 ohm-m, and we used 16 frequencies in the inversion: 320, 160, 80, 32, 20, 8, 5, 2.5, 1.25, 0.63, 0.31, 0.16, 0.08, 0.04, 0.02 and 0.01 Hz. We assigned error bars of ±5% to each impedance value, which corresponds to ±0.0217 in log apparent resistivity and ±1.43 degree in phase, with the exception of the 8 and 20 Hz data (which were assigned smaller error bars of ±2.5%, as they approximately coincide with the 1st and 3rd harmonics of the Schumann resonance) and the 80–320 Hz phase data (which were assigned larger error bars of ±20% due to increased contact resistance that sometimes caused the data to drift and fluctuate). Figure 6 shows the results of this 1-D inversion using the MT impedance data shown in Fig. 2. In this 1-D inversion, the final RMS approaches values of less than 1.0.

To elucidate temporal variations in resistivity structure, we attempted to invert the daily impedance. Figure 7 shows the variation in daily impedance at 8 and 20 Hz frequencies that correspond to the 1st and 3rd harmonics of the Schumann resonance, respectively. Because there is significant scatter in the arithmetic averages of impedance, the results of the 1-D inversion are also scattered. To address this problem, we smoothed the data before carrying out the inversions, which first involved removing the outliers (beyond ±10%) from the stacked impedance data (Fig. 2) at each frequency. This procedure essentially requires the assumption that impedance does not show any real temporal variations beyond ±10%. It should be noted that the percentage of outliers varies from approximately 1% at higher frequencies to around 50% at the lower frequencies, and reaches up to 90% at dead band values around 0.16 Hz. Second, a moving average with a time window of two weeks was applied to the daily impedance data. As a result, the smoothed data show approximately ±5% apparent resistivity changes and a ±1 degree phase change. This two-step procedure was applied to all frequencies, resulting in significant smoothing of the data, but even with this screening method the data corresponding to the dead band (0.31, 0.16, 0.08, and 0.04 Hz) were still scattered. Therefore, we assigned large error bars of ±50% to these dead-band data. The variation in smoothed daily impedance across the entire range of frequencies is shown as a color pseudo-section in Fig. 8; for simplicity, only $(Z_{xy} - Z_{yx})/2$ values are shown. The upper panel in this figure shows the change in apparent resistivity (in %), and the lower panel shows phase changes (in degrees). As noted in Subsection 2.1, the contact resistance between the soil and one electrode drifted to values of over 10,000 ohm-m; this drift in contact resistance was recovered by changing the electrodes in December 2011. Figure 8 shows that contact resistance has a significant phase effect at frequencies of 80–320 Hz, although the apparent resistivity across these frequencies was much less influenced. In addition, an apparent resistivity pseudo-section shows a negative correlation between temporal variations at higher frequencies (above 10 Hz) and lower frequencies (below 10 Hz).

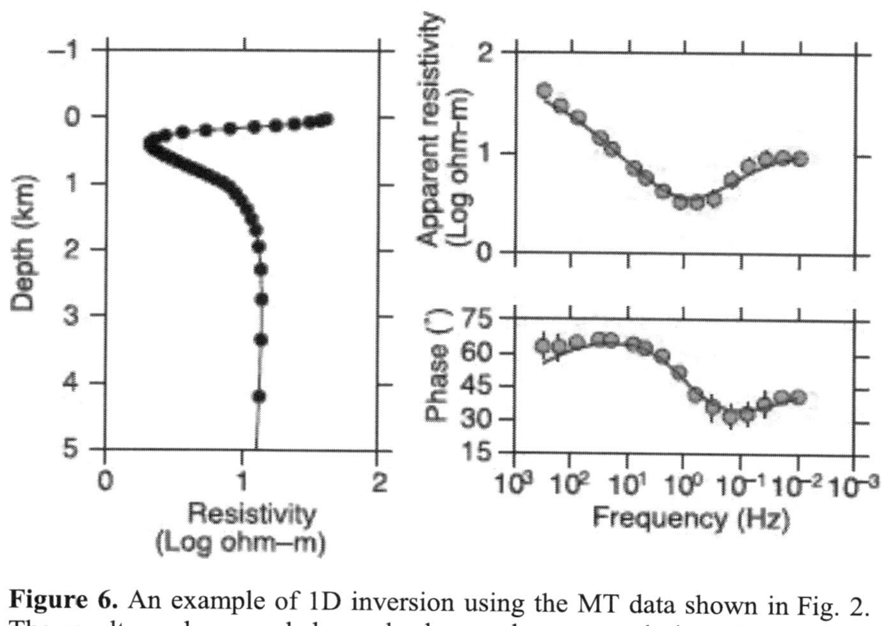

Figure 6. An example of 1D inversion using the MT data shown in Fig. 2. The result can be regarded as a background structure during observational period. Right panel shows the fitness of the resistivity model. Green circle and solid line indicate data and theoretical response, respectively.

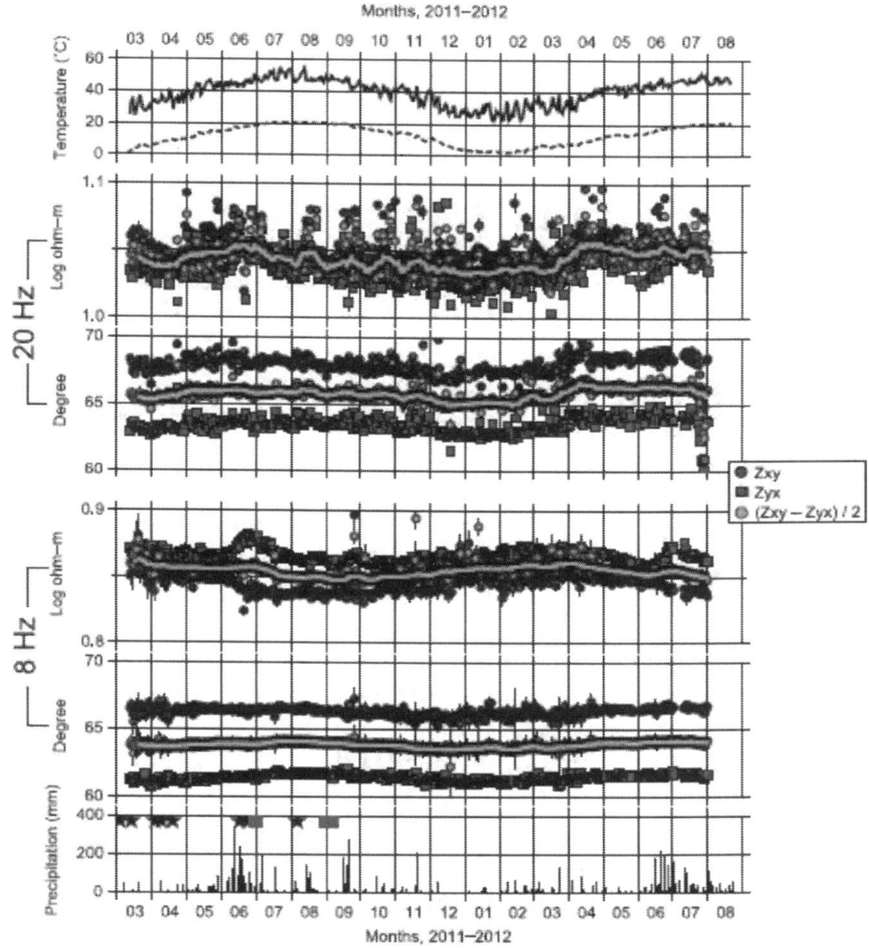

Figure 7. Apparent resistivity and phase variations in the frequency of 20 Hz at the MT monitoring site. Thick green line indicates moving average with a time window of 2 weeks.

By using the smoothed impedances, we were able to compute a stable convergence of each inversion to a target RMS of 1.0. Figure 9 shows the observed temporal variation of 1-D resistivity structure estimated in this study. To allow for direct comparisons, the temperature variations at site ASY (1.3 km from the MT monitoring site), and precipitation data for the region are shown alongside these 1-D resistivity structure data. Because 1-D inversion of the dead-band data is not plausible according to the two-step smoothing procedure outlined above, we also performed each inversion by using the data averaged over each one-month period, from which we confirmed the similarity of both results. The estimated resistivity changes are approximately ±5%, and the polarity differs between depths of 100 and 400 m (Fig. 9).

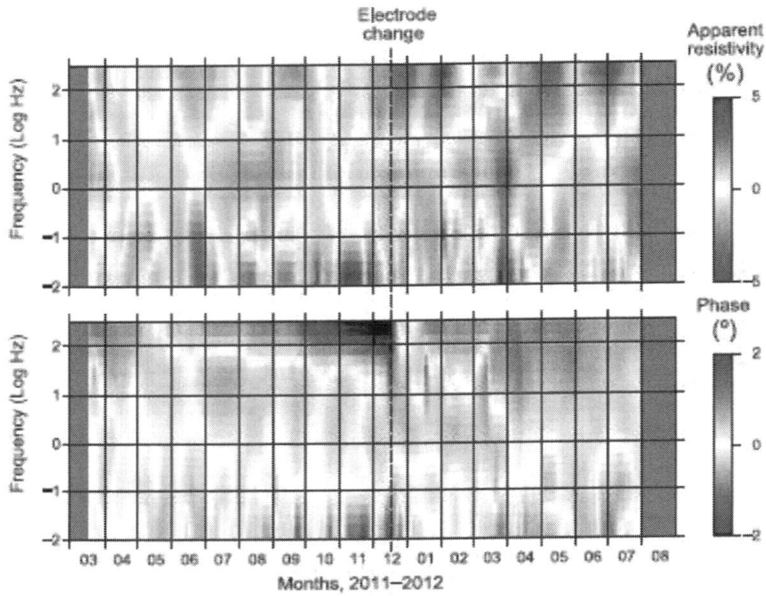

Figure 8. Temporal variations in the smoothed apparent resistivity and phase shown in a color pseudo-section. For simplicity, only $(Z_{xy} - Z_{yx})/2$ is shown. Upper panel shows change in apparent resistivity (%) while the lower panel shows change in phase (°); both are deviations from the average values of each frequency. The phase jump in December 2011 at 80–320 Hz is due to a change in electrodes; a detailed explanation is provided in the main text.

4.2 Assessing the significance of MT impedance variations

The fact that observed impedance and inferred resistivity variations are quite small means that evaluation of the causes of the temporal changes in impedance and inferred resistivity needs to be carefully undertaken. An initial assessment suggests that rainfall is the most likely cause of the resistivity changes; for instance, the 8 Hz data shown in Fig. 7 clearly indicate that Z_{xy} and Z_{yx} apparent resistivity values diverged during the rainy season in June 2011 and 2012, although with no corresponding phase changes (Fig. 7). These observations suggest that heavy and long-term rainfall caused local disturbances in surficial resistivity structures around the MT site with no corresponding temporal change in geomagnetic transfer functions (Fig. 4); these functions should theoretically not change during these sorts of local distortions. We also observe that rainfall-related impedance disturbances disappear when average impedance values are used (Fig. 7).

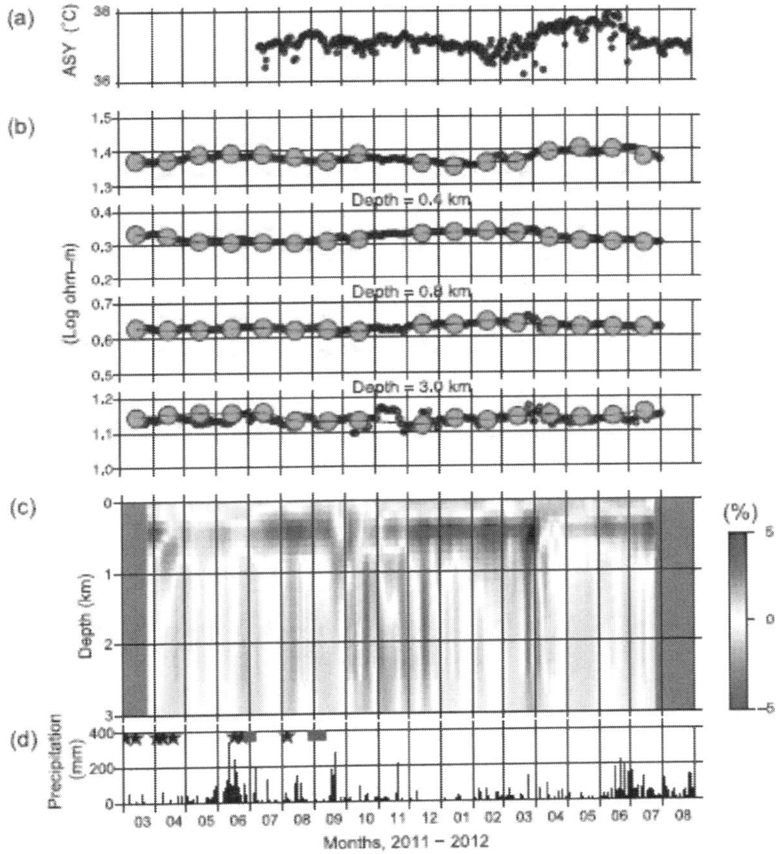

Figure 9. Temporal changes in resistivity structure within the study area compared with variations in temperature and precipitation. (a) Temperature variation at the ASY site, 2 km WSW of the MT monitoring sites (Fig. 5). (b) Temporal change in the resistivity structure estimated using a 1D Occam inversion; small and large green circles indicate results derived from smoothed 1 day and 1 month average data, respectively. Note that scatter associated with the 1 month average for November 2011 was caused by electrode problems, resulting in unreliable data. (c) Resistivity changes monitored by percentage deviation from the average value for each depth. (d) Timing of precipitation and volcanic eruptions in the study area; the notation used is as in Figs. 5 and 7.

Aside from rainfall, the actual change in temperature itself may affect the output of the measurement system, resulting in artificial variations of MT response functions. In particular, it is known that the induction coils in our MT monitoring instruments are sensitive to changes in the temperature of their surrounding environment, because the permeability of the core material varies systematically with changes in temperature. In comparison, Fig. 7 also shows temporal surface temperature variations, both in the MT logger and at a depth of 30 cm; note that the induction coils were installed at a depth of approximately 30 cm. It is immediately apparent that both sets of temperature data exhibit the same overall seasonal variability of approximately ±10°C. Here, we know that the output reading from the induction coil can vary by a value of 0.09%/1°C (at most), such that the observed artificial temperature increase results in a slight decrease in apparent resistivity. This temperature response corresponds to a value of ±1.8%/±10°C of the apparent resistivity change (Eqs. (1) and (3)). Although this estimate is equal to the observed temporal variations in impedance data (within one order of magnitude), both the polarity and the periods of time during which the peak variations of impedance occur (green curved lines in the 2nd and 4th plots from the top in Fig. 7) do not correlate well with the temperature monitoring curves (top in Fig. 7). Therefore, we conclude that such artifacts in the data, possibly relating to the effects of temperature change on the instruments, do not appear to have had significant effect on our data.

4.3 Limitations of the 1-D inversion analysis

In our 1-D inversion of subsurface resistivity structure, the largest uncertainties in the model arise due to the simple assumptions that must be made. The most significant issue in this regard is the assumption of a '1-D' structure. Figure 2 shows that our 1-D approximation breaks down in the deeper levels of the model (corresponding to the low-frequency data in Fig. 2). Furthermore, the data recorded on the induction vector (Fig. 4) show that temporal resistivity changes can occur not just beneath the site, but farther from the site. To address these issues, multi-site MT monitoring (Peacock et al., 2012) and 3-D modeling are necessary, especially with regards to evaluating variations in resistivity structure at great depth.

Another important uncertainty arises from the inversion algorithm itself. Since we applied the Occam approach (Constable et al., 1987), the results we obtained (Figs. 6 and 8) represent the smoothest available model among all possible candidates. Note that either a sharp contrast in resistivity or a sharp change in resistivity can explain our data. For example, the real subsurface geological structure may comprise three layers (resistive-conductive-resistive) rather than the smooth structure portrayed in Fig. 6. In that particular case, the amplitude of the deduced temporal resistivity change may increase.

5. DISCUSSION

5.1 Interpretation of background resistivity structure

Because the main characteristics of the inferred resistivity structure remained stable throughout the entire observation period (Fig. 9), it is possible to make meaningful geological interpretations of this resistivity structure. The inversion results (Figs. 6 and 9) reveal that at a depth of around 400 m, a conductive layer appears to be sandwiched between two resistive layers. The sounding curves (Fig. 2) indicate that the conductive layer corresponds to impedance around 1 Hz, where the apparent resistivity reaches a minimum. The resistivity of the conductive layer (a few ohm-m) is difficult to explain by the presence of water alone, as this would require an unrealistic porosity of greater than 30%, as estimated from Archie's law and assuming that the resistivity of the water in this area is identical to that of seawater (0.3 ohm-m). This suggests that this area must contain a large amount of conductive clay minerals, such as smectite, and that these clays are contributing to the low resistivity of this particular layer. Indeed, the results of exploration drilling during geothermal energy exploration some 6 km SW of our MT site (Uchida and Sasaki, 2006) identified a prominent smectite layer that causes a low-resistivity (conductive) zone beneath the Kirishima volcanoes. The fact that smectite breaks down at temperatures above 200°C (e.g., Ussher et al., 2000; Nurhasan et al., 2006; Uchida and Sasaki, 2006) means a high-temperature (~200°C) zone is likely to be present in the subsurface of the study area at a depth of >1 km, most probably beneath the MT site.

Conductive clay layers usually exhibit relatively low permeability (e.g., Jones and Dumas, 1993; Nurhasan et al., 2006; Aizawa et al., 2009a;Yamaya et al., 2009; Kanda et al., 2010). In particular, the top of such a clay layer can be sealed (i.e., impermeable), such that in the subsurface, any shallow, cold groundwater do not become well-mixed with deep, hot hydrothermal waters (Aizawa et al., 2009a). Although hot-spring waters are known to exist at a depth of 60 m near our MT site, the temperature of this water is actually quite low (~38°C), which provides one indication that the conductive layer that we modeled appears to act as a low-permeability layer. Drilling log data, obtained from drilling some 400 m west of ASY in 1988, are available to a depth of 350 m (Kagiyama et al., 1994) and indicate that temperatures are highest (60°C) at 100 m depth and gradually decrease to 36°C at a depth of 350 m, suggesting that the shallow hot-spring water did not significantly mix with any deeper hydrothermal waters. However, since a prominent fumarolic area existed on Iwo-Yama volcano before 1990, high-permeability fracture zones may exist locally within this putative clayey low-permeability layer. Therefore, the relatively low temperature (~38°C) of the hot-spring waters at site ASY may be interpreted as evidence of the mixing of shallow, cold groundwater with small amounts of hot hydrothermal water originating from depth (Fig. 10).

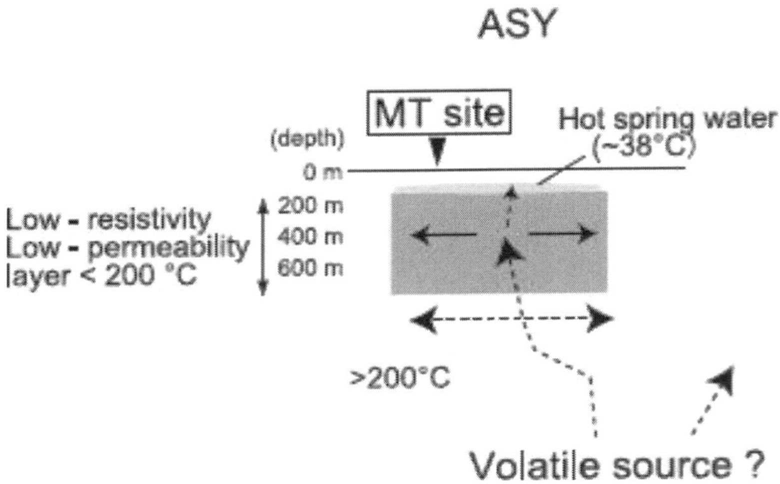

Figure 10. Schematic view of interpretation of resistivity and temperature change. Mixing of deep volatile and shallow groundwater causes the resistivity change at shallow level. Dashed and solid arrows indicate rapid and slow fluid movement, respectively.

5.2 Resistivity changes

Previous studies have shown that significant changes in subsurface resistivity may occur in association with large volcanic eruptions (Yukutake et al., 1990; Zlotnicki et al., 2003). In contrast, since the start of our monitoring study of temporal variations in subsurface resistivity structure and temperature, Shinmoe-dake volcano has gradually ceased activity and the observed MT impedance changes have only been very slight (±5%). However, based upon these data, we now suggest that significant changes in subsurface resistivity structure will occur prior to the next sub-Plinian eruption of this volcano. The geological scenario that we envisage is described as follows: 1) magma degassing occurs in association with the drop in pressure during ascent of the magma from deeper levels; 2) the resultant degassed volatiles reach shallow crustal levels faster than the ascending magma, to eventually become trapped near the surface by a shallow, low-permeability layer; (3) as a result, significant changes in resistivity structure occur in the subsurface (near the shallow conductive layer) before the next eruption ensues.

Although the above scenario may seem to be somewhat loosely constrained, there are two other independent studies that provide additional support for these hypotheses. The first arises from the fact that our MT site is located near Iwo-Yama volcano. Although fumaroles are not presently found at Iwo-Yama volcano, a fumarolic area existed there in 1990. At that time, $^3He/^4He$ ratios were determined for these fumarolic gases, yielding a value of 7.8 Ra (1 Ra corresponds to the atmospheric $^3He/^4He$ ratio), which demonstrates that magmatic gases were being supplied to the Iwo-Yama fumaroles (Sato et al., 1999). In another study on Kirishima volcanic group (Aizawa et al., submitted), the 3D resistivity structure was modeled to a depth of 20 km, indicating the presence of a sub-vertical conductive zone connecting a deep-seated conductive body (at a depth of 10 km) to a shallow conductive layer near the Iwo-Yama volcano (Fig. 1). This structure seems to suggest the existence of a volcanic fluid pathway between a deep subsurface magma chamber and Iwo-Yama volcano. In summary, we suggest that our MT monitoring site is located where deep-seated magmatic volatiles may be preferentially channeled to the surface.

The existence of such a volatiles pathway in the study area is also hinted at by the observed temporal variations in resistivity. Although the amplitudes are quite small, the resistivity changes recorded since March 2012 are the most remarkable in our data, both in terms of the observed MT impedance variations (Figs. 7–8) and the changes in inverted resistivity structure (Fig. 9). Notably, these changes appear to be synchronous with the temperature changes observed near the MT site (Fig. 9). In March of 2012, we observed concomitant temperature increases at two of the sites (ASY and SNY). At the ASY site, temperature was observed to gradually increase in March, lasting until late June. In contrast, at the SNY site, the temperature was suddenly observed to exceed the boiling point of water (97°C at the altitude of this site), and then gradually subside back to normal temperature levels. Significant rainfall was not observed at the time when these two temperature increases occurred at these sites. A more plausible interpretation of these observed temperature variations is that heat was transferred to these sites from deeper levels in the crust via volatiles. Faults and fracture networks are considered to be viable geological structures along which possible volatile pathways might occur in the subsurface (Uchida and Sasaki, 2006).

The MT inversion of the present study suggests that at shallow levels, resistivity structure changed at around March 2012, and accordingly, we interpret these changes in resistivity to be the result of the upward migration of volatiles, which may originated from magma degassing. For example, the ±5% resistivity change in Fig. 8 can be explained by a coincident ±2.5°C to

±5°C temperature change (e.g., Ussher et al., 2000), which is within an order of magnitude of the observations made at SNY. Furthermore, degassed volatiles can supply ions to groundwater, thereby reducing its resistivity, meaning that the resistivity decrease at a depth of 0.4 km during March 2012 can quantitatively be explained by an increase in volatile supply to the groundwater in this area. However, it is difficult to explain the observed increase in resistivity during March 2012 at a depth of 0.1 km; this depth roughly corresponds to the depth of low-temperature hot-spring waters in the site ASY area (Fig. 6). One possibility is that an increased supply of volatiles induced a widening of fractures in the low-permeability seal in this area, resulting in a depletion of shallow low-temperature groundwater. Alternatively, an increase in gas bubbles within the groundwater in this area may have caused a significant increase in bulk resistivity (Wang et al., 2010).

Because the timing of onset of the observed temperature increases was similar between sites ASY and SNY, we suggest that the volatiles inferred to have been introduced to both of these regions simultaneously, might have originated from a single, deep-seated volatile source (Fig. 9). The observed difference in duration of these temperature increases is explained as follows. At site ASY, an underground hot-spring exists at a depth of 60 m, below which the low-permeability clay layer exists, and we know that there are no fumaroles at the surface. Therefore, any volatiles present in that region are likely to be stored mostly at depth, and to have subsequently migrated gradually into shallow regions. In contrast, a high-temperature (~97°C) fumarolic zone exists at site SNY. Therefore, volatiles supplied to this site could escape to the surface, resulting in the observed sudden drop in the rate of temperature increase. Because the distance between the SNY and ASY sites is approximately 5 km, their volatile source could be located at relatively deep levels in the crust.

6. CONCLUSIONS

We conducted a MT and temperature monitoring study in a region NW of Shinmoe-dake volcano. From these data, we have come to the following conclusions. 1) By 1-D inversion, we have successfully inferred that slight (±5%) temporal variations in resistivity structure took place during the observation period. 2) We observed simultaneous temporal changes in resistivity and temperature. These observations suggest that volatiles are occasionally re-supplied to the shallower parts of the subsurface near the MT monitoring site, which is situated 2 km from a region where ground deformation has been observed.

The limitations of this MT monitoring mostly arise from the fact that we were only able to deploy 1 station, and it seems clear from our data that the subsurface resistivity structure at this site becomes increasingly three-dimensional at deeper levels. Furthermore, the temporal variations in MT response function suggest that resistivity changes may have occurred not only beneath the site, but also farther from the site. Ultimately, multi-site MT monitoring and the development of 3-D modeling in this area will be necessary to further explore the resistivity changes that take place in the future, especially at much deeper levels in the subsurface.

NOTES

Now at Institute of Seismology and Volcanology, Faculty of Sciences, Kyushu University, 2-5643-29 Shin'yama, Shimabara, Nagasaki 855–0843, Japan.

ACKNOWLEDGMENTS

We thank A. Watanabe and J. Oikawa for maintenance of the MT site. The observations were conducted under the supervision of the Japan Meteorological Agency. We thank landowners and regional forest offices (Kagoshima and Miyazaki prefectures) for permitting the observations. The geomagnetic data used for the remote-reference processing were provided by the Kakioka Geomagnetic Observatory, Japan Meteorological Agency. Comments by M. Unsworth and an anonymous reviewer significantly improved the manuscript. Helpful reviews and constructive editing by C. Newhall, the guest editor of this special volume, are also greatly appreciated. This work was supported by a Grant-in-Aid from MEXT (KAKENHI No. 22900001, and the Multidisciplinary Research Project for High Strain Rate Zones).

REFERENCES

1. Aizawa, K., Y. Ogawa, and T. Ishido, Groundwater flow and hydrothermal systems within volcanic edifices: Delineation by electric self-potential and magnetotellurics, J. Geophys. Res.—Solid Earth, **114**, B01208,

2. Aizawa, K., Y. Ogawa, M. Mishina, K. Takahashi, S. Nagaoka, N. Takagi, S. Sakanaka, and T. Miura, Structural controls on the 1998

volcanic unrest at Iwate volcano: Relationship between a shallow, electrically resistive body and the possible ascent route of magmatic fluid, J. Volcanol. Geotherm. Res., **187**(1–2), 131–139,

3. Aizawa, K., W. Kanda, Y. Ogawa, M. Iguchi, A. Yokoo, H. Yakiwara, and T. Sugano, Temporal changes in electrical resistivity at Sakurajima volcano from continuous magnetotelluric observations, J. Volcanol. Geotherm. Res., **199**(1–2), 165–175,

4. Aizawa, K., T. Koyama, H. Hase, M. Uyeshima, W. Kanda, M. Utsugi, R. Yoshimura, Y. Yamaya, T. Hashimoto, K. Yamazaki, S. Komatsu, A. Watanabe, K. Miyakawa, and Y. Ogawa, Three dimensional resistivity structure and magma pluming system of Kirishima volcanoes as inferred from broad-band magnetotelluric data, J. Geophys. Res., (submitted).

5. Brenguier, F., N. M. Shapiro, M. Campillo, V. Ferrazzini, Z. Duputel, O. Coutant, and A. Nercessian, Towards forecasting volcanic eruptions using seismic noise, Nat. Geosci., **1**(2), 126–130,

6. Chave, A. D. and D. J. Thomson, Bounded influence magne-totelluric response function estimation, Geophys. J. Int., **157**(3),

7. Constable, S. C., R. L. Parker, and C. G. Constable, Occams inversion—a practical algorithm for generating smooth models from electromagnetic sounding data, Geophysics, **52**(3), 289–300,

8. Gamble, T. D., J. Clarke, and W. M. Goubau, Magnetotellurics with a remote magnetic reference, Geophysics, **44**(1), 53–68, 1979.

9. Hanekop, O. and F. Simpson, Error propagation in electromagnetic transfer functions: What role for the magnetotelluric method in detecting earthquake precursors?, Geophys. J. Int., **165**, 763–774,

10. Hase, H., T. Hashimoto, S. Sakanaka, W. Kanda, and Y. Tanaka, Hydrothermal system beneath Aso volcano as inferred from self-potential mapping and resistivity structure, J. Volcanol. Geotherm. Res., **143**(4), 259–277,

11. Jones, A. G. and I. Dumas, Electromagnetic images of a volcanic zone, Phys. Earth Planet. Inter., **81**(1–4), 289–314, 1993.

12. Kagiyama, T., M. Yamaguchi, F. Masutani, and H. Utada, VLF, ELF-MT survey around Iwo-Yama, Kirishima volcanoes, Bull. Earthq. Res. Inst., **69**, 211–239, 1994.

13. Kanda, W., M. Utsugi, Y. Tanaka, T. Hashimoto, I. Fujii, T. Hasenaka, and N. Shigeno, A heating process of Kuchi-erabu-jima volcano, Japan,

as inferred from geomagnetic field variations and electrical structure, J. Volcanol. Geotherm. Res., **189**(1–2),

14. Kappler, K. N., H. F. Morrison, and G. D. Egbert, Long-term monitoring of ULF electromagnetic fields at Parkfield, California, J. Geophys. Res., **115**, B04406,

15. Kozono, T., H. Ueda, T. Ozawa, T. Koyaguchi, E. Fujita, A. Tomiya, and Y. J. Suzuki, Magma discharge variations during the 2011 eruptions of Shinmoe-dake volcano, Japan, revealed by geodetic and satellite observations, Bull. Volcanol., **75**, 695,

16. Miller, V. and M. Savage, Changes in seismic anisotropy after volcanic eruptions: Evidence from Mount Ruapehu, Science, **293**(5538), 2231–2233, 2001.

17. Nakada, S., M. Nagai, T. Kaneko, Y. Suzuki, and F. Maeno, The outline of the 2011 eruption at Shinmoe-dake (Kirishima), Japan, Earth Planets Space, **65**, this issue, 475–488,

18. Nakao, S., Y. Morita, K. Goto, H. Yakiwara, S. Hirano, J. Oikawa, H. Ueda, T. Kozono, Y. Hirata, H. Takahashi, M. Ichiyanagi, Y. Ohta, T. Matsushima, and M. Iguchi, Crustal deformation before and after eruption on Jan. 26, 2011 in Kirishima Shinmoedake, Japan, Abstract 2011 Fall Meeting of the Volcanological Society of Japan, p. 27, 2011 (in Japanese)

19. Nakao, S., Y. Morita, K. Goto, H. Yakiwara, S. Hirano, J. Oikawa, H. Ueda, T. Kozono, Y. Hirata, H. Takahashi, M. Ichiyanagi, Y. Ohta, T. Matsushima, and M. Iguchi, Crustal deformation after the eruption on January, 2011 by continuous GPS observation in Kirishima Volcano, Abstract 2012 JPGU meeting, 2012.

20. Nurhasan, Y. Ogawa, N. Ujihara, S. B. Tank, Y. Honkura, S. Onizawa, T. Mori, and M. Makino, Two electrical conductors beneath Kusatsu-Shirane volcano, Japan, imaged by audiomagnetotellurics, and their implications for the hydrothermal system, Earth Planets Space, **58**(8), 1053–1059, 2006.

21. Ogawa, Y, N. Matsushima, H. Oshima, S. Takakura, M. Utsugi, K. Hirano, M. Igarashi, and T. Doi, A resistivity cross-section of Usu volcano, Hokkaido, Japan, by audiomagnetotelluric soundings, Earth Planets Space, **50**(4), 339–346, 1998.

22. Ogawa, Y, M. Mishina, T. Goto, H. Satoh, N. Oshiman, T. Kasaya, Y Takahashi, T. Nishitani, S. Sakanaka, M. Uyeshima, Y Honkura, and M. Matsushima, Magnetotelluric imaging of fluids in intraplate earthquake

zones, NE Japan back arc, Geophys. Res. Lett., **28**(19), 3741–3744, 2001.

23. Patane, D., G. Barberi, O. Cocina, P. De Gori, and C. Chiarabba, Time-resolved seismic tomography detects magma intrusions at Mount Etna, Science, **313**(5788), 821–823,

24. Peacock, J. R., S. Thiel, P. Reid, and G. Heinson, Magnetotelluric monitoring of a fluid injection: Example from an enhanced geothermal system, Geophys. Res. Lett., **39**, L18403,

25. Sato, M., T. Mori, K. Notsu, and H. Wakita, Carbon and helium isotopic composition of fumarolic gases and hot spring gases from Kirishima volcanic area, Bull. Volcanol. Soc. Jpn., **44**, 279–283, 1999.

26. Suzuki, Y, M. Nagai, F. Maeno, A. Yasuda, N. Hokanishi, T. Shimano, M. Ichihara, T. Kaneko, and S. Nakada, Precursory activity and evolution of the 2011 eruption of Shinmoe-dake in Kirishima volcano-insights from ash samples, Earth Planets Space, **65**, this issue, 591–607,

27. Uchida, T. and Y Sasaki, Stable 3D inversion of MT data and its application to geothermal exploration, Explor. Geophys., **37**, 223–230, 2006.

28. Ueda, H., T. Kozono, E. Fujita, Y Kohno, M. Nagai, Y Miyagi, and T. Tanada, Crustal deformation associated with the 2011 Shinmoe-dake eruption as observed by tiltmeters and GPS, Earth Planets Space, **65**, this issue, 517–525,

29. Ussher, G., C. Harvey, R. Johnstone, and E. Anderson, Understanding the resistivities observed in Geothermal systems, Proceedings World Geothermal Congress, Kyushu, Japan, pp. 1915–1920, 2000.

30. Utada, H., Interpretation of time changes in the apparent resistivity observed prior to the 1986 eruption of Izu-Oshima volcano, Japan, J. Volcanol. Geotherm. Res., **126**, 97–107, 2003.

31. Uyeshima, M., Y Ogawa, Y Honkura, S. Koyama, N. Ujihara, T. Mogi, Y Yamaya, M. Harada, S. Yamaguchi, I. Shiozaki, T. Noguchi, Y Kuwaba, Y Tanaka, Y Mochido, N. Manabe, M. Nishihara, M. Saka, and M. Serizawa, Resistivity imaging across the source region of the 2004 Mid-Niigata Prefecture earthquake (M6.8), central Japan, Earth Planets Space, **57**(5), 441–446, 2005.

32. Wang, S., M. Lee, M. K. Park, and J.-M. Kim, Box experiments on monitoring the CO2 migration in a homogeneous medium using electrical resistivity survey, Geosci. J., **14**(1),

33. Yamaya, Y, T. Mogi, T. Hashimoto, and H. Ichihara, Hydrothermal system beneath the crater of Tarumai volcano, Japan: 3-D resistivity structure revealed using audio-magnetotellurics and induction vector, J. Volcanol. Geotherm. Res., **187**, 193–202,

34. Yamazaki, K., M. Teraishi, K. Ishihara, S. Komatsu, and K. Kato, Subtle changes 1 in strain prior to sub-Plinian eruptions recorded by vault-housed extensometers during the 2011 activity at Shinmoe-dake, Kirishima volcano, Japan, Earth Planets Space, (submitted).

35. Yoshimura, R. et al., Magnetotelluric transect across the Niigata-Kobe Tectonic Zone, central Japan: A clear correlation between strain accumulation and resistivity structure, Geophys. Res. Lett., **36**, L20311,

36. Yukutake, T., T. Yoshino, H. Utada, H. Watanabe, Y Hamano, and T. Shimomura, Changes in the electrical resistivity of the central cone, Mihara-yama, of Oshima volcano observed by a direct current method, J. Geomag. Geoelctr, **42**, 151–168, 1990.

37. Zlotnicki, J., Y Sasai, P. Yvetot, Y Nishida, M. Uyeshima, F. Fauquet, H. Utada, H. Takahashi, and G. Donnadieu, Resistivity and self-potential changes associated with volcanic activity: The July 8, 2000 Miyake-jima eruption (Japan), Earth Planet. Sci. Lett., **205**, 139–154, 2003.

Index

A

acoustic, 69, 70, 72, 73, 74, 75, 76, 77, 78, 84, 85
alpha particle, 190, 192, 195, 197, 200, 201, 202, 203, 205, 206, 207, 209
audio-frequency magnetotelluric, 147, 148, 151, 156, 165

B

Beam source, 194
biosphere, 91, 92, 93, 94, 95, 96, 97, 99, 100, 101, 102, 106, 107, 113, 115
Borehole heat exchanger system, 24
Borehole thermal energy store, 24, 25

C

complementary split-ring resonator, 46, 47
Concrete Bridge Deck, 84
Concrete Slab, 82
Conductive heat transfer, 216
conductivity structure, 148
Core-flood, 174
crack detection, 45, 46, 47, 52, 57, 59, 67
CT scanning, 173

D

Deep structure, 216
Density, 197, 216

E

Earth, 95
Endoliths, 101

G

gamma ray spectrometer, 190
Geometry, 193
Geothermal parameters, 24
Geothermal potential, 216
groundwater table, 2, 10

H

Heat and mass transport model, 24
heavy ion transportation, 190, 191

Hijima fault, 147, 148, 149, 150, 151, 152, 153, 155, 156, 161, 162, 163, 164, 165
Hydrogeology, 28
hydrothermal systems, 92, 104, 105, 106, 107, 112, 118, 119, 277

I

impact ejecta, 92, 94, 95, 107
India, 1, 2, 3, 4, 5, 13, 14, 15, 16, 17, 18, 19, 20, 149, 167

L

life on Mars, 92, 93, 102, 109, 110
lunar and planetary spectroscopy, 190

M

Magnetic and thermal modelling, 216
Magnetotellurics, 166, 256, 278
Mars, 103
MASS, 69, 70, 72, 76, 79, 80, 81, 82, 84, 85
metamaterials, 46
microstrip line technology, 46
mobile, 70, 72, 84, 85
Modeling, 17, 30
Moon, 190, 192, 205, 206, 210, 212, 214

N

near-field detection, 46, 63
Neutron, 189, 190, 192, 194, 197, 199, 206, 208

P

Pakistan, 2, 10, 11, 12, 14, 16, 17, 18, 19, 20
PHITS, 189, 190, 191, 192, 194, 196, 204, 206, 208, 209, 211, 212, 213, 214

R

radiating surface wave, 70, 72, 73, 74, 75, 76, 77, 78, 79, 81, 82, 83, 84, 86
resistivity monitoring, 256

S

Salinity, 2, 3, 13, 18
structural change, 256
subsurface, 1, 2, 3, 4, 5, 11, 12, 13, 14, 15, 16, 17, 18, 19, 20, 260, 266, 272, 273, 274, 275, 276, 277
Subsurface drainage, 1, 2, 4, 10, 14, 18, 19, 20
Subsurface temperature, 216

T

temperature monitoring, 255, 256, 261, 272, 276
Triassic carbonates, 23, 24, 40

W

Waterlogging, 2, 13
Well logs, 29

Y

Yamasaki fault system, 147, 148, 149, 150, 151, 162, 164, 165